Erlang Programming

Dear Chuck,
many thanks for all
your feedback and for being
part of this exciting project!

Francesco

Chuck — thanks so much for all your help — Simon

Francesco Cesarini and Simon Thompson

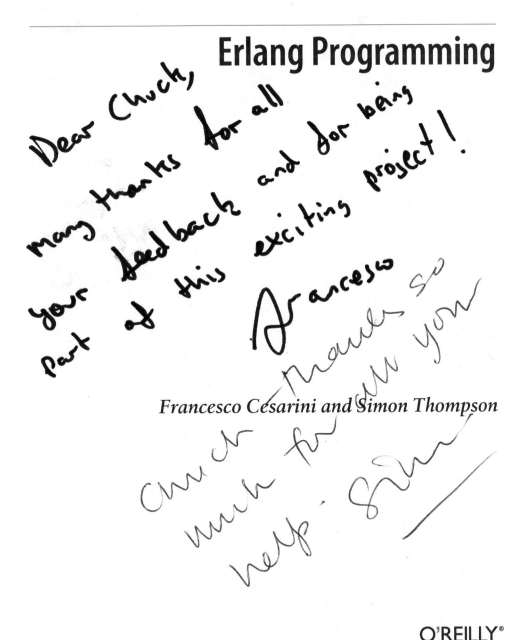

O'REILLY®

Beijing · Cambridge · Farnham · Köln · Sebastopol · Taipei · Tokyo

Erlang Programming

by Francesco Cesarini and Simon Thompson

Published by O'Reilly Media, Inc., 1005 Gravenstein Highway North, Sebastopol, CA 95472.

O'Reilly books may be purchased for educational, business, or sales promotional use. Online editions are also available for most titles (*http://my.safaribooksonline.com*). For more information, contact our corporate/institutional sales department: (800) 998-9938 or *corporate@oreilly.com*.

Editor: Mike Loukides

Production Editor: Sumita Mukherji

Copyeditor: Audrey Doyle

Proofreader: Sumita Mukherji

Indexer: Lucie Haskins

Cover Designer: Karen Montgomery

Interior Designer: David Futato

Illustrator: Robert Romano

Printing History:

June 2009:   First Edition.

RepKover™

This book uses RepKover™, a durable and flexible lay-flat binding.

ISBN: 978-0-596-51818-9

[M]

1244485946

Table of Contents

Foreword

Erlang is our solution to three problems regarding the development of highly concurrent, distributed "soft real-time systems":

- To be able to develop the software quickly and efficiently
- To have systems that are tolerant of software errors and hardware failures
- To be able to update the software on the fly, that is, without stopping execution

When we "invented" Erlang, we focused on telecommunication systems, but today these requirements are applicable to a large number of applications, and Erlang is used in applications as divergent as distributed databases, financial systems, and chat servers, among others. Recent interest in Erlang has been fueled by its suitability for use on multicore processors. While the world is struggling to find methods to facilitate porting applications to multicore processors, Erlang applications can be ported with virtually no changes.

Initially, Erlang was slow to spread; maybe it was too daring to introduce functional programming, lightweight concurrency, asynchronous message passing, and a unique method to handle failures, all in one go. It is easy to see why a language such as Java, which is only a small step away from C++, was easier for people to swallow. However, to achieve the goals I've just mentioned, we feel our approach has weathered the test of time. The use of Erlang is expanding rapidly.

This book is an excellent and practical introduction of Erlang, and is combined with a number of anecdotes explaining the ideas and background behind the development of Erlang.

Happy and, I trust, profitable reading.

—Mike Williams
Director of Traffic and Feature Software
Product Development Unit WCDMA, Ericsson AB
one of the inventors of Erlang

Preface

What made us start writing this book in the first place is the enthusiasm we share for Erlang. We wanted to help get the word out, giving back a little of what the community has given to us. Although we both got into Erlang for very different reasons, the end result was the same: lots of fun hours doing lots of fun stuff at a fraction of the effort it would have taken with other languages. And best of all, it is not a tool we use for hobby projects, but one we use on a daily basis in our real jobs!

Francesco: Why Erlang?

The year was 1994. While studying computer science at Uppsala University, one of the courses I took was on parallel programming. The lecturer held up the first edition of *Concurrent Programming in Erlang* (Prentice Hall) and said, "Read it." He then held up a handout and added, "These are the exercises, do them," after which Erlang barely got a mention; it was quickly overshadowed with the theory of threads, shared memory, semaphores, and deadlocks.

As the main exercise for this course, we had to implement a simulated world inhabited by carrots, rabbits, and wolves. Rabbits would roam this world eating carrots that grew in random patches. When they had eaten enough carrots, the rabbits would get fat and split in two. Wolves ran around eating up the rabbits; if they managed to catch and eat enough rabbits, they would also get fat and split. Rabbits and wolves within a certain distance of each other would broadcast information on food and predators. If a rabbit found a carrot patch, other rabbits would quickly join him. If a wolf found a rabbit, the pack would start chasing it.

The final result was amusingly fun to watch. The odd rabbit would run straight into a group of wolves, while others would run in other directions, sometimes stopping to grab a carrot en route. Every carrot patch, rabbit, and wolf was represented as an Erlang process communicating through message passing.

The exercise took me about 40 hours to solve. Although I enjoyed using Erlang and was positively surprised at the simplicity of its concurrency model and lack of OS threads for every process, I did not think that much of it right there and then. After all, it was one of the dozen or so languages I had to learn for my degree. Having used ML in my functional programming courses and ADA in my real-time programming courses, for me Erlang was just another language in the crowd. That changed a few months later when I started studying object-oriented programming.

In the object-oriented (OO) programming course, we were given the same simulated world lab but had to solve it with Eiffel, an OO language our new lecturer insisted was ideal for simulations. Although I had already solved the same problem and was able to reuse a good part of the algorithms, it took me and a fellow student 120 man-hours to solve.

This was the eye-opener that led me to believe the declarative and concurrent features in Erlang had to be the direction in which software development was heading. At the time, I was not sure whether the language that would lead the way in this paradigm shift was going to be Erlang, but I was certain that whatever language it was, it would be heavily influenced by Erlang and its ancestors. I picked up the phone and called Joe Armstrong, one of the inventors of Erlang. A week later, I visited the Ericsson Computer Science Lab for an interview, and I have never looked back.

Simon: Why Erlang?

I have worked in functional programming since the early 1980s, and have known about Erlang ever since it was first defined about 20 years ago. What I find most attractive about Erlang is that it's a language that was designed from the start to solve real and difficult problems, and to do it in an elegant and powerful way. That's why we've seen Erlang used in more and more systems in recent years.

It's also a small language, which makes writing tools for it much more practical than for a language such as Java, C++, or even Haskell. This, and the quality of the libraries we've been able to build on in our work, has helped the functional programming group at Kent to be very productive in implementing the Wrangler refactoring tool for Erlang.

Who Should Read This Book?

We have written this book to introduce you to programming in Erlang. We don't expect that you have programmed in Erlang before, nor do we assume that you are familiar with functional programming in other languages.

We do expect you to have programmed in Java, C, Ruby, or another mainstream language, and we've made sure that we point out to you where Erlang differs from what you're used to.

How to Read This Book

We wrote this book in two parts, the first to be read sequentially and the second can be read concurrently (or sequentially in whatever order you like), as the chapters are independent of each other.

The first 11 chapters of the book cover the core parts of Erlang:

- Chapter 1 gives a high-level introduction to the language, covering its key features for building high-availability, robust concurrent systems. In doing this, we also describe how Erlang came to be the way it is, and point out some of its high-profile success stories, which explain why you may want to adopt Erlang in one of your projects.

- The basics of sequential programming in Erlang are the subject of Chapters 2 and 3. In these chapters, we cover the central role of *recursion* in writing Erlang programs, as well as how *single assignment* in Erlang is quite different from the way variables are handled in other languages, such as C and Java.

- While covering sequential programming, we also introduce the *basic data types* of Erlang—numbers, atoms, strings, lists, and tuples—comparing them with similar types in other languages. Other types are covered later: records in Chapter 7, and function types and binaries in Chapter 9. Large-scale storage in ETS tables is the topic of Chapter 10.

- Erlang's distinctiveness comes to the fore in Chapters 4–6, which together cover the concurrent aspects of Erlang, embodied in *message passing* communication between concurrently executing *processes* running in separate memory spaces.

- It is possible to "hot-swap" code in a system, supporting *software upgrades* in running systems: this is the topic of Chapter 8.

- To conclude this part of the book, we cover *distributed programming* in Chapter 11. This allows different Erlang runtime systems (or *nodes*), which might be running on the same or different machines, to work together and interact as a distributed system.

In the remaining chapters, we cover a variety of different topics largely independent of each other. These include the following:

- The *Open Telecom Platform* (OTP) gives a set of libraries and design principles supporting the construction of robust, scalable systems in Erlang; this is the subject of Chapter 12.

- The Erlang distribution contains some standard computing applications: we cover the Mnesia *database* in Chapter 13 and the wxErlang *GUI programming* library in Chapter 14.

- Erlang distribution gives one mechanism for linking Erlang systems to each other. Chapter 15 shows how Erlang supports programming across the Internet using *sockets*, and Chapter 16 covers the various ways in which Erlang can *interwork* with systems written in C, Java, and Ruby, as well as many other languages.

- The standard Erlang distribution comes with a number of very useful tools, and we cover some of these next. Chapter 17 explains in depth how all aspects of Erlang systems can be traced without degrading their performance, and Chapter 18 covers tools for checking the correctness of programs, and for constructing documentation for Erlang systems. Unit testing, and how it is supported by EUnit, is the subject of Chapter 19.

- The last chapter, Chapter 20, looks at how to write programs that are elegant, readable, and efficient, and pulls together into one place much of the accumulated experience of the Erlang community.

The Appendix covers how to get started with Erlang, how to use the Erlang shell, popular tools for Erlang, and how to find out more about Erlang.

Each chapter is accompanied by a set of exercises, and you can download all the code in this book from its website:

http://www.erlangprogramming.org

The website also has references to further reading as well as links to the major sites supporting the Erlang community.

We wrote this book to be compatible with Erlang Release 13 (R13-B). Most of the features we describe will work with earlier releases; known incompatibilities with more recent earlier releases are detailed on our website.

Conventions Used in This Book

The following typographical conventions are used in this book:

Italic
Indicates new terms, URLs, filenames, file extensions, and occasionally, emphasis and keyword phrases.

`Constant width`
Indicates computer coding in a broad sense. This includes commands, options, variables, attributes, keys, requests, functions, methods, types, classes, modules, properties, parameters, values, objects, events, event handlers, XML and XHTML tags, macros, and keywords.

`Constant width bold`
Indicates commands or other text that the user should type literally.

Constant width italics

    Indicates text that should be replaced with user-supplied values or values determined by context.

    This icon signifies a tip, suggestion, or general note.

    This icon indicates a warning or caution.

Using Code Examples

This book is intended to help you write programs and systems in Erlang. In general, you may use the code in this book in your programs and documentation.

You do not need to contact the publisher for permission unless you are reproducing a significant portion of the code. For example, if you are writing a program that uses several chunks of code from this book you are not required to secure our permission. Answering a question by citing this book and quoting example code does not require permission.

Incorporating a significant amount of example code from this book into your product's documentation *does* require permission. Selling or distributing a CD-ROM of examples from O'Reilly books *does* require permission.

We appreciate, but do not require, attribution. An attribution usually includes the title, author, publisher, and ISBN. For example: "*Erlang Programming*, by Francesco Cesarini and Simon Thompson. Copyright © 2009 Francesco Cesarini and Simon Thompson, 978-0-596-51818-9."

If you feel your proposed use of code examples falls outside fair use or the permission given here, feel free to contact us as *permissions@oreilly.com*.

Safari® Books Online

Safari® When you see a Safari® Books Online icon on the cover of your favorite technology book, that means the book is available online through the O'Reilly Network Safari Bookshelf.

Safari offers a solution that's better than e-books. It's a virtual library that lets you easily search thousands of top tech books, cut and paste code samples, download chapters, and find quick answers when you need the most accurate, current information. Try it for free at *http://my.safaribooksonline.com*.

How to Contact Us

Please address comments and questions concerning this book to the publisher:

O'Reilly Media, Inc.
1005 Gravenstein Highway North
Sebastopol, CA 95472
800-998-9938 (in the United States or Canada)
707-829-0515 (international or local)
707-829-0104 (fax)

On the web page for this book we list errata, examples, and any additional information. You can access this page at:

http://www.oreilly.com/catalog/9780596518189

or at:

http://www.erlangprogramming.org

To comment or ask technical questions about this book, send email to:

bookquestions@oreilly.com

For more information about our books, conferences, Resource Centers, and the O'Reilly Network, see our website at:

http://www.oreilly.com/

Acknowledgments

In writing this book, we need to acknowledge everyone who made it possible. We start with Jan "Call Me Henry" Nyström, who helped jumpstart this project.

The team at O'Reilly Media provided us with endless support. In particular, our editor, Mike Loukides, patiently guided us through the process and provided encouragement, ensuring that the chapters kept on coming. Special thanks also go out to Audrey Doyle for the copyediting, and to Rachel Monaghan, Marlowe Shaeffer, Lucie Haskins, Sumita Mukherji, and everyone else on the production team.

We continue with the OTP team, and in particular, Bjorn Gustavsson, Sverker Eriksson, Dan Gudmundsson, Kenneth Lundin, Håkan Mattsson, Raimo Niskanen, and Patrik Nyblom, who helped us not only with the undocumented and unreleased features, ensuring that what is in print is in line with the latest release, but also with accuracy and correctness.

Other reviewers who deserve a special mention include Thomas Arts, Zvi Avraham, Franc Bozic, Richard Carlsson, Dale Harvey, Oscar Hellström, Steve Kirsch, Charles McKnight, Paul Oliver, Pierre Omidyar, Octavio Orozio, Rex Page, Michal Ptaszek, Corrado Santoro, Steve Vinoski, David Welton, Ulf Wiger, and Mike Williams.

Although we will not go into detail regarding what each of you did, it is important that you all know that your individual contributions had an influence in making this a better book. Thank you all!

Francesco needs to thank Alison for all her patience and support. I did not know what I was getting into when I agreed to write this book, and neither did you. Until the time to start working on the next book comes, I promise you laptop- and cell phone-free vacations. A thank you also goes to everyone at Erlang Training and Consulting for all the encouragement and to Simon for being such a great coauthor. We should all do it again sometime, as the result was worth it. But now, rest!

Simon wants to say a huge thank you to Jane, Alice, and Rory for their patience and support over the past few very busy months: without your encouragement, it just wouldn't have happened. Thanks, too, to Francesco for inviting me to join the project: it's been really enjoyable working together. I hope we get the chance to do it again, just not too soon....

Introduction

Why are we really excited about introducing you to Erlang? What do we feel is really special about the language? Its lightweight concurrency model with massive process scalability independent of the underlying operating system is second to none. With its approach that avoids shared data, Erlang is the perfect fit for multicore processors, in effect solving many of the synchronization problems and bottlenecks that arise with many conventional programming languages. Its declarative nature makes Erlang programs short and compact, and its built-in features make it ideal for fault-tolerant, soft real-time systems. Erlang also comes with very strong integration capabilities, so Erlang systems can be seamlessly incorporated into larger systems. This means that gradually bringing Erlang into a system and displacing less-capable conventional languages is not at all unusual.

Although Erlang might have been around for some time, the language itself, the virtual machine, and its libraries have been keeping pace with the rapidly changing requirements of the software industry. They are constantly being improved by a competent, enthusiastic, and dedicated team, aided by computer science researchers from universities around the world.

This introduction gives a high-level overview of the characteristics and features that have made Erlang so successful, providing insight into the context in which the language was designed, and how this influenced its current shape. Using case studies from commercial, research, and open source projects, we talk about how Erlang is used for real, comparing it with other languages and highlighting its strengths. We conclude by explaining the approaches that have worked best for us when running Erlang projects.

Why Should I Use Erlang?

What makes Erlang the best choice for your project? It depends on what you are looking to build. If you are looking into writing a number-crunching application, a graphics-intensive system, or client software running on a mobile handset, then sorry, you bought the wrong book. But if your target system is a high-level, concurrent, robust, soft real-time system that will scale in line with demand, make full use of multicore processors, and integrate with components written in other languages, Erlang should be your choice. As Tim Bray, director of Web Technologies at Sun Microsystems, expressed in his keynote at OSCON in July 2008:

> If somebody came to me and wanted to pay me a lot of money to build a large scale message handling system that really had to be up all the time, could never afford to go down for years at a time, I would unhesitatingly choose Erlang to build it in.

Many companies are using Erlang in their production systems:

- Amazon uses Erlang to implement SimpleDB, providing database services as a part of the Amazon Elastic Compute Cloud (EC2).
- Yahoo! uses it in its social bookmarking service, Delicious, which has more than 5 million users and 150 million bookmarked URLs.
- Facebook uses Erlang to power the backend of its chat service, handling more than 100 million active users.
- T-Mobile uses Erlang in its SMS and authentication systems.
- Motorola is using Erlang in call processing products in the public-safety industry.
- Ericsson uses Erlang in its support nodes, used in GPRS and 3G mobile networks worldwide.

The most popular open source Erlang applications include the following:

- The 3D subdivision modeler Wings 3D, used to model and texture polygon meshes.
- The Ejabberd system, which provides an Extensible Messaging and Presence Protocol (XMPP) based instant messaging (IM) application server.
- The CouchDB "schema-less" document-oriented database, providing scalability across multicore and multiserver clusters.
- The MochiWeb library that provides support for building lightweight HTTP servers. It is used to power services such as MochiBot and MochiAds, which serve dynamically generated content to millions of viewers daily.
- RabbitMQ, an AMQP messaging protocol implementation. AMQP is an emerging standard for high-performance enterprise messaging.

Although Uppsala University has for many years led the way with research on Erlang through the High Performance Erlang Project (HiPE), many other universities around the world are not far behind. They include the University of Kent in the United Kingdom and Eötvös Loránd University in Hungary, which are both working on refactoring tools. The Universidad Politécnica de Madrid of Spain together with Chalmers University of Technology and the IT University (both in Sweden) are working on Erlang property-based testing tools that are changing the way people verify Erlang programs.

With these companies, open source projects, and universities, we have just scratched the surface of what has today become a vibrant international community spread across six continents. Blogs, user groups, mailing lists, and dedicated sites are now helping to take the community to its next level.

The suitability of Erlang for server-side software has its roots in the history of the language, as it was originally developed to solve problems in a subset of this particular space, namely the telecom sector, and so it's worth looking back to the invention of Erlang in the 1980s.

The History of Erlang

In the mid-1980s, Ericsson's Computer Science Laboratory was given the task of investigating programming languages suitable for programming the next generation of telecom products. Joe Armstrong, Robert Virding, and Mike Williams—under the supervision of Bjarne Däcker—spent two years prototyping telecom applications with all of the available programming languages of the time. Their conclusion was that although many of the languages had interesting and relevant features, no single language encompassed them all. As a result, they decided to invent their own. Erlang was influenced by functional languages such as ML and Miranda, concurrent languages such as ADA, Modula, and Chill, as well as the Prolog logic programming language. The software upgrade properties of Smalltalk played a role, as did the Ericsson proprietary languages EriPascal and PLEX.

With a Prolog-based Erlang virtual machine (VM), the lab spent four years prototyping telecom applications with an evolving language that through trial and error became the Erlang we know today. In 1991, Mike Williams wrote the first C-based virtual machine, and a year later, the first commercial project with a small team of developers was launched. The project was a mobility server, allowing DECT cordless phone users to roam across private office networks. The product was successfully launched in 1994, providing valuable feedback on improvements and missing features that got integrated into the 1995 Erlang release.

Only then was the language deemed mature enough to use in major projects with hundreds of developers, including Ericsson's broadband, GPRS, and ATM switching solutions. In conjunction with these projects, the OTP framework was developed and released in 1996. OTP provides a framework to structure Erlang systems, offering robustness and fault tolerance together with a set of tools and libraries.

The history of Erlang is important in understanding its philosophy. Although many languages were developed before finding their niche, Erlang was developed to solve the "time-to-market" requirements of distributed, fault-tolerant, massively concurrent, soft real-time systems. The fact that web services, retail and commercial banking, computer telephony, messaging systems, and enterprise integration, to mention but a few, happen to share the same requirements as telecom systems explains why Erlang is gaining headway in these sectors.

Ericsson made the decision to release Erlang as open source in December 1998 using the EPL license, a derivative of the Mozilla Public License. This was done with no budget or press releases, nor with the help of the corporate marketing department. In January 1999, the erlang.org site had about 36,000 page impressions. Ten years later, this number had risen to 2.8 million. This rise is a reflection of an ever-growing community resulting from a combination of successful commercial, research, and open source projects, viral marketing, blogging, and books, all driven by the need to solve hard software problems in the domain for which Erlang had originally been created.

Erlang's Characteristics

Although Erlang on its own is an attractive programming language, its real strength becomes apparent when you put it together with the virtual machine (VM) and the OTP middleware and libraries. Each of them contributes to making software development in Erlang special. So, what are the features of Erlang that differentiate it from many of its peers?

High-Level Constructs

Erlang is a declarative language. Declarative languages work on the principle of trying to describe *what* should be computed, rather than saying *how* this value is calculated. A function definition—particularly one that uses *pattern matching* to select among different cases, and to extract components from complex data structures—will read like a set of equations:

```
area({square, Side})   ->    Side * Side ;
area({circle, Radius}) ->    math:pi() * Radius * Radius.
```

This definition takes a shape—here a square or a circle—and depending on which kind of shape it receives, it matches the correct function clause and returns the corresponding area.

In Erlang, you can pattern-match not only on high-level data but also on *bit sequences*, allowing a startlingly high-level description of protocol manipulation functions. Here is the start of a function to decode TCP segments:

```
decode(<< SourcePort:16, DestinationPort:16,
          SequenceNumber:32,
          AckNumber:32,
          DataOffset:4, _Reserved:4, Flags:8, WindowSize:16,
          Checksum:16, UrgentPointer:16,
          Payload/binary>>) when DataOffset>4 ...
```

In the preceding code, each numeric length, such as 4 in DataOffset:4, gives the number of *bits* to be matched to that variable. By comparison, think of how you would achieve the same effect in C or Java.

Another aspect of Erlang is that functions (or *closures*) are first-class data. They can be bound to a variable and can be treated just like any other data item: stored in a list, returned by a function, or communicated between processes.

List comprehensions, also taken from the functional programming paradigm, combine list generators and filters, returning a list containing the elements of the list generators after the filters have been applied. The following example of list comprehensions, which we explain fully in Chapter 9, is an implementation of the quicksort algorithm in a couple of lines of code:

```
qsort([]) -> [];
qsort([X|Xs]) ->
    qsort([Y || Y<-Xs, Y =< X]) ++ [X] ++ qsort([Y || Y<-Xs, Y > X]).
```

Concurrent Processes and Message Passing

Concurrency in Erlang is fundamental to its success. Rather than providing threads that share memory, each Erlang process executes in its own memory space and owns its own heap and stack. Processes can't interfere with each other inadvertently, as is all too easy in threading models, leading to deadlocks and other horrors.

Processes communicate with each other via *message passing*, where the message can be any Erlang data value at all. Message passing is *asynchronous*, so once a message is sent, the process can continue processing. Messages are retrieved from the process mailbox *selectively*, so it is not necessary to process messages in the order they are received. This makes the concurrency more robust, particularly when processes are distributed across different computers and the order in which messages are received will depend on ambient network conditions. Figure 1-1 shows an example, where an "area server" process calculates areas of shapes for a client, as we did earlier in "High-Level Constructs" on page 4.

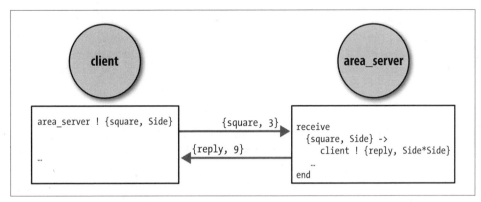

Figure 1-1. Communication between processes

Scalable, Safe, and Efficient Concurrency

Erlang concurrency is fast and scalable. Its processes are lightweight in that the Erlang virtual machine does not create an OS thread for every created process. They are created, scheduled, and handled in the VM, independent of the underlying operating system. As a result, process creation time is of the order of microseconds and independent of the number of concurrently existing processes. Compare this with Java and C#, where for every process an underlying OS thread is created: you will get some very competitive comparisons, with Erlang greatly outperforming both languages.

Erlang processes communicate with each other through message passing. Regardless of the number of concurrent processes in your system, exchanging messages within the system takes microseconds. All that is involved in message passing is the copying of data from the memory space of one process to the memory space of the other, all within the same virtual machine. This differs from Java and C#, which work with shared memory, semaphores, and OS threads. Even here, benchmarks show that Erlang manages to outperform these languages for the same reasons it outperforms them in process creation times.

You might think that comparing Erlang to C# and Java is unfair to these two languages, as we are comparing apples and oranges. Well, you are right. Our point is that if you want to build massively concurrent systems, you should be using the tool that is best for the job, regardless of the underlying concurrency mechanism. As a result, the concurrency model of an Erlang program would differ from that of languages where process creation and message passing times are not as small. We describe the Erlang way of dealing with concurrency in Chapters 4 through 6, and Chapter 12.

Soft Real-Time Properties

Even though Erlang is a high-level language, you can use it for tasks with soft real-time constraints. Storage management in Erlang is automated, with garbage collection implemented on a per-process basis. This gives system response times on the order of milliseconds even in the presence of garbage-collected memory. Because of this, Erlang can handle high loads with no degradation in throughput, even during sustained peaks.

Robustness

How do you build a robust system? Although Erlang might not solve all your problems, it will greatly facilitate your task at a fraction of the effort of other programming languages. Thanks to a set of simple but powerful error-handling mechanisms and exception monitoring constructs, very general library modules have been built, with robustness designed into their core. By programming for the correct case and letting these libraries handle the errors, not only are programs shorter and easier to understand, but they will usually contain fewer bugs.

The libraries are collectively known as the *OTP middleware*. What exception monitoring and error-handling mechanisms do they contain, and what libraries are built on top of them?

- Erlang processes can be *linked* together so that if one crashes, the other will be informed, and then can either handle the crash or choose to crash itself.

- OTP provides a number of *generic behaviors*, such as servers, finite state machines, and event handlers. These worker processes have built-in robustness, since they handle all the general (and therefore difficult) concurrent parts of these patterns; all the user needs to do is to program the specific behavior of the particular server, which is much more straightforward to program than the general behavior.

- These generic behaviors are linked to a *supervisor behavior* whose only task is to monitor and handle process termination. OTP puts the idea of links into a framework whereby a process *supervises* other workers and supervisors, and may itself be supervised by yet another process, all in a hierarchical structure. Figure 1-2 illustrates a typical supervision tree.

- Using this supervision and linking, Erlang programmers can concentrate on programming for the correct case, and can *let the process fail* in any other circumstances. This avoidance of defensive programming makes a programmer's task much easier, as well as making it more straightforward to understand how a program behaves.

Although in this book we concentrate on Erlang and its error-handling mechanisms and exception monitoring properties, we also provide an introduction to the OTP design patterns in Chapter 12.

Distributed Computation

Erlang has distribution incorporated into the language's syntax and semantics, allowing systems to be built with location transparency in mind. The default distribution mode is based on TCP/IP, allowing a node (or Erlang runtime system) on a heterogeneous network to connect to any other node running on any operating system. As long as these nodes are connected through a TCP/IP network and the firewall has been correctly configured, the result is a fully meshed network of nodes, where all the nodes can communicate with each other.

As Erlang clusters were designed to execute behind firewalls, security is based on secret cookies with very few restrictions on access rights. You have the ability to create more disparate networks of distributed Erlang nodes using gateways, and if necessary, make them communicate using secure Internet protocols such as SSL.

Erlang programs consist of processes that communicate via message passing. When you start programming in Erlang, these will all be on one node, but as the syntax of sending a message within the node is the same as sending it to a remote node, you can

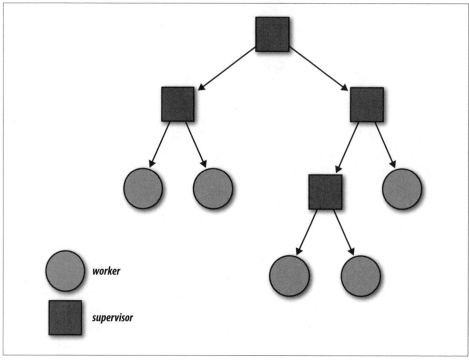

Figure 1-2. An example supervision tree

easily distribute your processes across a cluster of computers. With distribution built into the language, operations such as clustering, load balancing, the addition of hardware and nodes, communication, and reliability come with very little overhead and correspondingly little code.

Integration and Openness

You want to use the right tool for the right job. Erlang is an open language allowing you to integrate legacy code or new code where programming languages other than Erlang are more suitable for the job. As a result, there are mechanisms for interworking with C, Java, Ruby, and other programming languages, including Python, Perl, and Lisp.

High-level libraries allow Erlang nodes to communicate with nodes executing Java or C, making them appear and behave like distributed Erlang nodes. Other external languages can be tied in more tightly using drivers that are linked into the Erlang runtime system itself, as a device driver would be, and sockets can also be used for communication between Erlang nodes and systems written in other languages using popular protocols such as HTTP, SNMP, and IIOP.

The fact that distribution is built into Erlang means that integrating it with other systems is more natural than in other languages. The facilities for handling network data formats are an important part of the language and its libraries, rather than a bolted-on afterthought. The tracing and logging facilities also give you a clear picture of how the integration is working, enabling you to debug and tune systems much more effectively.

Erlang and Functional Programming

The recent success of Erlang is a success for functional programming, too, because it uses functional programming principles without making a big fuss about it: they are simply the right foundation on which to build a language with concurrency designed in from the start.

One of the prevalent myths in the community in the mid-1980s was that functional programming languages would be the only languages capable of working on the general-purpose parallel machines that were "just around the corner." It didn't turn out like that 20 years ago, but perhaps that's exactly what we are seeing now in the way that Erlang is being used to provide massive concurrency in server farms, cloud computing, and on the multicore processors inside all of our computers, from laptops on up.

Erlang and Multicore

The shift to multicore is inevitable. Parallelizing legacy C and Java code is very hard, and debugging parallelized C and Java is even harder...but what alternative is there?

The Erlang model for concurrency—separate processes with no shared memory communicating via message passing—naturally transfers to multicore processors in a way that is largely *transparent* to the programmer, so that you can run your Erlang programs on more powerful hardware without having to redesign them.

Symmetric multiprocessing (SMP) support in Erlang was first developed experimentally in the late 1990s, and is now an integral part of the standard release. The ethos of the Erlang/OTP development team at Ericsson is to make SMP work, measure its performance, find the bottlenecks, and optimize. Since releasing the first SMP-enabled version of Erlang, this has been their approach. Over recent releases, the virtual machine model has evolved from a single monolithic run queue—possibly with processes running on different processors—to a run queue for each processor, ensuring that the run queue is no longer a bottleneck for the system, as illustrated in Figure 1-3. As more complex processors emerge, the runtime system will be able to evolve with them.

The goal with Erlang's SMP is to hide the problems and awareness of SMP from the programmer. Programmers should develop and structure their code as they have always done, optimally using concurrency and without having to worry about the underlying operating system and hardware. As a result, Erlang programs should run perfectly well on any system, regardless of the number of cores or processors.

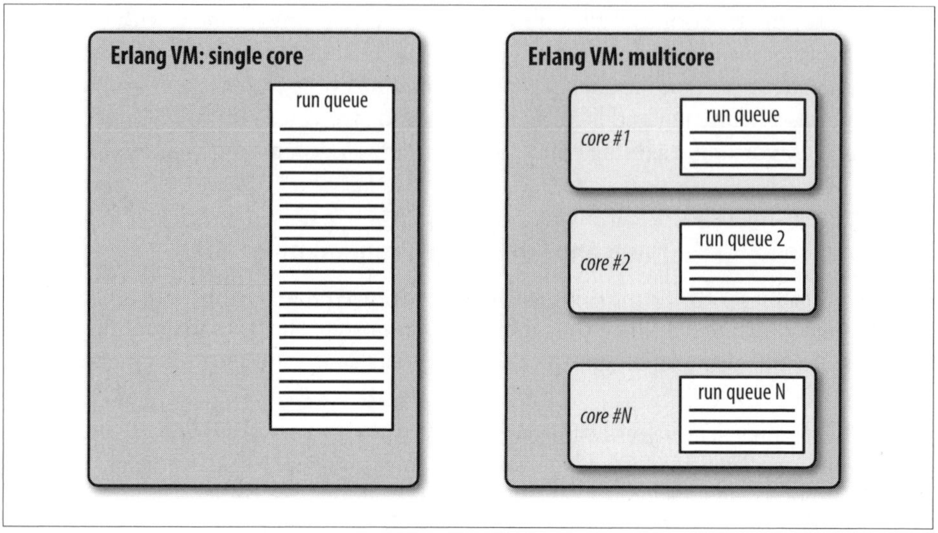

Figure 1-3. Run queues on a multicore processor

Case Studies

Let's start looking at how the features we just described have contributed to some of Erlang's successes. Ericsson's first major Erlang product was the AXD301 ATM switch; more recently, Erlang has been the key to implementing the CouchDB schema-free, document-oriented database. Finally, we report on a Motorola-based research project comparing the productivity of Erlang and C++ head on.

The AXD301 ATM Switch

The AXD301, a telephony-class 10–160 Gbps ATM switch, was designed and implemented from scratch in less than three years. At the heart of the AXD301 are more than 1.5 million lines of Erlang code, handling all the complex control logic, and overseeing operations and maintenance. This integrates with about half a million lines of C/C++ implementing low-level protocol and device drivers, much of it coming from third-party sources.

This ATM switch has been installed in networks all over the world, but the installation that shot to prominence was used by British Telecom to build what was at the time the largest "Voice over ATM" backbone in the world. According to an Ericsson press release issued at the end of the trial period, "Since cut-over of the first nodes in BT's network in January 2002 only one minor fault has occurred, resulting in 99.9999999% availability." The director of Ericsson's Next Generation Systems program, Bernt Nilsson, confirmed that "the network performance has been so reliable that there is almost a risk that our field engineers do not learn maintenance skills."

Experiences with the AXD301 suggest that "five nines" availability, downtime for software upgrades included, is a more realistic assessment. For nonstop operations, you need multiple computers, redundant power supplies, multiple network interfaces and reliable networks, cooling systems that never fail, and cables that system administrators cannot trip over, not to mention engineers who are well practiced in their maintenance skills. Considering that this target has been achieved at a fraction of the effort that would have been needed in a conventional programming language, it is still something to be very proud of.

How did Erlang contribute to the success of the AXD301? It supports incremental development, with the absence of side effects, making it easier to add or modify single components. Support for robustness and concurrency is built into the language and available from the start.

Erlang was very popular with the programming teams that found they were building much more compact code, thus dramatically improving their productivity. Experience from the project, although not scientifically documented, suggests that the Erlang code was 4 to 10 times shorter than similar systems written in C/C++, Java, and PLEX,[*] while the fault rate per thousand lines of code was the same.

Ericsson has gone on to use Erlang on other projects across the company, including a SIP telephony stack, control software for wireless base stations, telephony gateway controllers, media gateways, broadband solutions, and in GPRS and 3G data transmission. And these are just a few of the many we are allowed to talk about.

CouchDB

When Damien Katz decided to implement CouchDB, he wanted to be the one developing "cool stuff." He wanted to see whether he was good enough to develop something from scratch, pushing the code base to new levels. CouchDB is an open source database that provides a schema-less replicated document store, storing objects in JSON format and accessed through a RESTful interface.

He wrote the first version of CouchDB in C++. His system consisted of three components: a storage engine, a view engine, and a query language. The complexity of his components increased, and when he started hitting concurrency issues, he felt like he had hit a wall. He stumbled upon Erlang, downloaded it, and quickly came to the realization that it would solve his problems.

From the world Damien was coming from, Erlang initially sounded very complicated, and he believed it would be hard to learn. But when he got down to the details, what instead struck him was the simplicity of the language. Getting something to work with Erlang took extra effort compared to Java, as there were fewer tools and IDEs available,

---

[*] PLEX is a proprietary language developed by Ericsson and used extensively in the AXE-10 switches. Just like Erlang, many of its features were ahead of its time. It was never released to the public.

but to get something working *reliably* ended up taking much less talent and time than any of the other languages he knew.

Erlang gave Damien the features he needed for CouchDB at a fraction of the effort of using conventional languages. When migrating CouchDB to Erlang, he focused on the concurrency aspects and integrating it with his existing C++ components. He ended up replacing the entire C++ code base, as Erlang had all of the qualities he was looking for in a database application. They included support for intensive I/O, high reliability, and facilities for dealing with failure gracefully. The first benchmarks on the code, even before it was profiled, allowed in excess of 20,000 simultaneous connections. This compared pretty favorably with the 500 he expected to get on the C++ version!

Once it was released, CouchDB started getting lots of attention in the open source community. Damien made the decision to release the code under the Apache license, giving him the freedom he needed to continue development. Today, CouchDB is one of the best known Erlang open source applications currently being used in production systems worldwide.

What happened to Damien? He got a job with IBM, allowing him to continue developing CouchDB as an open source project. In Damien's words, now he is indeed "the guy who gets paid to work on cool stuff." You can read out more about CouchDB at *http://www.couchdb.org*.

Comparing Erlang to C++

Most experienced Erlang programmers will confirm that the Erlang programs they have written are substantially shorter than their counterparts in other mainstream programming languages used by the industry. Indeed, this was an urban legend among Ericsson programmers long before Erlang was released as open source. But until recently, there was very little scientific evidence to back up these claims. Quicksort (Chapter 9) using list comprehensions or remote procedure call server examples (Chapter 11), both of which we cover in this book, were used to argue the case. When comparing programming languages, however, you must benchmark whole systems in the application domain for which those languages were designed, not code snippets or simple functions.

Heriot-Watt University in the United Kingdom received an EPSRC[†] grant to study the impact of distributed functional programming languages in the telecom sector. When we first heard about this grant, our reaction was, why not speak with Ericsson and get it over with? We quickly changed our minds when we realized the research project was being done in cooperation with Motorola Labs, one of Ericsson's competitors. Although Heriot-Watt might have taken Ericsson's word that Erlang was suitable for programming telecom applications, Motorola wasn't having any of it.

[†] The Engineering and Physical Sciences Research Council provides U.K. government support for science research in universities.

The focus of the study consisted of two C++-based systems referred to as the Data Mobility (DM) component and the Dispatch Call Controller (DCC). These systems handled digital communication streams for pocket radio systems as used by emergency services. The DM been written with fault tolerance and reliability in mind. The implementation was done by good C++ programmers who based their development work on proprietary Motorola libraries. The DCC was an internal research prototype intended for evaluating the use of C++ and CORBA to gain scalability.

The Erlang rewrites were implemented by Jan Henry Nyström, an experienced Erlang programmer with an academic background. Two Erlang rewrites were done of the DM and only one of the DCC. The first DM implementation interfaced with Motorola's libraries, and the second was a pure Erlang implementation. The DCC was a pure Erlang implementation. Comparisons were made of the performance, robustness, productivity, and impact of the programming language constructs.

The interesting conclusions of this research came with the pure Erlang implementations. In the DM, there was an 85% reduction in code. This was explained by the fact that 27% of the C++ code consisted of defensive programming, 11% of memory management, and 23% of high-level communication, all features which in Erlang are part of the semantics of the language or are implemented in the OTP libraries. The DCC's code base was more in line with the folklore and urban legends, namely that it was about 70% smaller than its C++ counterpart.

The Erlang DM resulted in a 100% performance increase when compared to the C++ version, which crashed when severely overloaded. Although the throughput might sound surprising at first, it was a result of Erlang and its lightweight concurrency model being the right tool for the task. The mobility application in question had lots of concurrency, short messages, and little in terms of heavy processing and number crunching. The C++ implementation was never implemented to handle the loads it was subjected to, so as a result, the conclusion was that these load results might not be relevant and are certainly unfair to the C++ implementation. They do, however, demonstrate an important property of Erlang-based systems, which are stable under heavy loads and recover automatically when the load drops.

Although Erlang pioneers argued their case of shorter and more compact code based on experience, this study has finally provided empirical data to support the claims. A full report that fully confirms the "Erlang advantage" is available in "High-level distribution for the rapid production of robust telecoms software: comparing C++ and Erlang."[‡]

---

‡ Nyström, J.H., P.W. Trinder, and D.J. King. *Concurrency and Computation: Practice & Experience*, 20(8), 2008.

How Should I Use Erlang?

The philosophy used to develop Erlang fits equally well with the development of Erlang-based systems. Quoting Mike Williams, one of the three inventors of Erlang:

> Find the right methods—Design by Prototyping.

> It is not good enough to have ideas, you must also be able to implement them and know they work.

> Make mistakes on a small scale, not in a production project.

In line with these quotes, all successful Erlang projects should start with a *prototype*, and Erlang has all the support you need to get a prototype up and running quickly. Working prototypes usually cover a subset of the functionality and allow end-to-end tests of the system. If, for example, you were building an IM server (a recurring theme throughout this book), valid functionality to test could include the ability to sign on and send messages to a remote server without worrying about issues such as redundancy, persistency, and security.

Software development in Erlang is best achieved using an *agile* approach, incrementally delivering systems of increasing functionality over a short cycle period. Teams should be small in size, and, where possible, tests should be automated. The *tools* available with Erlang, discussed in the body of this book and in the section "Integration and Openness" on page 8, give excellent software development support. Testing is aided by EUnit for unit testing, and Common Test for system testing. Other tools include cover, providing coverage analysis, and Dialyzer, a static analysis tool that identifies software discrepancies such as type errors, dead code, and unsafe code.

If you're going to introduce Erlang to your organization, it can be a good strategy to *start small*, bringing in Erlang for a small project (or subsystem) where you can play to Erlang's strengths. This works particularly well for Erlang because it has distribution and *integration* designed in from the start, as we described in the section "Integration and Openness" on page 8, and virtually all production Erlang systems interwork with other languages and systems. Once you have achieved success on a small scale, you can start to think bigger!

The website for this book and the Appendix contain links about where you can go to learn more about Erlang itself, the tools that support program development, and the Erlang community. But now, it's time to get to work....

CHAPTER 2

Basic Erlang

This chapter is where we start covering the basics of Erlang. You may expect we'll just be covering things you have seen before in programming languages, but there will be some surprises, whether your background is in C/CC++, Java, Python, or functional programming. Erlang has assignment, but not as you know it from other imperative languages, because you can assign to each variable only *once*. Erlang has pattern matching, which not only determines control flow, but also binds variables and pulls apart complex data structures. Erlang pattern matching is different in subtle ways from other functional languages. So, you'll need to read carefully! We conclude the chapter by showing how to define Erlang functions and place them into modules to create programs, but we start by surveying the basic data types in Erlang.

Integers

Integers in Erlang are used to denote whole numbers. They can be positive or negative and expressed in bases other than 10. The notion of a maximum size of integers in Erlang does not exist, and so arbitrarily large whole numbers can be used in Erlang programming. When large integers do not fit in a word, they are internally converted to representation using an arbitrary number of words, more commonly known as *bignums*. While bignums give completely accurate calculation on arbitrary-size integers, this makes their implementation less efficient than fixed-size integers. The only limit on how large an integer can become depends on the physical constraints of the machine, namely the available memory. Some examples of integers include:

```
-234 0 10 100000000
```

To express integers in a base other than 10, the `Base#Value` notation is used. The base is an integer between 2 and 16, and the value is the number in that base; for example, `2#1010` denotes 10 in base 2, and `-16#EA` denotes −234 in base 16, since the letters `A` through `F` are used to denote the numbers 10 through 15 in base 16:

```
2#1010 -16#EA
```

To express characters as ASCII values, the $Character notation is used. $Character returns the ASCII value of Character. $a represents the integer 97 and $A represents the integer 65. The ASCII value representation of a newline, $\n, is 10:

```
$a $A $\n
```

The Erlang Shell

Start an Erlang shell by typing **erl** at the command prompt in a Unix shell, or in Windows by clicking the Erlang icon in the Start menu. More details about obtaining and running Erlang are given in the Appendix. When you get the Erlang command prompt (of the form number>), try typing some integers in their various notations. Do not forget to terminate your expression with a period or full stop (.), and then press the Enter key:

```
1> -234.
-234
2> 2#1010.
10
3> $A.
65
```

If you do not type a full stop at the end of your input, the Erlang shell will not evaluate what you have typed and will continue to collect input until you type a terminating full stop and press Enter:

```
4> 5-
4>
4> 4.
1
```

The 1>, 2>, and so on are the command prompts, which show that the Erlang shell is ready to receive an input expression. When you press Enter, and the line you typed in is terminated by a full stop, the shell will evaluate what you typed, and, if successful, will display the result. Note how the various integer notations are all translated and displayed in base 10 notation. If you type an invalid expression, you will get an error, as in:

```
4> 5-.
* 1: syntax error before: '.'
5> q().
```

Ignore errors for the time being, as we will cover them in Chapter 3. To recover from having made an error, just press the Enter key a few times, add a full stop, and terminate with a final press of the Enter key. If you want to exit the shell, just type **q()** followed by a full stop.

Floats

Floats in Erlang are used to represent real numbers. Some examples of floats include:

    17.368 -56.654 1.234E-10.

The `E-10` is a conventional floating-point notation stating that the decimal point has been moved 10 positions to the left: `1.234E-10` is the same as writing $1.234×10^{-10}$, namely 0.000000001234. The precision of the floats in Erlang is given by the 64-bit representation in the IEEE 754–1985 standard. Before going on to the next section, try typing a few floats in the Erlang shell.

 Soft real-time aspects of telecom applications rarely rely on floats. So historically, the implementation of efficient floating-point operations was a low priority for the Erlang virtual machine (VM). When Björn Gustavsson, one of the VM's maintainers, started working on a hobby project focused on modeling 3D graphics, Wings3D, he was not satisfied with the performance. Operations on real numbers suddenly became much more efficient. This was, of course, a great boon for anyone doing real number computations (e.g., graphics) in Erlang.

Mathematical Operators

Operations on integers and floats include addition, subtraction, multiplication, and division. As previously shown, + and − can be used as unary operators in front of expressions of the format *Op Expression*, such as `-12` or `+12.5`. Operations on integers alone always result in an integer, except in the case of floating-point division, where the result is a float. Using `div` will result in an integer without a remainder, which has to be computed separately using the `rem` operator. Table 2-1 lists the arithmetic operators.

Table 2-1. Arithmetic operators

Type	Description	Data type
+	Unary +	Integer | Float
−	Unary −	Integer | Float
*	Multiplication	Integer | Float
/	Floating-point division	Integer | Float
div	Integer division	Integer
rem	Integer remainder	Integer
+	Addition	Integer | Float
−	Subtraction	Integer | Float

All mathematical operators are left-associative. In Table 2-1, they are listed in order of precedence. The unary + and – have the highest precedence; multiplication, division, and remainder have the next highest precedence; and addition and subtraction have the lowest precedence.

So, for example, evaluating -2 + 3 / 3 will first divide 3 by 3, giving the float 1.0, and then will add -2 to it, resulting in the float -1.0. You can see here that it is possible to add an integer to a float: this is done by first *coercing* the integer to a float before performing the addition.

To override precedence, use parentheses: (-2 + 3) * 4 will evaluate to 4, whereas -2 + 3 * 4 gives the result 10 and -(2 + 3 * 4) evaluates to -14.

Now, let's use the Erlang shell as a glorified calculator[*] and test these operators. Note the results, especially when mixing floats and integers or dealing with floating-point division. Try out various combinations yourself:

```
1> +1.
1
2> -1.
-1
3> 11 div 5.
2
4> 11 rem 5.
1
5> (12 + 3) div 5.
3
6> (12+3)/5.
3.00000
7> 2*2*3.14.
12.5600
8> 1 + 2 + 3 + 5 + 8.
19
9> 2*2 + -3*3.
-5
10> 1/2 + (2/3 + (3/4 + (4/5))) - 1.
1.71667
```

Before going on to the next section, try typing **2.0 rem 3** in the shell:

```
13> 2.0 rem 3.
** exception error: bad argument in an arithmetic expression
     in operator  rem/2
         called as 2.0 rem 3
```

You are trying to execute an operation on a float and an integer when the Erlang runtime system is expecting two integers. The error you see is typical of errors returned by the runtime system. We will cover this and other errors in Chapter 3. If you come from a

---

[*] Based on personal experience and confirmed by threads on the Erlang-questions mailing list, this is more common than you might first believe.

C or a Java background, you might have noticed that there is no need to convert integers to floats before performing floating-point division.

Atoms

Atoms are constant literals that stand for themselves. Atoms serve much the same purpose as values in enumeration types in other languages; for the beginner, it sometimes helps to think of them as a huge enumeration type. To compare with other languages, the role of atoms is played by `#define` constants in C and C++, by "static final" values in Java, and by "enums" in Ruby.

The only operations on atoms are comparisons, which are implemented in a very efficient way in Erlang. The reason you use atoms in Erlang, as opposed to integers, is that they make the code clear yet efficient. Atoms remain in the object code, and as a result, debugging becomes easier; this is not the case in C or C++ where the definitions are only introduced by the preprocessor.

Atoms start with a lowercase letter or are delimited by single quotes. Letters, digits, the "at" symbol (@), the full stop (.), and underscores (_) are valid characters if the atom starts with a lowercase letter. Any character code is allowed within an atom if the atom is encapsulated by single quotes. Examples of atoms starting with a lowercase letter include:

```
january fooBar alfa21 start_with_lower_case node@ramone true false
```

When using quotes, examples include:

```
'January' 'a space' 'Anything inside quotes{}#@ \n\012'
'node@ramone.erlang-consulting.com'
```

 The concept of atoms in Erlang was originally inspired—as were a number of aspects of the language—by the logic programming language Prolog. They are, however, also commonly found in functional programming languages.

Now try typing some atoms in the shell. If at any point, the shell stops responding, you have probably opened a single quote and forgotten to close it. Type `'.` and press Enter to get back to the shell command line. Experiment with spaces, funny characters, and capital letters. Pay special attention to how the quotes are (and are not) displayed and how and where the expressions are terminated with a full stop:

```
1> abc.
abc
2> 'abc_123_CDE'.
abc_123_CDE
3> 'using spaces'.
'using spaces'
4> 'lowercaseQuote'.
```

```
lowercaseQuote
5> '\n\n'.
'\n\n'
6> '1
6> 2
6> 3
6> 4'.
'1\n2\n3\n4'
7> 'funny characters in quotes: !"£$%^&*()-='.
'funny characters in quotes: !"£$%^&*()-='
8> '1+2+3'.
'1+2+3'
9> 'missing a full stop.'
9> .
'missing a full stop.'
```

Booleans

There are no separate types of Boolean values or characters in Erlang. Instead of a Boolean type, the atoms true and false are used together with Boolean operators. They play the role of Booleans in being the results of tests, and in particular, comparisons:

```
1> 1==2.
false
2> 1<2.
true
3> a>z.
false
4> less<more.
true
```

Atoms are ordered in lexicographical order. We give more details of these comparisons later in this chapter. Erlang has a wide variety of built-in functions, usually called BIFs in the Erlang community, which can be used in your programs and in the shell. The built-in function is_boolean gives a test of whether an Erlang value is a Boolean:

```
5> is_boolean(9+6).
false
6> is_boolean(true).
true
```

Complex tests can be formed using the logical operators described in Table 2-2.

Table 2-2. Logical operators

Operator	Description
and	Returns true only if both arguments are true
andalso	Shortcut evaluation of and: returns false if the first argument is false, without evaluating the second
or	Returns true if either of the arguments is true
orelse	Shortcut evaluation of or: returns true if the first argument is true, without evaluating the second

Operator	Description
xor	"Exclusive or": returns true if one of its arguments is true and the other false
not	Unary negation operator: returns true if its argument is false, and vice versa

In the following code, the binary operators are *infixed*, or placed between their two arguments:

```
1> not((1<3) and (2==2)).
false
2> not((1<3) or (2==2)).
false
3> not((1<3) xor (2==2)).
true
```

Tuples

Tuples are a composite data type used to store a collection of items, which are Erlang data values but which do not have to all be the same type. Tuples are delimited by curly brackets, {...}, and their elements are separated by commas. Some examples of tuples include:

```
{123, bcd} {123, def, abc} {abc, {def, 123}, ghi} {}
{person, 'Joe', 'Armstrong'} {person, 'Mike', 'Williams'}
```

The tuple {123,bcd} has two elements: the integer 123 and the atom bcd. The tuple {abc, {def, 123}, ghi} has three elements, as the tuple {def, 123} counts as one element. The empty tuple {} has no elements. Tuples with one element, such as {123}, are also allowed, but because you could just use the element on its own "untupled," it's not a good idea in general to use them in your code.

In a tuple, when the first element is an atom, it is called a *tag*. This Erlang convention is used to represent different types of data, and will usually have a meaning in the program that uses it. For example, in the tuple {person, 'Joe', 'Armstrong'}, the atom person is the tag and might denote that the second field in the tuple is always the first name of the person, while the third is the surname.

The use of a first position tag is to differentiate between tuples used for different purposes in the code. This greatly helps in finding the cause of errors when the wrong tuple has been mistakenly passed as an argument or returned as a result of a function call. This is considered a best practice for Erlang.

A number of built-in functions are provided to set and retrieve elements as well as get the tuple size:

```
1> tuple_size({abc, {def, 123}, ghi}).
3
2> element(2,{abc, {def, 123}, ghi}).
{def,123}
3> setelement(2,{abc, {def, 123}, ghi},def).
```

```
{abc,def,ghi}
4> {1,2}<{1,3}.
true
5> {2,3}<{2,3}.
false
6> {1,2}=={2,3}.
false
```

In command 2 note that the elements of the tuple are indexed from 1 rather than zero. In the third example, the result is a *new* tuple, with a different value—def—in the second position, and the same values as the old tuple in the other positions. These functions are all *generic* in that they can be used over any kind of tuple, of any size.

Before starting to look at lists, make sure you experiment and get better acquainted with tuples and the tuple BIFs in the Erlang shell.

Lists

Lists and tuples are used to store collections of elements; in both cases, the elements can be of different types, and the collections can be of any size. Lists and tuples are very different, however, *in the way that they can be processed.* We begin by describing how lists are denoted in Erlang, and examine the way that strings are a special kind of list, before explaining in detail how lists can be processed.

Lists are delimited by square brackets, [...], and their elements are separated by commas. Elements in lists do not have to be of the same data type and, just like tuples, can be freely mixed. Some examples of lists include:

```
[january, february, march]
[123, def, abc]
[a,[b,[c,d,e],f], g]
[]
[{person, 'Joe', 'Armstrong'}, {person, 'Robert', 'Virding'},
 {person, 'Mike', 'Williams'}]
[72,101,108,108,111,32,87,111,114,108,100]
[$H,$e,$l,$l,$o,$ ,$W,$o,$r,$l,$d]
"Hello World"
```

The list [a,[b,[c,d,e],f], g] is said to have a length of 3. The first element is the atom a, the second is the list [b,[c,d,e],f], and the third is the atom g. The empty list is denoted by [], while [{person, 'Joe', 'Armstrong'}, {person, 'Robert', 'Virding'}, {person, 'Mike', 'Williams'}] is a list of tagged tuples.

Characters and Strings

Characters are represented by integers, and strings (of characters) are represented by lists of integers. The integer representation of a character is given by preceding the character with the $ symbol:

```
1> $A.
65
2> $A + 32.
97
3> $a.
97
```

There is no string data type in Erlang. Strings are denoted by lists of ASCII values and represented using the double quotes (") notation. So, the string "Hello World" is in fact the list [72,101,108,108,111,32,87,111,114,108,100]. And if you denote the integers using the ASCII integer notation $Character, you get [$H,$e,$l,$l,$o,$,$W,$o,$r,$l, $d]. The empty string "" is equivalent to the empty list []:

```
4> [65,66,67].
"ABC"
5> [67,$A+32,$A+51].
"Cat"
6> [72,101,108,108,111,32,87,111,114,108,100].
"Hello World"
7> [$H,$e,$l,$l,$o,$ ,$W,$o,$r,$l,$d].
"Hello World"
```

Strings and Binaries

Telecom applications do not rely on string operations, and as a result, strings were never included as a data type in Erlang. Every character in Erlang consumes 8 bytes in the 32-bit emulator (and 16 in the 64-bit emulator), ensuring that characters and strings are not stored in a memory-efficient way.

Erlang does include *binaries*, which we discuss in Chapter 9, and these are recommended for representing long strings, particularly if they are being transported by an application rather than being analyzed in any way. Recent releases of Erlang have improved the speed of binary processing, and it looks like this will continue.

This implementation has not stopped Erlang from making inroads in string-intensive applications, however. We have implemented websites that handle thousands of dynamic web pages per second and systems in which the XML API requires the parsing of thousands of SOAP requests per second. These systems run on small hardware clusters you can buy off eBay for a few hundred dollars.

As the spread of Erlang continues into new problem domains, the current implementation will probably be more of an issue, but so far very few production systems have suffered as a result of this implementation choice.

Atoms and Strings

What is the difference between atoms and strings? First, they can be processed in different ways: the only thing you can do with atoms is compare them, whereas you can process strings in a lot of different ways. The string "Hello World" can be split into its list of constituent words—["Hello", "World"], for instance; you can't do the same for the atom 'Hello World'.

You could use a string to play the role of an atom, that is, as a constant literal. However, another difference between atoms and strings is efficiency. Representation of a string takes up space proportional to the string's size, whereas atoms are represented in a system table and take a couple of bytes to reference regardless of their size. If a program is to compare two strings (or lists), it needs to compare the strings character by character while traversing both of them. When comparing atoms, however, the runtime system compares an internal identifier in a single operation.

Building and Processing Lists

As we said earlier, lists and tuples are processed in very different ways. A tuple can be processed only by extracting particular elements, whereas when working with lists, it is possible to break a list into a *head* and a *tail*, as long as the list is not empty. The head refers to the first element in the list, and the tail is another list that contains all the remaining elements; this list can itself be processed further. This is illustrated in Figure 2-1.

Just as a list can be split in this way, it is possible to build or construct a list from a list and an element. The new list is constructed like this—[Head|Tail], which is an example of a *cons*, short for *constructor*.

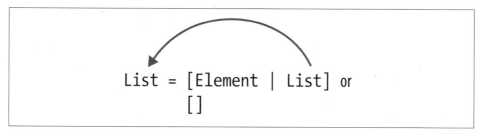

Figure 2-1. *The recursive definition of lists*

So, if you take the list [1,2,3], the head would be 1 and the tail would be [2,3]. Using the cons operator, the list can be represented as [1|[2,3]]. Breaking the tail further, you would get [1|[2|[3]]] and [1|[2|[3|[]]]]. A final valid notation for this list is of the format [1,2|[3|[]]], where you can have more than one element separated by commas before appending the tail with the cons operator. All of these lists are equivalent to the original list [1,2,3]. If the last tail term is the empty list, you have a *proper* or *well-formed* list.

When learning Erlang, the recursive definition of lists is the first hurdle that people can stumble on. So, just to be on the safe side, here is one more example where all of the lists are semantically equivalent:

```
[one, two, three, four]
[one, two, three, four|[]]
[one, two|[three, four]]
[one, two|[three|[four|[]]]]
[one|[two|[three|[four|[]]]]]
```

Note that you must have an element on the lefthand side of the cons operator and a list on the right, both within the square brackets, for the result to be a proper or well-formed list.

In fact, lists in Erlang do not have to be proper, meaning that the tail does not necessarily have to be a list. Try typing **[[1, 2]|3]** in the shell. What is the result? Expressions such as [1|2] and [1,2|foo] are syntactically valid Erlang data structures, but are of only limited value.[†] Nonproper lists can be useful in supporting demand-driven or lazy programming, and we talk about that in Chapter 9. Apart from this, it is one of the conventions of Erlang programming that use of nonproper lists should be avoided. That is because it is normally impossible to determine just by inspecting the code whether their use was intentional or an error. Writing [2|3] instead of [2|[3]], for example, results in a valid Erlang expression that compiles without any errors. It will, however, generate a runtime error when the tail of the list is treated as a list and not as an atom.

List Functions and Operations

Lists are one of the most useful data types in Erlang, and, especially in combination with tuples, they can be used to represent all sorts of complex data structures. In particular, lists are often used to represent collections of objects, which can be split into other collections, combined, and analyzed.

Many operations on lists are defined in the lists library module, and you can see some examples in the following shell session. These functions are not BIFs, and so are called by putting the module name in front of the function, separated by a colon (:) as in lists:split. The effect of the functions should be clear from the examples. You'll see how to define functions such as this in the next chapter:

```
1> lists:max([1,2,3]).
3
2> lists:reverse([1,2,3]).
[3,2,1]
3> lists:sort([2,1,3]).
[1,2,3]
4> lists:split(2,[3,4,10,7,9]).
{[3,4],[10,7,9]}
5> lists:sum([3,4,10,7,9]).
33
6> lists:zip([1,2,3],[5,6,7]).
[{1,5},{2,6},{3,7}]
7> lists:delete(2,[1,2,3,2,4,2]).
[1,3,2,4,2]
```

[†] Other than winding up people who are on a mission to find design flaws in Erlang.

```
8> lists:last([1,2,3]).
3
9> lists:member(5,[1,24]).
false
10> lists:member(24,[1,24]).
true
11> lists:nth(2,[3,4,10,7,9]).
4
12> lists:length([1,2,3]).
** exception error: undefined function lists:length/1
13> length([1,2,3]).
3
```

We said these are not BIFs: the exception to this is length, as you can see in commands 12 and 13.

There are also three operators on lists. You already saw the [...|...] operator, but there are also ++ and --, which join lists and "subtract" one list from another. Here are some examples:

```
1> [monday, tuesday, Wednesday].
[monday,tuesday,wednesday]
2>
2> [1|[2|[3|[]]]].
[1,2,3]
3> [a, mixed, "list", {with, 4}, 'data types'].
[a,mixed,"list",{with,4},'data types']
4> [1,2,3] ++ [4,5,6].
[1,2,3,4,5,6]
5> [1,2,2,3,4,4] -- [2,4].
[1,2,3,4]
6> "A long string I have split "
6> "across several lines.".
"A long string I have split across several lines."
```

The ++ operator takes two lists and joins them together into a new list. So, writing [1,2] ++ [3,4] will return [1,2,3,4].

The -- operator individually subtracts each element in the list on the righthand side from the list on the lefthand side. So, [1,1] -- [1] returns [1], whereas [1,2,3,4] -- [1,4] returns [2,3]. If you evaluate [1,2] -- [1,1,3], you get the list [2]. This is because if elements on the list on the righthand side of the operation do not match, they are ignored. Both ++ and -- are right-associative, and so an expression of the form [1,2,3]--[1,3]--[1,2] will be bracketed to the right:

```
7> [1,2,3]--[1,3]--[1,2].
[1,2]
8> ([1,2,3]--[1,3])--[1,2].
[]
```

Finally, writing "Hello " "Concurrent " "World" will result in the compiler appending the three strings together, returning "Hello Concurrent World".

If you want to add an element to the beginning of a list, you can do it in two ways:

- Using cons directly, as in [1|[2,3,4]].
- Using ++ instead, as in [1] ++ [2,3,4].

Both have the same effect, but ++ is less efficient and can lead to programs running substantially more slowly. So, when you want to add an element to the head of the list, you should always use cons ([...|...]) because it is more efficient.

The proplists module contains functions for working with property lists. *Property lists* are ordinary lists containing entries in the form of either tagged tuples, whose first elements are keys used for lookup and insertion, or atoms (such as blah), which is shorthand for the tuple {blah, true}.

To make sure you have grasped what's in this section, start the shell and test what you just learned about lists and strings:

- Pay particular attention to the way lists are built using [...|...], which some readers may struggle with the first time around. It is important that you understand how this works, as recursion, covered in Chapter 3, builds heavily on it.
- Look at lists that are not proper, because the next time you might come across them, they will probably be in the form of a bug.
- The append, subtract, and string concatenation operators will make your code more elegant, so make sure you spend some time getting acquainted with them as well.
- Remember that if the shell does not return the string you typed in, you probably forgot to close the double quotes. Type ". and press Enter a few times.
- Also, what happens when you type in a list of ASCII values? How does the shell display them?

We'll finish this section with a short discussion of the history of strings in the Erlang system.

Before the string concatenation construct was added to the language, programmers would make their strings span many lines. When the code became unreadable, they would often break the strings into manageable chunks and concatenate them using the append function in the lists library module.

When the ++ notation was added to the language, programmers went from using the append function to abusing the ++ operator. The ++ operator and the append function are expensive operations, as the list on the lefthand side of the expression has to be traversed. Not only are they expensive operations, but often they are redundant, as all I/O functions (including socket operations) in Erlang accept nonflat strings such as ["Hello ",["Concurrent "]|"World"].

Term Comparison

Term comparisons in Erlang take two expressions on either side of the comparison operator. The result of the expression is one of the Boolean atoms true or false. The equal (==) and not equal (/=) operators compare the values on either side of the operator without paying attention to the data types. Typing **1 == one** returns false, whereas **one == one** returns true.

Comparisons such as 1 == 1.0 will return true and 1 /= 1.0 will return false, as integers are converted to floats before being compared with them in such a comparison. You get around this by using the operators *exactly equal to* and *not exactly equal to*, as these operators compare not only the values on either side of the equation, but also their data types. So, for example, 1 =:= 1.0 and 1 =/= 1 will both return false, and 1 =/= 1.0 returns true.

As well as comparisons for (in)equality, you can compare the ordering between values, using < (less than), =< (less than or equal to), > (greater than), and >= (greater than or equal to). Table 2-3 lists the comparison operators.

Table 2-3. Comparison operators

Operator	Description
==	Equal to
/=	Not equal to
=:=	Exactly equal to
=/=	Exactly not equal to
=<	Less than or equal to
<	Less than
>=	Greater than or equal to
>	Greater than

If the expressions being compared are of different types, the following hierarchy is taken into consideration:

> number < atom < reference < fun < port < pid < tuple < list < binary

This means, for instance, that any number will be smaller than any atom and any tuple will be smaller than any list:

```
3> 11<ten.
true
4> {123,345}<[].
true
```

Lists are ordered *lexicographically*, like the words in a dictionary. The first elements are compared, and whichever is smaller indicates the smaller list: if they are the same, the second elements are compared, and so on. When one list is exhausted, that is the smaller list. So:

```
5> [boo,hoo]<[adder,zebra,bee].
false
6> [boo,hoo]<[boo,hoo,adder,zebra,bee].
true
```

On the other hand, when comparing tuples, the number of elements in the constructs is compared first, followed by comparisons of the individual values themselves:

```
7> {boo,hoo}<{adder,zebra,bee}.
true
8> {boo,hoo}<{boo,hoo,adder,zebra,bee}.
true
```

The ability to compare values from different data types allows you to write generic functions such as sort, where regardless of the heterogeneous contents of a list, the function will always be able to sort its elements. For the time being, do not worry about references, funs, ports, and binaries. We will cover these data types in Chapter 9 and Chapter 15.

Using the *exactly equal* and *not exactly equal* operators will provide the compiler and type tools with more information and result in more efficient code. Unfortunately, =:= and =/= are not the prettiest of operators and tend to make the code ugly. As a result, the *equal* and *not equal* operators are commonly used in programs, including many of the libraries that come with the Erlang runtime system.

Start the Erlang shell and try some of the comparison operators. Though not included in the following examples, try testing with different data types and comparing the results with the various equality operators. Make sure you also become acquainted with how the operators work with different data types:

```
1> 1.0 == 1.
true
2> 1.0 =:= 1.
false
3> {1,2} < [1,2].
true
4> 1 =< 1.2.
true
5> 1 =/= 1.0.
true
6> (1 < 2) < 3.
false
7> (1 > 2) == false.
true
```

Variables

Variables are used to store values of simple and composite data types. In Erlang, they always start with an uppercase letter,[‡] followed by upper- and lowercase letters, integers, and underscores. They may not contain other "special" characters. Examples of variables include the following:

```
A_long_variable_name Flag Name2 DbgFlag
```

Erlang variables differ from variables in most conventional programming languages. In the scope of a function, including the Erlang shell process, *once you've bound a variable, you cannot change its value*. This is called *single assignment*. So, if you need to do a computation and manipulate the value of a variable, you need to store the results in a new variable. For example, writing the following:

```
Double = 2,
Double = Double * Double
```

would result in a runtime error, because `Double` is already bound to the integer `2`. Trying to bind it to the integer `4` fails as it is already bound. As mentioned, the way around this feature is to bind the results in a fresh variable:

```
Double = 2,
NewDouble = Double * Double
```

Single assignment of variables might feel awkward at first, but you'll get used to it very quickly. It encourages you to write shorter functions and puts in place a discipline that often results in code with fewer errors. It also makes debugging of errors related to incorrect values easy, as tracing the source of the error to the place where the value was bound can lead to only one place.

All calls with variables in Erlang are *call by value*: all arguments to a function call are evaluated before the body of the function is evaluated. The concept of call by reference does not exist, removing one way in which side effects can be caused.[§] All variables in Erlang are considered local to the function in which they are bound. Global variables do not exist, making it easier to debug Erlang programs and reduce the risk of errors and bad programming practices.

Another useful feature of Erlang variables is that there is no need to declare them: you just use them. Programmers coming from a functional programming background are used to this, whereas those coming from a C or Java background will quickly learn to appreciate it. The reason for not having to declare variables is that Erlang has a *dynamic type system*.[‖] Types are determined at runtime, as is the viability of the operation you

---

[‡] Variables can also begin with an underscore; these play a role in pattern matching and are discussed in the section "Pattern Matching" on page 33.

[§] We cover side effects and destructive operations later in the book.

[‖] Other languages can avoid variable declarations for other reasons. Haskell, for instance, uses a *type inference* algorithm to deduce types of variables.

are trying to execute on the variable. The following code attempting to multiply an atom by an integer will compile (with compiler warnings), but will result in a runtime error when you try to execute it:

```
Var = one,
Double = Var * 2
```

At first, using variables that start with capital letters might feel counterintuitive, but you'll get used to it quickly. After a few years of programming in Erlang, when reading C code, don't be surprised if you react over the fact that whoever wrote the code used atoms instead of variables. It has happened to us!

Before using variables, remember: *variables can be bound only once!* This might be a problem in the Erlang shell, as programs are meant to run nonstop for many years and the same shell is used to interact with them. Two operations can be used as a workaround to this problem. Using f() forgets all variable bindings, whereas f(Variable) will unbind a specific Variable. You can use these operations only in the shell. Attempts to include them in your programs are futile and will result in a compiler error:

```
1> A = (1+2)*3.
9
2> A + A.
18
3> B = A + 1.
10
4> A = A + 1.
** exception error: no match of right hand side value 10
5> f(A).
ok
6> A.
** 1: variable 'A' is unbound **
```

In fact, what you see here with assignment to variables is a special case of pattern matching, which we'll discuss shortly.

The Erlang Type System

The reason for not having a more elaborate type system is that none of the Erlang inventors knew how to write one, so it never got done. The advantage of a static type system is that errors can be predicted at compile time rather than at runtime, therefore allowing faults to be detected earlier and fixed at a lower cost.

A number of people have tried to build a static type system for Erlang. Unfortunately, due to design decisions taken when Erlang was invented, no project has been able to write a comprehensive type system, since with hot code loading, this is intrinsically difficult. To quote Joe Armstrong in one of the many type system flame wars, "It seems like it should be 'easy'—and indeed, a few weeks programming can make a type system that handles 95% of the language. Several man-years of work [by some of the brightest minds in computer science] have gone into trying to fix up the other 5%—but this is really difficult."

If the right tools are available, it's possible to detect many of the faults in a program without running it. An excellent tool that resulted from research related to an Erlang type system by Uppsala University is TypEr, which can infer types of Erlang functions. TypEr, taken together with the Dialyzer tool, which also came out of Uppsala University and can pick up other faults at compile time, results in a powerful mechanism for finding faults in Erlang programs. We discuss these tools in Chapter 18.

Complex Data Structures

When we refer to Erlang *terms*, we mean legal *data structures*. Erlang terms can be simple data values, but we often use the expression to describe arbitrarily complex data structures.

In Erlang, complex data structures are created by nesting composite data types together. These data structures may contain bound variables or the simple and composite values themselves. An example of a list containing tuples of type person (tagged with the atom person) with the first name, surname, and a list of attributes would look like this:

```
[{person,"Joe","Armstrong",
    [ {shoeSize,42},
      {pets,[{cat,zorro},{cat,daisy}]},
      {children,[{thomas,21},{claire,17}]}]
    },
  {person,"Mike","Williams",
    [ {shoeSize,41},
      {likes,[boats,wine]}]
    }
]
```

Or, if we were to write it in a few steps using variables, we would do it like this. Note how, for readability, we named the variables with their data types:

```
1> JoeAttributeList = [{shoeSize,42},  {pets,[{cat, zorro},{cat,daisy}]},
1>                        {children,[{thomas,21},{claire,17}]}].
 [{shoeSize,42},
  {pets,[{cat,zorro},{cat,daisy}]},
  {children,[{thomas,21},{claire,17}]}]
2> JoeTuple = {person,"Joe","Armstrong",JoeAttributeList}.
{person,"Joe","Armstrong",
        [{shoeSize,42},
         {pets,[{cat,zorro},{cat,daisy}]},
         {children,[{thomas,21},{claire,17}]}]}
3> MikeAttributeList = [{shoeSize,41},{likes,[boats,wine]}].
[{shoeSize,41},{likes,[boats,wine]}]
4> MikeTuple = {person,"Mike","Williams",MikeAttributeList}.
{person,"Mike","Williams",
        [{shoeSize,41},{likes,[boats,wine]}]}
5> People = [JoeTuple,MikeTuple].
[{person,"Joe","Armstrong",
        [{shoeSize,42},
         {pets,[{cat,zorro},{cat,daisy}]},
```

```
        {children,[{thomas,21},{claire,17}]}]},
    {person,"Mike","Williams",
          [{shoeSize,41},{likes,[boats,wine]}]}]
```

One of the beauties of Erlang is the fact that there is no explicit need for memory allocation and deallocation. For C programmers, this means no more sleepless nights hunting for pointer errors or memory leakages. Memory to store the complex data types is allocated by the runtime system when needed, and deallocated automatically by the garbage collector when the structure is no longer referenced.

Memory Management in Erlang

When the first Erlang-based products were being developed in 1993, critics said it was madness to use a language compiling for a VM with a garbage collector (just like Java!) for soft real-time systems. The VM automatically handles the task of allocating memory for the system, and more importantly, it recycles that memory when it is no longer needed (hence the term "garbage collection"). It is thanks to the design of the garbage collector, however, that the soft real-time properties of these systems are not affected.

The current implementation of the Erlang VM uses a copying, generational garbage collector. The garbage collection is done *separately for each concurrent process*: when no more memory is available for a particular process to store values, a garbage collection will be triggered.

A *copying* garbage collector works by having two separate areas (heaps) for storing data. When garbage collection takes place, the active memory is copied to the other heap, and the garbage left behind is overwritten in the other heap.

The garbage collector is also *generational*, meaning that it has several generations of the heap (in Erlang's case, two). A garbage collection can be shallow or deep. A shallow garbage collection looks only at data in the youngest generation; all data that survives three shallow garbage collections will be moved to the old generation. A deep garbage collection will occur only when a shallow collection fails to recycle enough memory or after a (VM version dependent) number of shallow collections.

Pattern Matching

Pattern matching in Erlang is used to:

- Assign values to variables
- Control the execution flow of programs
- Extract values from compound data types

The combination of these features allows you to write concise, readable yet powerful programs, particularly when pattern matching is used to handle the arguments of a function you're defining. A pattern match is written like this:

```
Pattern = Expression
```

And as we said earlier, it's a generalization of what you already saw when we talked about variables.

The `Pattern` consists of data structures that can contain both bound and unbound variables, as well as literal values (such as atoms, integers, or strings). A *bound* variable is a variable which already has a value, and an *unbound* variable is one that has not yet been bound to a value. Examples of patterns include:

```
Double
{Double, 34}
{Double, Double}
[true, Double, 23, {34, Treble}]
```

The `Expression` consists of data structures, bound variables, mathematical operations, and function calls. It may not contain unbound values.

What happens when a pattern match is *executed*? Two results are possible:

- The pattern match can *succeed*, and this results in the unbound variables becoming bound (and the value of the expression being returned).
- The pattern match can *fail*, and no variables become bound as a result.

What determines whether the pattern match succeeds? The `Expression` on the right-hand side of the = operator is first evaluated and then its value is compared to the `Pattern`:

- The expression and the pattern need to be of the same shape: a tuple of three elements can match only with a tuple of three elements, a list of the form [X|Xs] can match only with a nonempty list, and so on.
- The *literals* in the pattern have to be equal to the values in the corresponding place in the value of the expression.
- The *unbound* variables are bound to the corresponding value in the `Expression` if the pattern match succeeds.
- The *bound* variables also must have the same value as the corresponding place in the value of the expression.

Taking a concrete example, writing Sum = 1+2 where the variable Sum is unbound would result in the sum of 1 and 2 being calculated and compared to Sum. If Sum is unbound, pattern matching succeeds and Sum is bound to 3. Just to be clear, this would *not* bind 1 to Sum and then add 2 to it. If Sum is already bound, pattern matching will succeed if (and only if) Sum is already bound to 3.

Let's look at some examples in the shell:

```
1> List = [1,2,3,4].
[1,2,3,4]
```

In command 1, the pattern match succeeds, and binds the list [1,2,3,4] to the List variable:

```
2> [Head|Tail] = List.
[1,2,3,4]
3> Head.
1
4> Tail.
[2,3,4]
```

In command 2, the pattern match succeeds, because the List is nonempty, so it has a head and a tail which are bound to the variables Head and Tail. You can see this in commands 3 and 4.

```
5> [Head|Tail] = [1].
** exception error: no match of right hand side value [1]
6> [Head|Tail] = [1,2,3,4].
[1,2,3,4]
7> [Head1|Tail1] = [1].
[1]
8> Tail1.
[]
```

What goes wrong in command 5? It looks as though this should succeed, but the variables Head and Tail are bound already, so this pattern match becomes a *test* of whether the expression is in fact [1,2,3,4]; you can see in command 6 that this would succeed.

If you want to extract the head and tail of the list [1], you need to use variables that are not yet bound, and commands 7 and 8 show that this is now successful.

```
9> {Element, Element, X} = {1,1,2}.
{1,1,2}
10> {Element, Element, X} = {1,2,3}.
** exception error: no match of right hand side value {1,2,3}
```

What happens if a variable is repeated in a pattern, as in command 9? The first occurrence is unbound, and results in a binding: here to the value 1. The next occurrence is bound, and will succeed only if the corresponding value is 1. You can see that this is the case in command 9, but not in command 10, hence the "no match" error.

```
11> {Element, Element, _} = {1,1,2}.
{1,1,2}
```

As well as using variables, it is possible to use a wildcard symbol, _, in a pattern. This will match with anything, and produces no bindings.

```
12> {person, Name, Surname} = {person, "Jan-Henry", "Nystrom"}.
{person,"Jan-Henry","Nystrom"}
13> [1,2,3] = [1,2,3,4].
** exception error: no match of right hand side value [1,2,3,4]
```

Why do we use pattern matching? Take assignment of variables as an example. In Erlang, the expression Int = 1 is used to compare the contents of the variable Int to the integer 1. If Int is unbound, it gets bound to whatever the righthand side of the equation evaluates to, in this case 1. That is how variable assignment actually works. We are not assigning variables, but in fact pattern-matching them. If we now write Int = 1 followed by Int = 1+0, the first expression will (assuming Int is unbound) bind the variable Int to the integer 1. The second expression will add 1 to 0 and compare it to the contents of the variable Int, currently bound to 1. As the result is the same, the pattern matching will be successful. If we instead wrote Int = Int + 1, the expression on the righthand side would evaluate to 2. Attempting to compare it to the contents of Int would fail, as it is bound to 1.

Pattern matching is also used to pick the execution flow in a program. Later in this and in the following chapters, we will cover case statements, receive statements, and function clauses. In each of these constructs, pattern matching is used to determine which of the clauses has to be evaluated. In effect, we are testing a pattern match that either succeeds or fails. For example, the following pattern match fails:

```
{A, A, B} = {abc, def, 123}
```

The first comparison is to ensure that the data type on the righthand side of the expression is the same as the data type on the left, and that their size is the same. Both are tuples with three elements, so thus far, the pattern matching is successful. Tests are now done on the individual elements of the tuple. The first A is unbound and gets bound to the atom abc. The second A is now also bound to abc, so comparing it to the atom def will fail because the values differ.

Pattern matching [A,B,C,D] = [1,2,3] fails. Even if both are lists, the list on the lefthand side has four elements and the one on the right has only three. A common misconception is that D can be set to the empty list and the pattern matching succeeds. In this example, that would not be possible, as the separator between C and D is a comma and not the cons operator. [A,B, C|D] = [1,2,3] will pattern-match successfully, with the variables A, B, and C being bound to the integers 1, 2, and 3, and the variable D being bound to the tail, namely the empty list. If we write [A,B|C] = [1,2,3,4,5,6,7], A and B will be bound to 1 and 2, and C will be bound to the list containing [3,4,5,6,7]. Finally, [H|T] = [] will also fail, as [H|T] implies that the list has at least one element, when we are in fact matching against an empty list.

The last use of pattern matching is to extract values from compound data types. For example:

```
{A, _, [B|_], {B}} = {abc, 23, [22, 23], {22}}
```

will successfully extract the first element of the tuple, the atom abc, and bind it to the variable A. It will also extract the first element of the list stored in the third element of the tuple and bind it to the variable B.

In the following example

```
14> Var = {person, "Francesco", "Cesarini"}.
{person, "Francesco", "Cesarini"}
15 {person, Name, Surname} = Var.
{person, "Francesco", "Cesarini"}
```

we are binding a tuple of type `person` to the variable `Var` in the first clause and extracting the first name and the surname in the second one. This will succeed, with the variable `Name` being bound to the string `"Francesco"` and the variable `Surname` to `"Cesarini"`.

We mentioned earlier that variables can start with an underscore; these denote *"don't care" variables*, which are placeholders for values the program does not need. "Don't care" variables behave just like normal variables—their values can be inspected, used, and compared. The only difference is that compiler warnings are generated if the value of the normal variable is never used. Using "don't care" variables is considered good programming practice, informing whoever is reading the code that this value is ignored. To increase readability and maintainability, one often includes the value or type in the name of a "don't care" variable. The underscore on its own is also a "don't care" variable, but its contents cannot be accessed: its values are ignored and never bound.

When pattern matching, note the use of the "don't care" variables, more specifically in the following example:

```
{A, _, [B|_], {B}} = {abc, 23, [22, 23], {22}}
```

As _ is never bound, it does not matter whether the values you are matching against are different. But writing:

```
{A, _int, [B|_int], {B}} = {abc, 23, [22, 23], {22}}
```

completely changes the semantics of the program. The variable `_int` will be bound to the integer 23 and is later compared to the list containing the integer 23. This will cause the pattern match to fail.

 Using variables that start with an underscore makes the code more legible, but inserts potential bugs in the code when they are mistakenly reused in other clauses in the same function. Since the introduction of compiler warnings for singleton variables (variables that appear once in the function), programmers mechanically add an underscore, but tend to forget about the single assignment rule and about the fact that these variables are actually bound to values. So, use them because they increase code legibility and maintainability, but use them with care, ensuring that you do not introduce bugs.

You can see from what we have said that pattern matching is a powerful mechanism, with some subtleties in its behavior that allow you to do some amazing things in one or two lines of code, combining tests, assignment, and control.

At the risk of sounding repetitive, try pattern matching in the shell. You can experiment with defining lists to be really sure you master the concept, and use pattern matching to deconstruct the lists you have built. Make pattern-matching clauses fail and inspect the errors that are returned.# When you do so, experiment with both bound and

---

Different versions of the Erlang runtime system will format errors differently.

unbound variables. As pattern matching holds the key to writing compact and elegant programs, understanding it before continuing will allow you to make the most of Erlang as you progress.

Functions

Now that we've covered data types, variables, and pattern matching, how do you use them? In programs, of course. Erlang programs consist of functions that call each other. Functions are grouped together and defined within modules. The name of the function is an atom. The head of a function clause consists of the name, followed by a pair of parentheses containing zero or more formal parameters. In Erlang, the number of parameters in a function is called its *arity*. The arrow (->) separates the head of the clause from its body.

Before we go any further, *do not* try to type functions directly in the shell. You can if you want to, but all you will get is a syntax error. Functions have to be defined in modules and compiled separately. We will cover writing, compiling, and running functions in the next section.

Example 2-1 shows an Erlang function used to calculate the area of a shape.[*] Erlang functions are defined as a collection of clauses separated by semicolons and terminated by a full stop. Each clause has a head specifying the expected argument patterns and a function body consisting of one or more comma-separated expressions. These are evaluated in turn, and the return value of a function is the result of the last expression executed.

Example 2-1. An Erlang function to calculate the area of a shape

```
area({square, Side}) ->
  Side * Side ;
area({circle, Radius}) ->
  math:pi() * Radius * Radius;
area({triangle, A, B, C}) ->
  S = (A + B + C)/2,
  math:sqrt(S*(S-A)*(S-B)*(S-C));
area(Other) ->
  {error, invalid_object}.
```

When a function is called, its clauses are checked sequentially by pattern matching the arguments passed in the call to the patterns defined in the function heads. If the pattern match is successful, variables are bound and the body of the clause is executed. If it is not, the next clause is selected and matched. When defining a function, it is a good practice to make sure that for every argument there is one clause that succeeds; this is

---

[*] In the case of the triangle, the area is calculated using Heron's formula where `math:sqrt/1` is used to give the square root of a float.

often done by making the final clause a *catch-all* clause that matches all (remaining) cases.

In Example 2-1, `area` is a function call that will calculate the area of a square, a circle, or a triangle or return the tuple `{error, invalid_object}`. Let's take an example call to the area function:

```
area({circle, 2})
```

Pattern matching will fail in the first clause, because even though we have a tuple of size 2 in the argument and parameter, the atoms `square` and `circle` will not match. The second clause is chosen and the pattern match is successful, resulting in the variable `Radius` being bound to the integer 2. The return value of the function will be the result of the last expression in that clause, `math:pi()*2*2`, namely `12.57` (rounded). When a clause is matched, the remaining ones are not executed.

The last clause with the function head `area(Other) ->` is a catch-all clause. As `Other` is unbound, it will match any call to the function `area` when pattern matching fails in the first three clauses. It will return an expression signifying an error: `{error, invalid_object}`.

A common error is shadowing clauses that will never match. The `flatten` function defined in the following example will always return `{error, unknown_shape}`, because pattern matching against `Other` will always succeed, and so `Other` will be bound to any argument passed to `flatten`, including `cube` and `sphere`:

```
flatten(Other)  -> {error, unknown_shape};
flatten(cube)   -> square;
flatten(sphere) -> circle.
```

Let's look at an example of the factorial function:

```
factorial(0) -> 1;
factorial(N) ->
  N * factorial(N-1).
```

If we call `factorial(3)`, pattern matching in the first clause will fail, as 3 does not match 0. The runtime system tries the second clause, and as N is unbound, N will successfully be bound to 3. This clause returns `3 * factorial(2)`. The runtime system is unable to return any value until it has executed `factorial(2)` and is able to multiply its value by 3. Calling `factorial(2)` results in the second clause matching, and in this call N is bound to 2 and returns the value `2 * factorial(1)`, which in turn results in the call `1 * factorial(0)`. The call `factorial(0)` matches in the first clause, returning 1 as a result. This means `1*factorial(0)` in level 3 returns 1, in level 2 returns 2*1, and in level 1 returns the result of `factorial(3)`, namely 6:

```
factorial(3).
Level 1: 3 * factorial(3 - 1)        (returns 6)
Level 2:       2 * factorial(2 - 1)  (returns 2)
Level 3:             1 * factorial(1 - 1) (returns 1)
Level 4:                   1         (returns 1)
```

As we mentioned earlier, pattern matching occurs in the function head, and an instance of the variable N is bound after a successful match. Variables are local to each clause. There is no need to allocate or deallocate them; the Erlang runtime system handles that automatically.

Modules

Functions are grouped together in *modules*. A program will often be spread across several modules, each containing functions that are logically grouped together. Modules consist of files with the *.erl* suffix, where the file and module names have to be the same. Modules are named using the –module(Name) directive, so in Example 2-2, the demo module would be stored in a file called *demo.erl*.

Example 2-2. A module example

```
-module(demo).
-export([double/1]).

% This is a comment.
% Everything on a line after % is ignored.

double(Value) ->
  times(Value, 2).
times(X,Y) ->
  X*Y.
```

The export directive contains a list of exported functions of the format Function/Arity. These functions are *global*, meaning they can be called from outside the module. And finally, comments in Erlang start with the percent symbol (%) and span to the end of the line. Make sure you use them everywhere in your code!

Global calls, also called fully qualified function calls, are made by prefixing the module name to the function. So, in Example 2-2, calling demo:double(2) would return 4. Local functions can be called only from within the module. Calling them by prefixing the call with the module name will result in a runtime error. If you were wondering what math:sqrt/1 did in Example 2-1, it calls the sqrt (square root) function from the math module, which comes as part of the Erlang standard distribution.

Functions in Erlang are uniquely identified by their name, their arity, and the module in which they are defined. Two functions in the same module might have the same name but a different arity. If so, they are different functions and are considered unrelated. There is no need to declare functions before they are called, as long as they are defined in the module.

Compilation and the Erlang Virtual Machine

To run functions exported from a module, you have to compile your code, which results in a *module.beam* file being written in the same directory as the module:

- If you are using a Unix derivative, start the Erlang shell in the same directory as the source code.
- In Windows environments, one way to start a werl shell in the right directory is to right-click a *.beam* file, and in the pop-up window select the Open With option and choose "werl". This will from now on allow you to get an Erlang shell in the right directory just by double-clicking any *.beam* file in the same location where you have placed your source code.

With both operating systems, you can otherwise move to the directory by using the cd(Directory) command in the Erlang shell. Once in the directory, you compile the code using c(Module) in the Erlang shell, omitting the erl suffix from the module name. If the code contained no errors, the compilation will succeed.

Large Erlang systems consist of loosely coupled Erlang modules, all compiled on a standalone basis. Once you have compiled your code, look at the source code directory and you will find a file with the same name as the module, but with the *.beam* suffix. This file contains the byte code that you can call from any other function. The *.beam* suffix stands for Björn's Erlang Abstract Machine, an abstract machine on which the compiled code runs.

Once compiled, you need to make a fully qualified function call to run your functions. This is because you are calling the function from outside the module. Calling non-exported functions will result in a runtime error:

```
1> cd("/home/francesco/examples").
/home/francesco/examples
ok
2> c(demo).
{ok,demo}
3> demo:double(10).
20
4> demo:times(1,2).
** exception error: undefined function demo:times/2
```

Module Directives

Every module has a list of attributes of the format *–attribute(Value)*. They are usually placed at the beginning of the module, and are recognizable by the – sign in front of the attribute name and the full stop termination. The module attribute is mandatory, and describes the module name. Another attribute we have come across is the export attribute, which takes a list of function/arity definitions.

A useful directive when programming is the –compile(export_all) directive, which at compile time will export all functions defined in the module. Another way of doing this is to specify an option on compiling the file:

```
c(Mod,[export_all]).
```

This directive should be used only for testing purposes. Do not do like many others and forget to replace it with an **export** directive before your code goes into production! The compile directive takes on other options that are useful only in special conditions. If you are curious and want to read more about them, check out the manual page for the compile module.

Another directive is −import(Module, [Function/Arity,...]). It allows you to import functions from other modules and call them locally. Going back to the function area example, including −import(math,[sqrt/1]) as a directive in your module would allow you to rewrite the function clause calculating the area of a triangle call. As a reminder, do not forget to terminate your directives with a full stop:

```
-import(math, [sqrt/1]).

area({triangle, A, B, C})  ->
  S = (A + B + C)/2,
  sqrt(S*(S-A)*(S-B)*(S-C));
```

Using the import directive can make your code hard to follow. Someone trying to understand it may at first glance believe **sqrt/1** is a local function and unsuccessfully search for it in the module; on the other hand, she can check the directives at the head of the file to see that it is indeed imported. That being said, it's a convention in the Erlang community to use import sparingly, if at all.

You can make up your own module attributes. Common examples include -author(Name) and -date(Date). User-defined attributes can have only one argument (unlike some of the built-in attributes).

All attributes and other module information can be retrieved by calling Mod:module_info/0 or selectively calling the Mod:module_info/1 function. From the shell, you can use the m(Module) command:

```
5> demo:module_info().
[{exports,[{double,1},{module_info,0},{module_info,1}]},
 {imports,[]},
 {attributes,[{vsn,[74024422977681734035664295266840124102]}]},
 {compile,[{options,[]},
           {version,"4.5.1"},
           {time,{2008,2,25,18,0,28}},
           {source,"/home/francesco/examples/demo.erl"}]}]
6> m(demo).
Module demo compiled: Date: February 25 2008, Time: 18.01
Compiler options:  []
Object file: /home/francesco/examples/demo.beam
Exports:
        double/1
        module_info/0
        module_info/1
ok
```

If you read through the Erlang libraries, other attributes you will come across will include -behaviour(Behaviour) (U.K. English spelling), -record(Name, Fields), and -vsn(Version). Note that we did not have any vsn attribute in the demo module, but one appeared in the preceding example. When vsn is not defined, the compiler sets it to the MD5 of the module. Note also that the module_info functions appear in the list of exported functions, as they are meant to be accessible outside the module. Do not worry about records, vsn, and behaviour for now, as we cover them in Chapters 7 and 12.

We covered the basics of Erlang in this chapter, and you saw some of its peculiarities: you can assign values to variables, but only once; you can pattern-match against a variable and it may turn into a test for equality with that variable. Other features of the language, such as the module system and the basic types it contains, are more familiar.

We'll build on this in Chapter 3, where we talk about the details of sequential programming, and then in Chapter 4, where we'll introduce you to concurrency in Erlang—probably the single most important feature of the language.

Exercises

Exercise 2-1: The Shell

Type in the following Erlang expressions in the shell and study the results. They will show the principles of pattern matching and single-variable assignment described in this chapter. What happens when they execute? What values do the expressions return, and why?

A. Erlang expressions

```
1 + 1.
[1|[2|[3|[]]]].
```

B. Assigning through pattern matching

```
A = 1.
B = 2.
A + B.
A = A + 1.
```

C. Recursive list definitions

```
L = [A|[2,3]].
[[3,2]|1].
[H|T] = L.
```

D. Flow of execution through pattern matching

```
B = 2.
B = 2.
2 = B.
B = C.
C = B.
B = C.
```

E. Extracting values in composite data types through pattern matching

```
Person = {person, "Mike", "Williams", [1,2,3,4]}.
{person, Name, Surname, Phone} = Person.
Name.
```

Exercise 2-2: Modules and Functions

Copy the demo module from the example in this chapter. Compile it and try to run it from the shell. What happens when you call demo:times(3,5)? What about double(6) when omitting the module name?

Create a new module called shapes and copy the area function in it. Do not forget to include all the module and export directives. Compile it and run the area function from the shell. When you compile it, why do you get a warning that variable Other is unused? What happens if you rename the variable to _Other?

Exercise 2-3: Simple Pattern Matching

Write a module boolean.erl that takes logical expressions and Boolean values (represented as the atoms true and false) and returns their Boolean result. The functions you write should include b_not/1, b_and/2, b_or/2, and b_nand/2. You should not use the logical constructs and, or, and not, but instead use pattern matching to achieve your goal.

Test your module from the shell. Some examples of calling the exported functions in your module include:

```
bool:b_not(false)  ⇒ true
bool:b_and(false, true)  ⇒ false
bool:b_and(bool:b_not(bool:b_and(true, false)), true)  ⇒ true
```

The notation foo(X) ⇒ Y means that calling the function foo with parameter X will result in the value Y being returned. Keep in mind that and, or, and not are reserved words in Erlang, so you must prefix the function names with b_.

Hint: implement b_nand/2 using b_not/1 and b_and/2.

CHAPTER 3

Sequential Erlang

Erlang's design was heavily influenced by functional and logic programming languages. When dealing with sequential programs, those familiar with languages such as Prolog, ML, or Haskell will recognize the influence they have had on Erlang's constructs and development techniques. When working in functional programming languages, you replace iterative constructs such as `while` and `for` loops with recursive programming techniques.

Recursion is the most useful and powerful of all the techniques in a functional programmer's armory. It allows a programmer to traverse a data structure via successive calls to the same function, with the patterns of function calls mirroring the structure of the data itself. The resulting programs are more compact and easier to understand and maintain. Functional programs are, importantly, *side-effect-free*, unless side effects are specifically needed for printing or for access to external storage.

You control recursion using various conditional constructs that enhance the expressive power of pattern matching; in the example of data structure traversal, different patterns correspond to different traversals: bottom-up, top-down, breadth-first, and so forth.

This chapter also introduces other features directly related to sequential programming. The absence of a strong type system and the flexibility and dynamic nature of some of Erlang's constructs result in runtime errors that, although rare, have to be handled. Through exception-handling mechanisms, programs can recover and continue execution.

The Erlang distribution comes with an extensive set of modules which contain libraries, tools, and utilities, as well as complete applications. New modules are being added in every release, and existing libraries are often enhanced with new functionality. Some of the libraries consist of what we in the Erlang world call *built-in functions* (BIFs), because they are part of the Erlang runtime system. They do things that are either not programmable in Erlang itself or would be slow to execute were they to be defined by the user.

Conditional Evaluations

Erlang has three forms of conditional evaluation that are (at least partially) inter-changeable. The first form you encountered in Chapter 2: the choice of a function clause to evaluate through pattern matching over the arguments to the function. The second is the case statement, which works in a similar way to function clause selection. The third form is the if construct, which you can view as a simplified form of the case construct. Let's start with the case construct.

The case Construct

The case construct relies on pattern matching to choose what code to evaluate, in a strikingly similar manner to the selection of function clauses by pattern matching. Instead of matching actual parameters against patterns for the formal parameters to the function, case evaluates an expression and matches the result against a list of pattern-matching clauses, separated by semicolons.

The general case expression has the following form:

```
case conditional-expression of
  Pattern1 -> expression1, expression2, .. ;,
  Pattern2 -> expression1, expression2, .. ;
  ... ;
  Patternn -> expression1, expression2, ..
end
```

The keywords used are case, of, and end. The *conditional-expression* is evaluated and matched against *Pattern1*, ..., *Patternn* in turn until the first clause with a pattern that matches is found. The -> separates the pattern or *head* of the clause from the *body* which consists of a comma-separated list of expressions. Once a pattern has been matched, the selected clause expressions are evaluated in order and the result of the case construct is the result of the last expression evaluated.

In the following case construct, the list List is examined using the member function from the lists library module to determine whether the atom foo is a member of the List, in which case the atom ok is returned. Otherwise, the tuple {error, unknown_element} is returned:

```
case lists:member(foo, List) of
  true -> ok;
  false -> {error, unknown_element}
end
```

Case expressions always return a value, so nothing is stopping you from binding the return value to a variable. It is possible, although rare, to type case clauses directly in the Erlang shell. From now on, the constructs are getting complex enough that it makes sense for you to try what you learn by typing trial functions in a module, compiling the modules, and running their functions.

As with function definitions, the result of **case** expression evaluation should match one of the patterns; otherwise, you will get a runtime error. If you have _ or an unbound variable as the last pattern, it will match any Erlang term and act as a catch-all clause (discussed in Chapter 2). It is not mandatory to have a catch-all clause; in fact, it is discouraged if it is used as a form of defensive programming (see the sidebar "Defensive Programming" for more information on catch-all clauses).

Defensive Programming

Assume your program has to map to an integer an atom representing a day of the week. A defensive programming approach with a catch-all clause would look like this:

```
convert(Day) ->
  case Day of
    monday    -> 1;
    tuesday   -> 2;
    wednesday -> 3;
    thursday  -> 4;
    friday    -> 5;
    saturday  -> 6;
    sunday    -> 7;
    Other     -> {error, unknown_day}
  end.
```

We recommend strongly that you *don't* take this approach. It is better to make your program terminate in the **convert** function with a *clause error* (meaning no clause was matched), because then *the error is apparent at the point where it occurred*.

The alternative is to handle the error by returning an *error value*, but then *every* function that calls **convert/1** will have to handle the error, or risk there being an arithmetical error, due to the result of **convert** being a tuple rather than an integer.

In the past, defensive programming like this has left us searching in 2-million-line code bases looking for the function that returned the error tuple that caused a bad match error in a completely different part of the system.

Function definitions and **case** expressions have a lot in common. Take the following simple example of a one-argument function to calculate the length of a list:

```
listlen([])     -> 0;
listlen([_|Xs]) -> 1 + listlen(Xs).
```

You can rewrite this directly using a **case** expression:

```
listlen(Y) ->
  case Y of
    []     -> 0;
    [_|Xs] -> 1 + listlen(Xs)
  end.
```

In a multiple-argument function, it is possible to pattern-match simultaneously on all the arguments:

```
index(0,[X|_])         -> X;
index(N,[_|Xs]) when N>0 -> index(N-1,Xs).
```

A `case` expression, on the other hand, matches a single expression. Making the arguments into a tuple allows a `case` to be used to define `index`, too:

```
index(X,Y) ->
    index({X,Y}).

index(Z) ->
    case Z of
        {0,[X|_]}         -> X;
        {N,[_|Xs]} when N>0 -> index(N-1,Xs)
    end.
```

Alternatively, the pattern matching could be performed by nested `case` expressions, matching separately on the two arguments:

```
index(X,Y) ->
    case X of
        0 ->
            case Y of
                [Z|_]   -> Z
            end;
        N when N>0 ->
            case Y of
                [_|Zs]  -> index(N-1,Zs)
            end
    end.
```

So, pattern matching in function definitions can be more compact than using a `case` expression, but remember that a `case` expression can be used *anywhere* in a function definition and not just in the head, so each kind of pattern match has a part to play in Erlang. A case statement with only one clause is considered bad practice in Erlang. In our example, we do it to demonstrate a point. You should however avoid it in your code, and instead use pattern matching on its own.

Variable Scope

The *scope* of a variable is the region of the program in which that variable can be used. The same variable *name* may be used in many places in the program; some uses will refer to the same variable and others to a different variable, which happens to have the same name. In the following example:

```
f(X) -> Y=X+1,Y*X.
```

the scope of X, which is introduced in the head of the function clause, is the whole of the clause—that is, Y+X, Y*X—whereas the scope of Y is the remainder of the body after it is introduced, which here is the single expression Y*X.

In the next example, there are two separate variables with the name Y. The first is defined in the function f/1 and is used in the final expression of the body of f. The second is defined in the head of the second clause for g/1 and its scope is the entire body of this clause:

```
f(X)       -> Y=X+1,Y*X.
g([0|Xs]) -> g(Xs);
g([Y|Xs]) -> Y+g(Xs);
g([])      -> 0.
```

As we said, in Erlang the scope of a variable is any position in the same function clause after it has been bound, either by an explicit match using = or as part of a pattern. This creates a problem when variables are bound in only some of the clauses of a case or if construct, and are later used within the same function body. The small code example that follows demonstrates this: what would be the result of calling unsafe(one) and unsafe(two)? In fact, there is nothing to worry about because the compiler will not let you compile modules with "unsafe" variables in them, that is, variables defined in only one case or if clause and used outside that clause:

```
unsafe(X) ->
  case X of
    one -> Y = true;
    _   -> Z = two
  end,
  Y.
```

You can use a variable safely only if it is bound in all the clauses of a case or if construct. This is considered bad coding practice, as it tends to make the code much harder to read and understand. Here is an example of a safe use in the preferred style, where there is a single binding to the variable in question, with a value determined by a case expression:

```
safe(X) ->                    preferred(X) ->
  case X of                     Y =   case X of
    one -> Y = 12;                      one -> 12;
    _   -> Y = 196                      _   -> 196
  end,                                end,
  X+Y.                          X+Y.
```

The if Construct

The if construct looks like a case without the *conditional-expression* and the of keyword:

```
if
  Guard1 -> expression11, expression12, .. ;
  Guard2 -> expression21, expression22, .. ;
  ... ;
  Guardn -> expressionn1, expressionn2, ..
end
```

The guard expressions *Guard1*, ..., *Guardn* are evaluated in turn one after another until one evaluates to true. If this is *Guardi*, the body of the following clause is evaluated:

```
expressioni1, expressioni2,... expressionin
```

The result of the complete `if` expression is the result of this sequence; that is, the result of evaluating the last expression in the body of the clause that was executed.

The guard expressions are a *subset* of the Erlang Boolean expressions that can only contain calls to a restricted set of functions together with comparisons and arithmetic operations. We describe exactly what the guards may contain in the next section.

If none of the guards evaluates to the atom `true`, a runtime error is generated. To get a catch-all clause, you can allow the last clause to have the atom `true` as the guard; it is not mandatory to have such a catch-all clause.

In the following example, the variable X is examined to determine whether it is smaller than, larger than, or equal to 1:

```
if
  X < 1 -> smaller;
  X > 1 -> greater;
  X == 1 -> equal
end
```

This could be written equally well with a catch-all clause, which makes it clear that it will always return a result:

```
if
  X < 1 -> smaller;
  X > 1 -> greater;
  true  -> equal
end
```

Erlang novices, especially those coming from an imperative background, tend to overuse `if` statements when the same result can be achieved more elegantly using pattern matching in `case` statements. The following example demonstrates how you can rewrite an `if` statement using a `case` expression, when the `if` expression has a sequence of guards that test the value of a particular expression:

```
if                          case X rem 2 of
  X rem 2 == 1 -> odd;        1 -> odd;
  X rem 2 == 0 -> even        0 -> even
end                         end
```

To test your understanding of `if` and `case` expressions, try evaluating some in the shell and in simple programs that you can compile and run. Pay special attention to the return value of these clauses, trying to bind the value to a variable and using it in later computations. When testing, write a program with an unsafe variable (defined in some clauses and not others) and try to compile it.

Guards

Guards are additional constraints that can be placed in a function clause—either a `case` or a `receive` clause (we will cover `receive` expressions in Chapter 4). Guards are placed before the `->` separating the clause head from the body.

A guard consists of a when keyword followed by a *guard expression*. The clause will be selected only if the pattern matches and the guard expression evaluate to the atom true.

Let's rewrite the factorial example from Chapter 2:

```
factorial(0) -> 1;
factorial(N) ->
  N * factorial(N-1).
```

this time using guards:

```
factorial(N) when N > 0 ->
  N * factorial(N - 1);
factorial(0) -> 1.
```

We reordered the clauses in the factorial function. In the previous version, we had to have factorial(0) as the first clause to ensure that the function terminates. Now we select the recursive clause (i.e., the one that calls the factorial function in its body) only if the parameter N is larger than 0.

If the pattern matching together with the guards uniquely identifies what clause should be selected, their order becomes irrelevant, and that is the case here. A last thing to note concerning the rewrite of the factorial function is that if factorial(-1) is called, a runtime error is generated, as no clause can be selected, since –1 is less than 0 and not equal to 0. In the previous version of the factorial function, the function would never have returned a value, since factorial(-1) calls factorial(-2), and so on. This would eventually cause the Erlang runtime system to run out of memory and terminate.

The individual guard expressions can be built using the following constructs:

- Bound variables
- *Literal* Erlang terms denoting data values including numbers, atoms, tuples, lists, and so forth
- Type tests, such as is_binary, is_atom, is_boolean, is_tuple, and so on
- Term comparisons using ==, =/=, <, >, and so on, as listed in Chapter 2
- Arithmetic expressions built using the arithmetical operators given in Chapter 2
- Boolean expressions as described in Chapter 2
- Guard built-in functions

Guard subexpressions resulting in a runtime error are treated as returning false.

As an example, we can write:

```
guard(X,Y) when not(((X>Y) or not(is_atom(X)) ) and (is_atom(Y) or (X==3.4))) ->
  X+Y.
```

which shows that guards can be complex combinations of tests, but do not allow reference to any *user-defined functions*.

The reason for not allowing developers to implement their own guard functions, limiting them to permitted operations, is to ensure that guards are *free of side effects*. The guards are executed for all clauses up to the successful clause, meaning that if you had an io:format call in a guard that fails, you would still see the printout even if the clause was not selected for evaluation.

 The Erlang language also has legacy versions of the type BIFs that are simply called by the type name: atom/1, integer/1, and so forth. The use of these BIFs is discouraged, as they are deprecated and are available only for backward compatibility reasons. The new guards are is_atom/1, is_integer/1, and so on.

Erlang allows simple logical combinations of guards to be written in a different way:

- Separating individual guard expressions with a comma (,) gives their *conjunction*, so that such a sequence evaluates to true only if *all* expressions in the sequence evaluate to true.

- Separating individual expressions (or indeed, comma-separated conjunctions) with a semicolon (;) gives their *disjunction*, where the sequence evaluates to true if *any* expression evaluates to true.

As an example of "... ; ...,..." notation, we can rewrite the guard function (after a couple of applications of De Morgan's Laws) to the following:

```
guard2(X,Y) when not(X>Y) , is_atom(X) ; not(is_atom(Y)) , X=/=3.4 ->
    X+Y.
```

Simple combinations with only commas or semicolons are fine; we would not recommend using semicolons and commas together in practice, as it's too easy to get the logic wrong.

To conclude this section, copy the following example and run it from the shell. The function even will take the remainder of an integer when divided by 2; if the result is 0, it returns the atom true, and if it is 1 the integer is not even, so it returns false. Try calling the function even with a float or an atom. What happens? The second function, number, returns the atom float or integer depending on the argument passed to number/1. If anything other than a number is passed, the function returns false:

```
-module(examples).
-export([even/1, number/1]).

even(Int) when Int rem 2 == 0 -> true;
even(Int) when Int rem 2 == 1 -> false.

number(Num) when is_integer(Num) -> integer;
number(Num) when is_float(Num)   -> float;
number(_Other)                   -> false.
```

Built-in Functions

The following subsections will familiarize you with a few of the more commonly used *built-in functions* grouped according to the type of function, with examples illustrating their use. We will refer to built-in functions as *BIFs*, a practice almost universal in the Erlang community. Standard and nonstandard BIFs are listed in the manual page of the `erlang` module.

BIFs are usually written in C and integrated into the virtual machine (VM), and can be used to manipulate, inspect, and retrieve data as well as interact with the operating system. An example of a data manipulation function is the conversion of an atom to a string: `atom_to_list/1`. Other BIFs, such as `length/1`, which returns the length of a list, are implemented in the runtime system for efficiency.

Originally, all built-in functions were considered to belong to the module `erlang`, but they have made their way to other modules for practicality and efficiency reasons. Among the modules that contain built-in functions are `ets` and `lists`.

Although most built-in functions are seen as being an integral part of Erlang, others are VM-dependent and do not necessarily exist in other VM implementations or even in specific OS ports of the existing VM. Standard built-in functions are *auto-imported*, so you can call them without the module prefix. Nonstandard BIFs, however, have to be prefixed with the `erlang` module prefix, as in `erlang:function`. Examples of nonstandard built-in functions include `erlang:hash(Term, Range)`, which returns the hash of the `Term` in the specified range; and `erlang:display(Term)`, which prints the term to standard output and is mainly used for debugging purposes.

Object Access and Examination

A large number of BIFs deal with built-in types such as lists and tuples:

`hd/1`
> Returns the first element of a list

`tl/1`
> Returns the remaining elements when the first element has been removed

`length/1`
> Returns the length of a list

`tuple_size/1`
> Returns the number of elements in a tuple

`element/2`
> Returns the *n*th element of a tuple

`setelement/3`
> Replaces an element in a tuple, returning the new tuple

```
erlang:append_element/2
```
Adds an element to the tuple, as the final element

These functions are shown in action in the following code:

```
1> List = [one,two,three,four,five].
[one,two,three,four,five]
2> hd(List).
one
3> tl(List).
[two,three,four,five]
4> length(List).
5
5> hd(tl(List)).
two
6> Tuple = {1,2,3,4,5}.
{1,2,3,4,5}
7> tuple_size(Tuple).
5
8> element(2, Tuple).
2
9> setelement(3, Tuple, three).
{1,2,three,4,5}
10> erlang:append_element(Tuple, 6).
{1,2,3,4,5,6}
```

Type Conversion

Type conversions have to be BIFs since they change the underlying representation of the data. This would be impossible to do efficiently within the language, even if it were possible to do at all. There are numerous type conversion functions, not only to change numerical types, but also to convert the basic types to and from a printable representation (i.e., string). When changing a float to an integer, you can choose between rounding and truncating it:

atom_to_list/1, list_to_atom/1, list_to_existing_atom/1
All convert atoms to strings and back. If the atom was not previously used by the runtime system in the current session, calling the function list_to_existing_atom/1 will fail.

list_to_tuple/1, tuple_to_list/1
Both convert between the two data types.

float/1, list_to_float/1
Both create a float, one with an integer parameter and the other from a string.

float_to_list/1, integer_to_list/1
Both return strings.

round/1, trunc/1, list_to_integer/1
All return integers.

Here they are in action:

```
1> atom_to_list(monday).
"monday"
2> list_to_existing_atom("tuesday").
** exception error: bad argument
     in function  list_to_existing_atom/1
        called as list_to_existing_atom("tuesday")
3> list_to_existing_atom("monday").
monday
4> list_to_tuple(tuple_to_list({one,two,three})).
{one,two,three}
5> float(1).
1.00000
6> round(10.5).
11
7> trunc(10.5).
10
```

Process Dictionary

There is a set of BIFs that allow functions to store values associated with a key and later retrieve them in other parts of the program; this set of BIFs is called the *process dictionary*. The retrieval and manipulation of these values unfortunately introduces global variables into Erlang.

Using the process dictionary might provide the programmer with a quick win while developing the program, but the result is code that is very hard to debug and maintain. As most Erlang functions are side effect free, the parameters passed to the function included in the crash report usually contain enough information to solve the bug. Introducing the process dictionary greatly complicates this task, as the state of the process dictionary is lost when the program crashes. We will not cover these BIFs in this book, as we do not want to be seen as encouraging bad practices. If you are desperate to write ugly, hard-to-debug programs or are putting together a submission to the obfuscated Erlang competition, you can read about these BIFs in the documentation that comes with the Erlang distribution. At least you will not be able to say you picked up this bad habit from us.

Meta Programming

One often refers to the ability of a function to determine what other function to call at runtime as *meta programming*, that is, programs that create other programs and run them. For this use, we have the apply/3 function that takes three arguments, namely a module name, an exported function name, and a list of arguments. When called, it executes the named function on the specified arguments and returns its result.

The beauty of apply/3 is that the module, function, and arguments do not have to be known at compile time. They can be passed to the BIF as variables. So, in the following example, the call to `apply/3` returns what `examples:even(10)` returns: `true`. This ability to dynamically determine the function to run is essential when writing generic code:

```
1> Module = examples.
examples
2> Function = even.
even
3> Arguments = [10].
[10]
4> apply(Module, Function, Arguments).
true
```

A common pitfall for beginners is to forget to put the arguments (even if there is just one) into a list when using `apply`. Suppose we take our definition of `listlen` from earlier in the chapter, and we forget to put the argument in a list when using `apply`:

```
5> apply(sequential, listlen, [2,3,4]).
** exception error: undefined function sequential:listlen/3
```

The correct use of `apply` gives us the following:

```
6> apply(sequential, listlen, [[2,3,4]]).
3
```

which is the answer we expect.

If the number of arguments is known at compile time, you can use the following notation (if there are two arguments):

```
Mod:Fun(Arg1, Arg2)
```

instead of the more general `apply(Mod,Fun,[Arg1,Arg2])`. We will look at other ways that functions can be created dynamically when we look at higher-order functions in Chapter 9.

Process, Port, Distribution, and System Information

In the chapters dealing with concurrency, we will cover several BIFs directly related to processes, process inspection, and error handling. The same applies to port handling and distribution. We will mention these BIFs and others in their relevant chapters throughout the book. There is a variety of information concerning the system that we might want to know, and of course, all the access functions for this information have to be BIFs. The information includes low-level system information, trace stacks, as well as the current time and date. The list is long, but they are all documented in the `erlang` module.

The `date/0` function returns the current date as a tuple of `{Year, Month, Day}`, and the `time/0` function returns the current time as a tuple of `{Hour, Minute, Second}`. The `now/0` function returns a tuple of `{MegaSeconds, Seconds, MicroSeconds}` that have passed since midnight, January 1, 1970. The `now/1` BIF will always return a unique value in a particular Erlang node, even if called more than once in the same microsecond. As a result, it can be used as a unique identifier.

Input and Output

The io module provides input and output from an Erlang program. In this section, we describe the main functions that read from standard input and write to standard output. Each function can take a file handle (of type io_device()) as an additional (first) argument: file operations are defined in the file module.

To read a line from standard input, use io:get_line/1, which takes a prompt string (or atom) as its input:

```
1> io:get_line("gissa line>").
gissa line>lkdsjfljasdkjflkajsdf.
"lkdsjfljasdkjflkajsdf.\n"
```

It is also possible to read a specified number of characters:

```
2> io:get_chars("tell me> ",2).
tell me> er
"er"
```

The most useful input function is io:read/1, which reads an *Erlang term* from standard input:

```
3> io:read("ok, then>>").
ok, then>>atom.
{ok,atom}
4> io:read("ok, then>>").
ok, then>>{2,tue,{mon,"weds"}}.
{ok,{2,tue,{mon,"weds"}}}
5> io:read("ok, then>>").
ok, then>>2+3.
{error,{1,erl_parse,"bad term"}}
```

As command 5 reminds us, a term is a *fully evaluated* value, and not an arbitrary Erlang expression such as 2+3.

Output in Erlang is provided by io:write/1, which will print an Erlang term, but the function most commonly used is io:format/2, which provides *formatted output*. io:format takes the following:

- A *formatting string* (or binary) that controls the formatting of the arguments
- A list of values to be printed

The formatting string contains characters that are printed as they are with *control sequences* for formatting.

Control sequences begin with a tilde (~), and the simplest form is a single character, indicating the following:

~c

An ASCII code to be printed as a character.

~f

A float to be printed with six decimal places.

~e

A float to be printed in scientific notation, showing six digits in all.

~w

Writes any term in *standard syntax*.

~p

Writes data as ~w, but in "pretty printing" mode, breaking lines in appropriate places, indenting sensibly, and outputting lists as strings where possible.

~W, ~P

Behave as ~w, ~p, but eliding structure at a depth of 3. These take an extra argument in the data list indicating the maximum depth for printing terms.

~B

Shows an integer to base 10.

Here they are in action:

```
1> List = [2,3,math:pi()].
[2,3,3.141592653589793]
2> Sum = lists:sum(List).
8.141592653589793
3> io:format("hello, world!~n",[]).
hello, world!
ok
4> io:format("the sum of ~w is ~w.~n", [[2,3,4],ioExs:sum([2,3,4])]).
the sum of [2,3,4] is 9.
ok
5> io:format("the sum of ~w is ~w.~n", [List,Sum]).
the sum of [2,3,3.141592653589793] is 8.141592653589793.
ok
6> io:format("the sum of ~W is ~w.~n", [List,3,Sum]).
the sum of [2,3|...] is 8.141592653589793.
ok
7> io:format("the sum of ~W is ~f.~n", [List,3,Sum]).
the sum of [2,3|...] is 8.141593.
ok
```

The full control sequence has the form ~F.P.PadC, where F is the field width of the printed argument, P is its precision, Pad is the padding character, and C is the control character. The full details of these are available in the documentation for the io module; for now, the next two examples illustrate the point:

```
8> io:format("the sum of ~W is ~.2f.~n", [List,3,Sum]).
the sum of [2,3|...] is 8.14.
ok
9> io:format("~40p~n", [{apply, io, format, ["the
 sum of ~W is ~.2f.~n", [[2,3,math:pi()],3,ioExs:sum([2,3,math:pi()])]]}]).
{apply,io,format,
        ["the sum of ~W is ~.2f.~n",
         [[2,3,3.141592653589793],
          3,8.141592653589793]]}
ok
```

To see how much ~p prettifies the output, it is worth trying the last command, but with ~w replacing ~40p in the formatting string.

When printing lists of integers, sometimes the output formatted with ~p will be confusing. The pretty printing mode tries to figure out what you are trying to print, and formats it accordingly. But if your list of integers happens to be valid ASCII values instead of a list of integers, you will get a string. If they are integers you need to print out, use ~w instead:

```
1> List = [72,101,108,108,111,32,87,111,114].
"Hello Wor"
2> io:format("~p~n",[List]).
"Hello Wor"
ok
3> io:format("~w~n",[List]).
[72,101,108,108,111,32,87,111,114]
ok
```

Recursion

The best way to tackle programming problems is to use the well-tested strategy of divide and conquer to break the problem into a number of simpler subproblems. By joining together the solutions of several simple problems, you solve the bigger one without even realizing it! Let's try this approach by taking a list of integers and adding 1 to every element in the list. Because Erlang is a single assignment language, we have to create a new list, in which we will store the result.

We will name the function bump/1 and divide the problems into smaller tasks that are easier to solve, implementing one clause at a time. If the old list is empty, the new one should also be empty. The following function clause takes care of this case:

```
bump([]) -> [];
```

The second possibility is that the list contains at least one element. If so, we split the list to a head and a tail. We take the head and create a new list whose head is the head of the old list incremented by 1:

```
bump([Head | Tail]) -> [Head + 1 | ?].
```

Now, the question is how do we proceed with the rest of list? We want to construct a new list where all the elements are one larger than in the old list. But that is exactly what the bump function is supposed to do! The solution is to call the function we are defining recursively using the tail of the list:

```
bump([Head | Tail]) -> [Head + 1 | bump(Tail)].
```

This provides us with what we are looking for. We recurse on the tail, and ensure that bump/1 returns a well-formed list that can be the tail of the new list we just created. So, the solution would be:

```
bump([]) -> [];
bump([Head | Tail]) -> [Head + 1 | bump(Tail)].
```

Does the function really work? Let's try working through the call bump([1,2,3]):

```
bump([1, 2, 3] => [1 + 1 | bump([2, 3])
    1 + 1 => 2
    bump([2, 3]) => [2 + 1 | bump([3])
        2 + 1 => 3
        bump([3]) => [3 + 1 | bump([])
            3 + 1 => 4
            bump([]) => []
            [4 | []] => [4]
        [4] <=
        [3 | [4]] => [3, 4]
    [3, 4] <=
    [2 | [3, 4]] => [2, 3, 4]
[2, 3, 4] <=
```

In the bump example, we exposed two important issues. The first one is the common technique of tackling the problem piecemeal, breaking it up into smaller problems. This resulted in a very common recursive programming pattern in Erlang. How does this piece of magic work? We are calling the same function, reusing variables. Are they not already bound? No, they are not. The important thing to remember is that the variables are unique to every call and considered fresh in every iteration. For each call to a function, a frame is created on the call stack with information regarding where to return, together with the parameters to the function and its local variables. This is important, even if it's hidden from you in the runtime system, and so we will get back to it later in the chapter when we talk about tail-recursive functions.

We are now going to look at a more elaborate example, revisiting a similar type of problem and, indeed, a solution. We want to compute the average of a list of numbers. So, let's call the function average. What is the average? It is the sum of the elements divided by the length of the list. We can thus define average to be as follows:

```
average(List) -> sum(List) / len(List).
```

And this has solved the problem! All we need to do is to define the sum and len functions, and we are done. To compute the sum, we do a similar case analysis to what we did for bump, breaking the problem into smaller problems.

Let's start with the sum. If the list is empty, the sum of its elements is obviously zero:

```
sum([]) -> 0;
```

If the list contains at least one element, we break the list into a head and a tail, and add the head (the first element) to the sum of the tail (the rest of the list). And as you may remember from the bump example, since we are already defining a function to solve this problem, let's use it:

```
sum([Head | Tail]) -> Head + sum(Tail).
```

The next step is to write the len/1 function. We are calling the length function len to avoid clashing with the built-in function. Here we want to add one for each element of the list, so it is almost identical to the sum/1 code, with two small exceptions: we do not care about the value of the elements, and we add one instead, giving us the following:

```
len([_ | Tail]) -> 1 + len(Tail).
```

We used the "don't care" variable (_) to signify that we are not interested in the value of the head of the list. The case for the empty list, of course, returns zero. Putting it all together we get the following:*

```
average(List) -> sum(List) / len(List).

sum([]) -> 0;
sum([Head | Tail]) -> Head + sum(Tail).

len([]) -> 0;
len([_ | Tail]) -> 1 + len(Tail).
```

In practice, we would have used the length/1 BIF rather than redefining the function for ourselves, because it's better to reuse code if you can, and also because it's more efficiently implemented as a BIF, but it's useful to see it defined recursively here.

Let's take a closer look now at how **average** works in an example:

```
average([1, 2, 3]) => sum([1, 2, 3]) / len([1, 2, 3])
    sum([1, 2, 3]) => 1 + sum([2, 3])
        sum([2, 3]) => 2 + sum([3])
            sum([3]) => 3 + sum([])
                sum([]) => 0
                3 + 0 => 3
            3 <=
            2 + 3 => 5
        5 <=
        1 + 5 => 6
    6 <=
    len([1, 2, 3]) => 1 + len([2, 3])
        len([2, 3]) => 1 + len([3])
            len([3]) => 1 + len([])
                len([]) => 0
                1 + 0 => 1
            1 <=
            1 + 1 => 2
        2 <=
        1 + 2 => 3
    3 <=
    6 / 3 => 2.0
2.0 <=
```

---

* It would be better if the function were not to cause a "division by zero" error on an empty list; how would you modify the definition so that **average** returns zero on an empty list?

The most striking point of this example is how similar the sum/1 and len/1 functions are. It is a pattern that, with variations, is by far the most common in Erlang code. We will now continue with an example that is a variation of this pattern. We will traverse a list, filtering out the elements that are not even. We now have three cases to take care of. In the first, the list is empty, meaning there are no elements to examine. This base case will return the empty list:

```erlang
even([]) -> [];
```

If the first element of the list is even, we want to include it in the list we are constructing, and to make the rest of the list consist of those elements in the Tail that are even:

```erlang
even([Head | Tail]) when Head rem 2 == 0 -> [Head | even(Tail)];
```

Note that we are using a guard to determine whether the head is even by checking whether the remainder of the division by two is equal to zero. Finally, the third case is when the first element of the list is odd. If so, we want to drop that element by applying the recursive call to the tail without appending the first element to the front:

```erlang
even([_ | Tail]) -> even(Tail).
```

Note that we have no guard in the last clause ensuring that the head is not even. We know this already, as the previous clause selects all the even numbers, leaving only the odd ones. Not needing to check the head of the list, we use a "don't care" variable:

```erlang
even([]) -> [];
even([Head | Tail]) when Head rem 2 == 0 -> [Head | even(Tail)];
even([_ | Tail]) -> even(Tail).
```

Let's take a closer look at how this evaluates:

```erlang
even([10, 11, 12]) => [10 | even([11, 12])] (10 rem 2 == 0)
    even([11, 12]) => even([12])             (11 rem 2 == 1)
        even([12]) => [12 | even([])]        (12 rem 2 == 0)
            even([]) => []
            [12 | []] => [12]
        [12] <=
    [12] <=
    [10 | [12]] => [10, 12]
[10, 12] <=
```

In all of these examples, we traversed the entire list, either constructing a new list or calculating a value. We will conclude these examples with one where the condition for termination is not necessarily the empty list. A condition that terminates the recursive calls is referred to as a *base case*. We will write a member/2 function which, given an element and a list, traverses the list, returning true if the element is a member of the list and false if not.

No element is a member of the empty list. So, if the empty list is sent to this function call, we return false:

```erlang
member(_, []) -> false;
```

At this point, we do not care what element we were looking for, as it isn't there, so we use the "don't care" variable. Our second case is if the list contains at least one element, and we check whether the first element is the one we are looking for. We break the list into a head and a tail, and pattern-match the element with the head. If they are equal, we return true:

```
member(H, [H | _]) -> true;
```

We establish that the element and the head are the same by matching them with a common variable; the rest of the list is no longer of importance. Finally, the list contains at least one element, but since we have not selected the previous clause, the head obviously does not match what we are looking for, meaning we have to recurse through the rest of the list looking for the element:

```
member(H, [_ | T]) -> member(H, T).
```

Pulling the three clauses together we have the following:

```
member(_, [])        -> false;
member(H, [H | _]) -> true;
member(H, [_ | T]) -> member(H, T).
```

Using pen and paper, try working step by step through the following examples. Pick out the base case and understand how the lists are broken up into a head and tail using the recursive definition of lists:

```
1> c(recursion).
{ok,recursion}
2> recursion:member(friday, [monday, tuesday, wednesday, thursday, friday]).
true
3> recursion:member(sunday, [monday, tuesday, wednesday, thursday, friday]).
false
```

Recursion is one of the most fundamental tools not only in Erlang, but also functional programming. You would benefit from copying the recursive examples we went through into a module called **recursion** and testing them with different parameters.

Tail-Recursive Functions

In writing the sum function earlier:

```
sum([])           -> 0;
sum([Head | Tail]) -> Head + sum(Tail).
```

we used a *direct recursion* style. This style means you can read the function definition as a *description* of the sum of a list, as in "the result of sum of the list [2,3,4] is equal to 2 added to the sum of [3,4]". Or you can read it as an *equation*:

```
sum([2,3,4] = 2 + sum([3,4])
```

Another approach to defining sum uses an additional function parameter, called an *accumulating parameter*, to hold the value of the sum as it is calculated.

If you call your function sum_acc, and the second parameter holds the "sum so far," how should you define the function? The first case is when the list is empty—you must then return the "sum so far":

```
sum_acc([],Sum) -> Sum;
```

If, on the other hand, the list is not empty, you take off the Head and add it to the Sum, and call sum_acc on the Tail and the new "sum so far":

```
sum_acc([Head|Tail], Sum) -> sum_acc(Tail, Head+Sum).
```

How do you call the function to sum a list? You start off with the list and a "sum so far" of zero:

```
sum(List) -> sum_acc(List,0).
```

To be clear about how this works, let's write down the evaluation of an example (hiding the calculation of the arithmetic):

```
sum([2,3,4])
=> sum_acc([2,3,4],0)
=> sum_acc([3,4],2)
=> sum_acc([4],5)
=> sum_acc([],9)
=> 9
```

The definition of sum_acc is called *tail-recursive* because the body of the function is a *call to the function itself*. In general, a function f is tail-recursive when the only calls to f occur as the last expression (i.e., *tail*) in the bodies of the clauses of f. What is the difference between the two definitions of sum?

- The *direct* definition is easier to understand: you can read the definition as a *direct description* of the sum of a list.

- The *tail-recursive* definition is more like a program written in C or Java: you have to understand how the program will evolve through its execution to see that the final value of the second variable is in fact the sum of the list. On the other hand, in some circumstances, a tail-recursive definition can be *more efficient* memory wise.

 It is one of the principal *Myths of Erlang Performance*[†] that tail recursion is much more efficient than direct recursion in Erlang. Perhaps this was true in the early days of the language, but optimizations applied between releases 7 and 12 have meant that it's no longer true that tail recursion will give you a more efficient program.

The advice of the developers of the system is "*The choice is now mostly a matter of taste*. If you really do need the utmost speed, you must measure. *You can no longer be absolutely sure that a tail-recursive function will be the fastest in all circumstances.*"

[†] To be found as a part of the *Efficiency Guide* in the Erlang documentation.

Why might sum_acc be more efficient than the original sum? The clue is in the evaluation we gave earlier, where you can see that its evaluation looks like a loop, which simply changes the values of the two parameters until the base case is hit. In other words, it might use less memory in its implementation and as a result be more space-efficient. On the other hand, optimizations in the compiler may well permit efficient implementation of a nontail-recursive function, too.

Another example is a reimplementation of the bump/1 function. For the new bump/1, we will add an accumulator that contains the list we are constructing: as we said earlier, this parameter looks rather like a variable in an imperative language, as its value will change in each iteration!

We do not want to change the interface to bump/1, so we define a new helper function, bump_acc/2, which takes two arguments: the list we are "bumping" and the new list we are constructing.

In the initial call to bump_acc/2, we supply the original list as well as an empty list. This empty list is the starting value of our accumulator, which in its first iteration will be empty, as we have not constructed anything:

```
bump(L) -> bump_acc(L, []).
```

Now we do the same case analysis as in the earlier examples, but we construct the functions differently. If the old list is empty, we are done and the newly constructed list should be the result:

```
bump_acc([], Acc) -> Acc;
```

If the old list contains at least one element, we break it up into a head and a tail, increment the head, and insert it into the list we are building, calling bump_acc/2 with the new accumulator and the tail of the list:

```
bump_acc([H | T], Acc) -> bump_acc(T, [H + 1 | Acc]).
```

Before reading on, try copying the preceding code and test it or step through it with pen and paper. Do you notice anything wrong? You are traversing the list and adding elements to the beginning of the new list. But that will mean the first element added to the list will end up at the back, resulting in a *reversed* list. You can remedy this by reversing the list you get as a result from the base case. The resulting code will be:

```
bump(List) -> bump_acc(List, []).

bump_acc([], Acc)             -> reverse(Acc);
bump_acc([Head | Tail], Acc) -> bump_acc(Tail, [Head + 1 | Acc]).
```

To understand how this version differs from the original, let's look at an example evaluation:

```
bump([1, 2, 3])
 => bump_acc([1, 2, 3], [])
 => bump_acc([2, 3] , [2])
 => bump_acc([3], [3, 2])
 => bump_acc([], [4, 3, 2])
```

```
=> reverse([4, 3, 2])
=> [2,3,4]
```

When writing bump_acc/2, you saw that using the accumulator reversed the elements of the list. You use the same principle and do nothing to the elements to get an accumulator-based reverse function:

```
reverse(List) -> reverse_acc(List, []).

reverse_acc([], Acc) -> Acc;
reverse_acc([H | T], Acc) -> reverse_acc(T, [H | Acc]).
```

Tail-Call Recursion Optimization

Recall that in general, a function f is tail-recursive when the only calls to f occur as the last expression (i.e., *tail*) in the bodies of the clauses of f. Now, let's get to the optimization this enables.

Looking at the evaluation of bump_acc earlier, it is apparent that it could be implemented by overwriting the information in the arguments—which is held in the stack frame for the original function call—and then jumping to the instructions for the tail-call function: in this case the same function. This is done without allocating a new stack frame.

This same optimization is possible for functions that are indirectly tail-recursive. For example, the following function merges two lists (of the same length) by interleaving their values:

```
merge(Xs,Ys) ->
  lists:reverse(mergeL(Xs,Ys,[])).

mergeL([X|Xs],Ys,Zs) ->
  mergeR(Xs,Ys,[X|Zs]);
mergeL([],[],Zs) ->
  Zs.

mergeR(Xs,[Y|Ys],Zs) ->
  mergeL(Xs,Ys,[Y|Zs]);
mergeR([],[],Zs) ->
  Zs.
```

This tail-call optimization, and therefore a tail-recursive definition, is most important for functions that run forever: these will form the bodies of concurrent processes, which we introduce in the next chapter.

Two accumulators example

We will now show you a more elaborate example of a tail-recursive function to ensure that you really understand the concept. We will convert the sum/1 and len/1 functions to be tail-recursive. There is no point in traversing the list twice, once to compute the sum and a second time to get the length, when you can get away with traversing it only once. To do so, we'll add one helper function that does *both* computations.

This helper function takes two accumulators, one used to store the sum and the other to store the average. The call to the helper function, where we initialize the accumulators to zero, is as follows:

```erlang
average(List) -> average_acc(List, 0, 0).
```

We now use our divide-and-conquer technique to solve the problem. If the list is empty, the accumulators contain the sum and length, so we perform the division:

```erlang
average_acc([], Sum, Length) -> Sum/Length;
```

Our second case is that if the list contains at least one element, we add the element to the sum accumulator, increment the length accumulator, and recursively call the helper function with the accumulators and the tail of the list:

```erlang
average_acc([H | T], Sum, Length) -> average_acc(T, Sum + H, Length + 1).
```

The code for the average is faster, uses less space, and—once you get used to accumulators—is more readable than the original version. Before going on to the next section, use pen and paper to work your way through a few examples of the tail-recursive version of **average**. You should also think about how to modify the **average** function so that it does not return an error on the empty list:

```erlang
average(List) -> average_acc(List, 0,0).

average_acc([], Sum, Length) ->
  Sum / Length;
average_acc([H | T], Sum, Length) ->
  average_acc(T, Sum + H, Length + 1).
```

Iterations Versus Recursive Functions

We said we iterate using recursion, but to clarify what we mean, we will show you how iteration in C can be rewritten in Erlang, tying these concepts to each other.

We will start with a simple iterative function in C that sums the integers from 1 to the boundary passed as an argument. The C function has two local integer variables containing the iterator (or *index variable*) and the sum so far. The summation is performed using a **for** loop, a typical iterative construct:

```c
int sum(int boundary) {
  int i, sum = 0;

  for(i = 1; i <= boundary; i++)
    sum += i;
  return sum;
}
```

The Erlang version uses a helper function with an accumulator, resulting in a similar pattern. The **sum** function mimics the C **sum** function by calling the helper function with the initial values. The values correspond to the initialization of the **sum** variable when it is declared and the **for** loop initialization of the **i** variable.

The base case of the helper function corresponds to when the for loop exits and the explicit return of the value of the variable sum is made.

The recursive call corresponds to the body of the for loop, where the loop destructively increments sum. The Erlang function does the same by a recursive call where the argument Sum is increased by the value of the Index argument. Finally, the increase of the iterator variable i is mimicked by the Index argument to the recursive call and is incremented by one:

```erlang
sum(Boundary) -> sum_acc(1, Boundary, 0).

sum_acc(Index, Boundary, Sum) when Index =< Boundary ->
  sum_acc(Index + 1, Boundary, Sum + Index);
sum_acc(_I, _B, Sum)->
  Sum.
```

Runtime Errors

For you to become productive in Erlang, you should know the runtime errors that can occur and how they are reported in the shell. The runtime errors in Erlang are exceptions that are thrown by the system. They are presented in the upcoming list, together with an example of a function that generates them and how that will actually look in the shell when you call that function.

In the examples, we use a module test which exports the functions factorial/1, test1/1, and test2/1. In some of the last cases, we can induce the runtime error directly in the shell. With each runtime error, we also give a brief description of its cause:

function_clause
> This is returned when none of the existing function patterns match in the called function. This error normally occurs when you have either forgotten a case in your case analysis or inadvertently called the function with the wrong argument:
>
> ```erlang
> factorial(N) when N > 0 ->
>     N * factorial(N - 1);
> factorial(0) -> 1.
> ```
>
> ```erlang
> 1> test:factorial(-1).
> ** exception error: no function clause matching test:factorial(-1)
> ```

case_clause
> This is returned when none of the existing patterns in the case construct match. The most common reason for this is that you have forgotten one or more possible cases:
>
> ```erlang
> test1(N) ->
>   case N of
>     -1 -> false;
>     1  -> true
>   end.
> ```

```
1> test:test1(0).
** exception error: no case clause matching 0
     in function   test:test1/1
```

if_clause

This is returned when none of the existing expressions in the `if` construct evaluate to `true`. As this is really a simplified `case` construct, the error is typically caused by a missing pattern:

```
test2(N) ->
  if
    N < 0 -> false;
    N > 0 -> true
  end.
```

```
1> test:test2(0).
** exception error: no true branch found when evaluating an if expression
     in function   foo:test2/1
```

badmatch

Errors occur in situations when pattern matching fails and there are no other alternative clauses to choose from. For the `badmatch` exception, it is very hard to point to a single cause, but one recurrent cause is when you inadvertently *try to bind a variable that is already bound*, as in the following:

```
1> N=45.
45
2> {N,M}={23,45}.
** exception error: no match of right hand side value {23,45}
```

This fails to bind `N` to `23`, as `N` is already bound to `45`.

Another relatively common cause is when you match to retrieve parts of a result from a function call. For example, it is common when searching for a tuple in a list of tuples to use the library function `lists: keysearch/3`, which returns on success the tuple {`value, Tuple`}, where `Tuple` is the sought tuple. Now, this function is quite often called in the following way:

```
{value, Tuple} = lists:keysearch(Key, Pos, List)
```

because we want to immediately use the retrieved tuple. But the function returns `false` when no tuple with a matching key can be found, thereby resulting in a `badmatch`:

```
1> Tuple = {1, two, 3}.
{1,two,3}
2> {1, two, 3, Four} = Tuple.
** exception error: no match of right hand side value {1,two,3}
```

badarg

This is returned when a BIF is called with the wrong arguments. In the following example, `length` requires a list, but is called with an atom:

```
1> length(helloWorld).
** exception error: bad argument
      in function  length/1
         called as length(helloWorld)
```

undef

> This is returned if the global function being called is not defined or exported. The cause of this exception is often that you have misspelled the function name, or have called the function without prepending the module name to the function call:

```
1> test:hello().
** exception error: undefined function test:hello/0
```

badarith

> This is returned when arithmetical operations are executed with an inappropriate argument, such as nonintegers or floats or trying to divide by zero:

```
1> 1+a.
** exception error: bad argument in an arithmetic expression
      in operator  +/2
         called as 1 + a
```

There are a few more error types that we will cover in later chapters of this book as they become relevant.

Handling Errors

In the preceding section, you saw some of the errors that can occur in an Erlang system, together with diagnoses of the potential causes.

When executing an expression and a runtime error occurs, you might want to catch the exception and prevent the thread of execution from terminating. Alternatively, you might want to let it fail and for some other part of the system deal with recovery: this latter option, which involves process linking and supervision, will be covered in Chapters 6 and 12. In the meantime, in this section we will discuss how errors can be caught and handled, using the try ... catch construct.

Using try ... catch

The idea behind the try ... catch construct is to evaluate an expression, and providing ways of handling the normal result of the expression as well as abnormal termination. What's more, the construct allows you to differentiate between different return values that arise as a result of Erlang's different exception-handling mechanisms, and to handle them in different ways.

Before the expression to be evaluated, you insert the reserved word try. You pattern-match the (normal) result as you would have done in a case statement, but instead of terminating the clauses with an end, you replace it with a catch followed by clauses to handle exceptions. These clauses have an *exception type* (also called *classes*) and *exception patterns* at their head, and corresponding return expressions.

The `try ... catch` construct has the following form:

```
try Exprs of
   Pattern1 [when Guard1] ->
               ExpressionBody1;
   Pattern2 [when Guard2] ->
               ExpressionBody2
catch
    [Class1:]ExceptionPattern1
               [when  ExceptionGuardSeq1] ->
               ExceptionBody1;
    [ClassN:]ExceptionPatternN
               [when ExceptionGuardSeqN] ->
               ExceptionBodyN
end
```

Here is an example of this in action, shown through a series of commands in the Erlang shell. In the first command, X is bound to 2, so any subsequent attempt to bind X to 3 will fail with a `badmatch` error:

```
1> X=2.
2
2> try (X=3) of
2>    Val -> {normal, Val}
2> catch
2>    _:_ -> 43
2> end.
43
```

In the second command, all error patterns in all classes (matching `_:_`) are mapped to 43; this result is returned here, as it is when only patterns in the `error` class (matching `error:_`) are matched in the third command:

```
3> try (X=3) of
3>    Val -> {normal, Val}
3> catch
3>    error:_ -> 43
3> end.
43
```

In the fourth command, we return the *error type* as a part of the result, and we can see that it is indeed a `badmatch` of the expression 3 (with X):

```
4> try (X=3) of
4>    Val -> {normal, Val}
4> catch
4>    error:Error -> {error,Error}
4> end.
{error,{badmatch,3}}
```

Finally, `throw` allows us to execute a nonnormal return within a `try ... catch` statement:

```
5> try (throw(non_normal_return)) of
5>   Val -> {normal, Val}
5> catch
5>   throw:Error -> {throw, Error}
5> end.
{throw,non_normal_return}
```

You can use the throw/1 BIF within the context of a try ... catch in what is called a *nonlocal return*. The return value of the expression passed to throw is returned by the try ... catch expression, bypassing the call stack.

Imagine you are parsing a very large and deeply nested XML structure. By only handling positive cases where the structure is correctly parsed, you do not need to check the return value of each recursive call for an error, and instead can concentrate on returning the parsed structure. Should you come across a parse error, the function would throw an exception. This exception is intercepted by try ... catch, which bypasses the whole recursive call stack to become the return value of the expression.

You should avoid using throw, as nonlocal *returns* make your code very hard to follow and debug. The only exceptions to this guideline are examples such as the earlier parser example, where a throw allows an exit from a deeply nested structure on an error condition. If you really have to use catch and throw, have pity on those who will be maintaining your code and ensure that they are both called in the same module. Trying to figure out which catch in which module handles a throw defined elsewhere is not for those with little patience.

What error classes are there?

error
> This is the principal class of errors, and you saw the various types of runtime errors in the preceding section; error can also be raised by calling the BIF erlang:error(Term).

throw
> This is the class that is generated by an explicit call to throw an exception, which will be caught by an enclosing try ... catch expression. Use of throw in Erlang is discouraged, because it makes understanding program behavior substantially more difficult.

exit
> This can be raised by calling the exit/1 BIF, invoked with a reason for termination; exits can also be produced by an exit signal, which we cover in more detail in Chapter 6.

To see these in action, let's take the function return_error, defined here:

```
-module(exception).
-export([return/1]).

return_error(X) when X < 0 ->
  throw({'EXIT', {badarith,
```

```
                     [{exception,return_error,1},
                      {erl_eval,do_apply,5},
                      {shell,exprs,6},
                      {shell,eval_exprs,6},
                      {shell,eval_loop,3}]}}});
return_error(X) when X == 0 ->
    1/X;
return_error(X) when X > 0->
    {'EXIT', {badarith, [{exception,return_error,1},
                         {erl_eval,do_apply,5},
                         {shell,exprs,6},
                         {shell,eval_exprs,6},
                         {shell,eval_loop,3}]}}.
```

This function will produce three different kinds of behavior depending on whether its argument is positive, negative, or zero. We define **try_return** to catch the errors:

```
try_return(X) when is_integer(X) ->
  try return_error(X) of
    Val -> {normal, Val}
  catch
    exit:Reason -> {exit, Reason};
    throw:Throw -> {throw, Throw};
    error:Error -> {error, Error}
  end.
```

```
4> exception:try_return(1).
{normal,{'EXIT',{badarith,[{exception,return_error,1},
                           {erl_eval,do_apply,5},
                           {shell,exprs,6},
                           {shell,eval_exprs,6},
                           {shell,eval_loop,3}]}}}
5> exception:try_return(0).
{error,badarith}
6> exception:try_return(-1).
{throw,{'EXIT',{badarith,[{exception,return_error,1},
                          {erl_eval,do_apply,5},
                          {shell,exprs,6},
                          {shell,eval_exprs,6},
                          {shell,eval_loop,3}]}}}
```

You can use wildcards in your **try ... catch** expressions, and if you are not pattern matching on the return value, you can omit the **of**. Type in the following example and use it to experiment with various pattern matches of exceptions and wildcards. Don't forget to export the function, but ignore the warnings generated by the last two clauses of the **catch** clause. These clauses are included to demonstrate the syntax and allow you to experiment, but they will never execute, as all exceptions will be handled in the previous clauses:

```
try_wildcard(X) when is_integer(X) ->
  try return_error(X)
    catch
    throw:Throw -> {throw, Throw};
    error:_     -> error;
```

```
      Type:Error  -> {Type, Error};
           _        -> other;               %% Will never be returned
       _:_          -> other                %% Will never be returned
    end.

7> exception:try_wildcard(-1).
{throw,{'EXIT',{badarith,[{exception,return_error,1},
                          {erl_eval,do_apply,5},
                          {shell,exprs,6},
                          {shell,eval_exprs,6},
                          {shell,eval_loop,3}]}}}
8> exception:try_wildcard(0).
error
9> exception:try_wildcard(1).
{'EXIT',{badarith,[{exception,return_error,1},
                   {erl_eval,do_apply,5},
                   {shell,exprs,6},
                   {shell,eval_exprs,6},
                   {shell,eval_loop,3}]}}
```

Before moving off this example, don't think that all you can do with caught errors is to pass them on in some form. Instead, it's possible to return values that show no trace that an error has been raised; this is illustrated in a final version of **try_return**:

```
try_return(X) when is_integer(X) ->
  try return_error(X) of
    Val -> {normal, Val}
  catch
    exit:_  -> 34;
    throw:_ -> 99;
    error:_ -> 678
  end.
```

Try examples of this; you will see that for positive values the 'EXIT' is not trapped (we return to this in Chapter 6).

Using catch

The original Erlang mechanism for exception handling was the catch. Because of its somewhat peculiar behavior, Richard Carlson, a member of the High Performance Erlang Team at Uppsala University, suggested a review of Erlang's exception handling, resulting in the try ... catch expression. He got the buy-in from the OTP team, which included this construct as a documented and permanent feature of the Erlang R10B release. We include a discussion of catch here because of the amount of legacy code that will use this construct. Be warned, however, that it's not as elegant as try ... catch.

The catch expression allows you to trap when runtime errors occur. The format of this is catch *expression*, where if the expression evaluates correctly it returns the value of the expression. But if a runtime error occurs, it returns the tuple {'EXIT', Error}, where Error contains information on the runtime error.

Let's try catching exceptions by using the `list_to_integer/1` BIF, calling it with a numeric string rather than a number:

```
1> list_to_integer("one").
** exception error: bad argument
     in function  list_to_integer/1
        called as list_to_integer("one")
2> catch list_to_integer("one").
{'EXIT',{badarg,[{erlang,list_to_integer,["one"]},
                 {erl_eval,do_apply,5},
                 {erl_eval,expr,5},
                 {shell,exprs,6},
                 {shell,eval_exprs,6},
                 {shell,eval_loop,3}]}}
```

Look at the result of the two calls. In the first call, we generate a runtime error that is printed out by the shell. If this were to occur in your code, the program would terminate abnormally. In the second call, we execute the expression within the scope of a `catch`. As a result, instead of generating a runtime error, the expression returns a tuple of the format {'EXIT', {Reason, Stack}}. `Reason` is an atom describing the error type—here it's a `badarg`—whereas `Stack` is the function call stack, allowing you to locate where the BIF was called with the incorrect arguments.

As the following example demonstrates, precedence in Erlang can at times be counterintuitive. If you are binding the return value of an expression encapsulated in a `catch`, you need to encapsulate the `catch` expression in parentheses, giving it a higher precedence than the assignment. If you don't, the compiler will return a syntax error:

```
3> catch 1/0.
{'EXIT',{badarith,[{erlang,'/',[1,0]},
                   {erl_eval,do_apply,5},
                   {erl_eval,expr,5},
                   {shell,exprs,6},
                   {shell,eval_exprs,6},
                   {shell,eval_loop,3}]}}
4> X = catch 1/0.
* 1: syntax error before: 'catch'
4> X = (catch 1/0).
{'EXIT',{badarith,[{erlang,'/',[1,0]},
                   {erl_eval,do_apply,5},
                   {erl_eval,expr,5},
                   {erl_eval,expr,5},
                   {shell,exprs,6},
                   {shell,eval_exprs,6},
                   {shell,eval_loop,3}]}}
5> X.
{'EXIT',{badarith,[{erlang,'/',[1,0]},
                   {erl_eval,do_apply,5},
                   {erl_eval,expr,5},
                   {erl_eval,expr,5},
                   {shell,exprs,6},
                   {shell,eval_exprs,6},
                   {shell,eval_loop,3}]}}
```

The throw/1 BIF will also work with catch. Look at the following example:

```erlang
-module(math).
-export([add/2]).

add(X,Y) ->
  test_int(X),
  test_int(Y),
  X + Y.

test_int(Int) when is_integer(Int) -> true;
test_int(Int) -> throw({error, {non_integer, Int}})
```

Let's interact with this module in the Erlang shell:

```erlang
1> math:add(1,1).
2
2> math:add(one, 1).
** exception throw: {error,{non_integer,one}}
     in function  math:test_int/1
3> catch math:add(one, 1).
{error,{non_integer,one}}
```

Calling the function add/2 within the scope of a catch results in the tuple {error, Reason} being returned. Calling the same function outside the scope of the catch results in an *exception throw* runtime error.

Another problem with catch is that it does not differentiate semantically between a runtime error, a throw, an exit, or the return value of a function, and instead treats them equally. There is no way to determine whether the {'EXIT', Error} returned when executing a call encapsulated within a catch was the result of a throw({'EXIT', {Reason, Stack}}), the result of a runtime error, a call to the exit/1 BIF, or merely an expression returning the tuple {'EXIT', Error}.

This is best illustrated by returning to the example of return_error defined earlier:

```erlang
-module(exception).
-export([return/1]).

return(X) when is_integer(X) ->
  catch return_error(X).
```

Interacting with this, you can see:

```erlang
1> exception:return(-1).
{'EXIT',{badarith,[{exception,return_error,1},
                   {erl_eval,do_apply,5},
                   {shell,exprs,6},
                   {shell,eval_exprs,6},
                   {shell,eval_loop,3}]}}
2> exception:return(0).
{'EXIT',{badarith,[{exception,return_error,1},
                   {erl_eval,do_apply,5},
                   {shell,exprs,6},
                   {shell,eval_exprs,6},
                   {shell,eval_loop,3}]}}
```

```
3> exception:return(1).
{'EXIT',{badarith,[{exception,return_error,1},
                   {erl_eval,do_apply,5},
                   {shell,exprs,6},
                   {shell,eval_exprs,6},
                   {shell,eval_loop,3}]]}}
```

You think you've seen it all? Try typing the following in the Erlang shell:

```
catch exit({badarith, [{exception, return_error, 1}, {erl_eval, do_apply, 5},
                        {shell, exprs, 6}, {shell, eval_exprs, 6},
                        {shell, eval_loop, 3}]})
```

You'll get a fourth, syntactically identical but semantically different way of generating the error in the example.

Erlang programmers were not too bothered by this, and for well more than a decade they made do by avoiding `throw` and not returning tuples that contained the `'EXIT'` atom. But although programmers might not have been too bothered, the proposal of the `try ... catch` expression solved these problems for new programs—the problem is working with legacy code written before the Erlang R10B release.

Library Modules

A large number of library modules are distributed with the Erlang runtime system. It always pays to spend some time reviewing what is available and making a mental note of the library modules that might contain functionality you might find useful. In between releases, it makes sense to read the release notes, which contain the major changes you should be aware of in the libraries, as well as provide pointers to new modules. If you cannot find a module providing you with a generic solution to your problem, you should search for it in the open source community, as the chance that someone has written and released it under a friendly open source license is good. This section deals with the most commonly used libraries in Erlang systems. But before looking at these libraries, let's look at how you access the documentation associated with them.

Documentation

The Erlang distribution comes with documentation, available in both HTML and Unix manpage formats. It is often bundled with the Erlang release, but can also be downloaded separately or accessed online at *http://erlang.org* (see Figure 3-1). You can access the root page of the HTML documentation by opening the *file:///<erl_root_dir>/doc/index.html* page, where the *doc* subdirectory is located in the root directory of the Erlang installation. In Windows, a shortcut to the documentation is included in the Erlang/OTP installation directory within the Program Files menu.

If you are using a Unix-based system, `erl –man Module` is a convenient way of accessing the manual pages. If that command does not work, the manual pages have not been

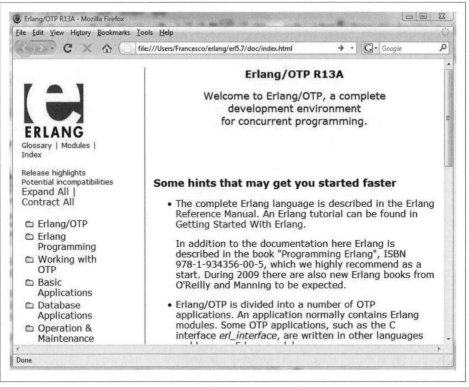

Figure 3-1. The home page of the Erlang online documentation

installed. Finally, most editors with an Erlang mode should have direct hooks into the manual pages.

You access the documentation using the menu on the lefthand side of the page. At the top there are links to the following:

Glossary
A list of the terminology in common use in the Erlang community.

Modules
An alphabetical list of the modules in the Erlang distribution, with web documentation for each of them. This includes the `erlang` and `erl` modules. Many modules have appropriately descriptive titles, and so a search of this page in your browser can lead you to the functionality you are looking for.

Index
A permuted index of Erlang/OTP functions and commands. Browser search is again useful here, and once you have found something of interest, links are provided in the index to the module identified.

The remaining links in the left column can be expanded to submenus containing a wealth of information about various aspects of the system. For example, browsing these will give you information about tools, higher-level descriptions of the architecture of parts of Erlang/OTP, and getting started with the system.

Useful Modules

To get an idea of what functionality and which modules are available in the Erlang distribution, open the main HTML documentation page as described earlier. In the top left, you should find a link to a page listing all existing modules, and the documentation there describes every function exported by the module, as well as their types and any related type definitions.

The most important ones you should be aware of are listed here. Spend some time browsing their manual pages and experiment with them in the shell:

array
> The array module contains an abstract data type for functional, extensible arrays. They can have a fixed size, or grow as needed. The module contains functionality to set and inspect values as well as to define recursions over them.

calendar
> The calendar module provides functions to retrieve local and universal times as well as providing time conversions for the day of the week, date, and time. Time intervals can be computed, ranging from dates down to a microsecond granularity. The calendar module is based on the Gregorian calendar and the now/0 BIF.

dict
> The dict module is a simple key value dictionary, which allows you to store, retrieve, and delete elements, merge dictionaries, and traverse them.

erlang
> All BIFs are considered to be implemented in the erlang module. The manual page for this module lists all of the Erlang BIFs, differentiating between the generic ones and the ones that are specific to the VM, and therefore those that are auto-imported and those that are not.

file
> The file module provides an interface to the filesystem, allowing you to read, manipulate, and delete files.

filename
> The filename module allows you to write generic file manipulation and inspection functions that will work regardless of the file notation used by the underlying operating system.

io
> The io library module encapsulates the standard I/O server interface functions, allowing you to read and write strings to I/O devices, including stdout.

lists

> The lists list-manipulation module is without a doubt the most used library module in all major Erlang systems. It provides functions for inspecting, manipulating, and processing lists.

math

> All of the standard mathematical functions, including pi/0, sin/1, cos/1, and tan/1, are implemented in the math library module.

queue

> The queue module implements an abstract data type for FIFO queues.

random

> The random module, given a seed, provides a pseudorandom number generator.

string

> The string module contains an array of string processing functions. It differentiates itself from the lists module in that it takes into consideration the fact that the contents of the lists are ASCII characters.

timer

> The timer module contains functions that relate to time, including generation of events and conversion of various time formats to milliseconds, the main unit used by this module.

The Debugger

The Erlang debugger is a graphical tool providing mechanisms to debug sequential code and influence program execution. It allows the user to step through programs while inspecting and manipulating variables. You can set breakpoints that stop the execution as well as inspect the recursive stack and variable bindings at each level. This section should serve as only a brief introduction to the debugger, and should be enough to get you started in using it. There are many features and details we have not covered, all of which are documented in the Debugger User's Guide, available in the online documentation.

You start the debugger by typing **debugger:start()** and a monitor window appears. This window displays a list of trace-compiled modules, attached (traced) processes, and other debug-related settings (see Figure 3-2).

To trace a module, you first need to compile it with the debug_info flag. In your Unix shell, you do that using the following command:

```
erlc +debug_info Module.erl
```

From the Erlang shell, use either of the following two commands:

```
c(Module, [debug_info]).
```

```
compile:file(exception, [debug_info]).
```

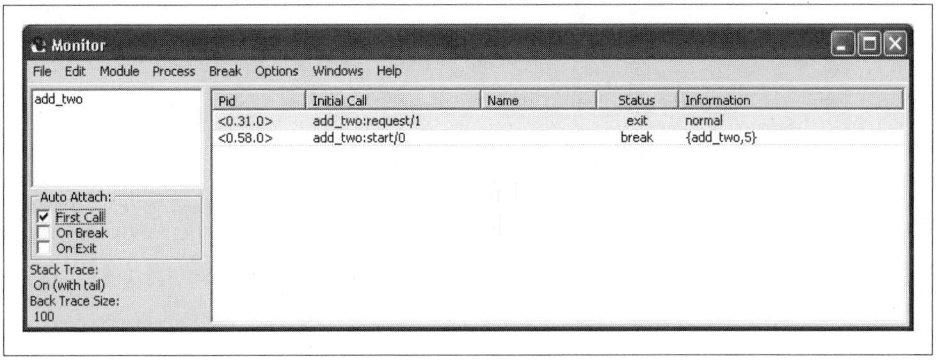

Figure 3-2. The monitor window

You then pick the module in the debugger by opening an interpret dialog window in the module menu. Trace-compiled modules will be listed in this window alongside nontrace-compiled ones (the latter will be shown in parentheses). Click the module you want to trace and a * will appear next to it. As soon as a process starts executing in the traced module, an entry will appear in the monitor window. You can double-click it, opening what we call an *attach window*. This window (see Figure 3-3) allows you to step through the code, view and manipulate variables, as well as inspect the recursive stack. Another way of opening an attach window is by preselecting one of the attach options in the monitor window. The options include hitting a breakpoint in the code, running an interpreted module and exiting, or simply calling an interpreted module for the first time.

You can insert breakpoints in the code by choosing an appropriate entry in the break menu or by clicking the line in the monitor or module window. Breakpoints can be set only in executable expressions, so breakpoints in function heads, patterns, or clause delimiters have no effect. Breakpoints can have a status of active or inactive. When an active breakpoint is reached, the breakpoint can, through triggers, be deleted, deactivated, or kept active.

Now that we are now done with sequential programming, it is time to move on to concurrency. But before doing so, make sure you grasp the key concepts of recursion and its various patterns, as concurrency builds on tail-recursive functions stop iterating only when we stop the process. Also spend some time browsing through the available library modules in the documentation, to be aware of what is available.

The following exercises will get you familiar with recursion and its different uses. Pay special attention to the different recursive patterns that we covered in this chapter. If you are having problems finding bugs or following the recursion, try using the debugger.

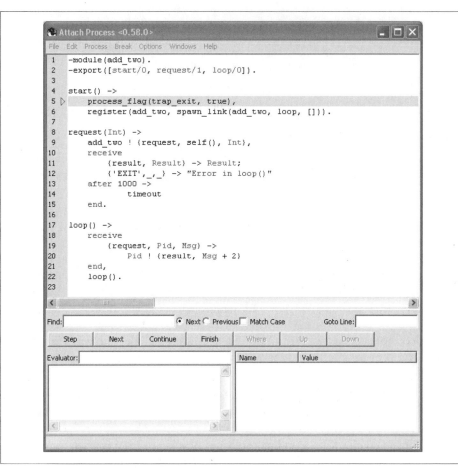

```
   Attach Process <0.58.0>                              _ □ ✕
 File  Edit  Process  Break  Options  Windows  Help
 1   -module(add_two).
 2   -export([start/0, request/1, loop/0]).
 3
 4   start() ->
 5 ▷     process_flag(trap_exit, true),
 6       register(add_two, spawn_link(add_two, loop, [])).
 7
 8   request(Int) ->
 9       add_two ! {request, self(), Int},
10       receive
11           {result, Result} -> Result;
12           {'EXIT',_,_} -> "Error in loop()"
13       after 1000 ->
14               timeout
15       end.
16
17   loop() ->
18       receive
19           {request, Pid, Msg} ->
20               Pid ! {result, Msg + 2}
21       end,
22       loop().
23
```

Find: ⦿ Next ○ Previous □ Match Case   Goto Line:
  ┌──────┐ ┌──────┐ ┌─────────┐ ┌──────┐ ┌──────┐ ┌──────┐ ┌──────┐
  │ Step │ │ Next │ │ Continue│ │Finish│ │ Where│ │ Up │ │ Down │
  └──────┘ └──────┘ └─────────┘ └──────┘ └──────┘ └──────┘ └──────┘
 Evaluator: Name   Value

Figure 3-3. The attach window

Exercises

Exercise 3-1: Evaluating Expressions

Write a function sum/1 which, given a positive integer N, will return the sum of all the integers between 1 and N.

Example:

   sum(5) ⇒ 15.

Write a function sum/2 which, given two integers N and M, where N =< M, will return the sum of the interval between N and M. If N > M, you want your process to terminate abnormally.

Example:

```
sum(1,3) ⇒ 6.
sum(6,6) ⇒ 6.
```

Exercise 3-2: Creating Lists

Write a function that returns a list of the format [1,2,..,N-1,N].

Example:

```
create(3) ⇒ [1,2,3].
```

Write a function that returns a list of the format [N, N-1,..,2,1].

Example:

```
reverse_create(3) ⇒ [3,2,1].
```

Exercise 3-3: Side Effects

Write a function that prints out the integers between 1 and N.

Hint: use io:format("Number:~p~n",[N]).

Write a function that prints out the even integers between 1 and N.

Hint: use guards.

Exercise 3-4: Database Handling Using Lists

Write a module db.erl that creates a database and is able to store, retrieve, and delete elements in it. The destroy/1 function will delete the database. Considering that Erlang has garbage collection, you do not need to do anything. Had the db module stored everything on file, however, you would delete the file. We are including the destroy function to make the interface consistent. You may *not* use the lists library module, and you have to implement all the recursive functions yourself.

Hint: use lists and tuples as your main data structures. When testing your program, remember that Erlang variables are single-assignment:

Interface:

```
db:new()                        ⇒ Db.
db:destroy(Db)                  ⇒ ok.
db:write(Key, Element, Db)      ⇒ NewDb.
db:delete(Key, Db)             ⇒ NewDb.
db:read(Key, Db)               ⇒{ok, Element} | {error, instance}.
db:match(Element, Db)          ⇒ [Key1, ..., KeyN].
```

Example:

```
1> c(db).
{ok,db}
```

```
2>  Db = db:new().
[]
3>  Db1 = db:write(francesco, london, Db).
[{francesco,london}]
4>  Db2 = db:write(lelle, stockholm, Db1).
[{lelle,stockholm},{francesco,london}]
5>  db:read(francesco, Db2).
{ok,london}
6>  Db3 = db:write(joern, stockholm, Db2).
[{joern,stockholm},{lelle,stockholm},{francesco,london}]
7>  db:read(ola, Db3).
{error,instance}
8>  db:match(stockholm, Db3).
[joern,lelle]
9>  Db4 = db:delete(lelle, Db3).
[{joern,stockholm},{francesco,london}]
10> db:match(stockholm, Db4).
[joern]
11>
```

 Due to single assignment of variables in Erlang, you need to assign the updated database to a new variable every time. Use f() to forget existing variable bindings in the shell.

Exercise 3-5: Manipulating Lists

Write a function that, given a list of integers and an integer, will return all integers smaller than or equal to that integer.

Example:

    filter([1,2,3,4,5], 3) ⇒ [1,2,3].

Write a function that, given a list, will reverse the order of the elements.

Example:

    reverse([1,2,3]) ⇒ [3,2,1].

Write a function that, given a list of lists, will concatenate them.

Example:

    concatenate([[1,2,3], [], [4, five]]) ⇒ [1,2,3,4,five].

Hint: you will have to use a help function and concatenate the lists in several steps.

Write a function that, given a list of nested lists, will return a flat list.

Example:

    flatten([[1,[2,[3],[]]], [[[4]]], [5,6]]) ⇒ [1,2,3,4,5,6].

Hint: use concatenate to solve flatten.

Exercise 3-6: Sorting Lists

Implement the following sort algorithms over lists:

Quicksort
> The head of the list is taken as the pivot; the list is then split according to those elements smaller than the pivot and the rest. These two lists are then recursively sorted by `quicksort`, and joined together, with the pivot between them.

Merge sort
> The list is split into two lists of (almost) equal length. These are then sorted separately and their results merged in order.

Exercise 3-7: Using Library Modules

Implement the database-handling list in Exercise 3-4 using the `lists` module library functions. Maintain the same interface to the `db` module, allowing your two modules to be interchangeable. How much shorter is your solution?

Exercise 3-8: Evaluating and Compiling Expressions

This exercise asks you to build a collection of functions that manipulate arithmetical expressions. Start with an expression such as the following:

```
((2+3)-4)    4    ~((2*3)+(3*4))
```

which is fully bracketed and where you use a tilde (~) for unary minus.

First, write a *parser* for these, turning them into Erlang representations, such as the following:

```
{minus, {plus, {num, 2}, {num,3}}, {num, 4}}
```

which represents `((2+3)-4)`. We call these *exps*. Now, write a number of functions:

- An *evaluator*, which takes an exp and returns its value
- A *pretty printer*, which will turn an exp into a string representation
- A *compiler*, which transforms an exp into a sequence of code for a stack machine to evaluate the exp
- A *simulator* which will implement expressions for the stack machine
- A *simplifier*, which will simplify an expression so that 0*e is transformed to 0, 1*e to e, and so on (there are quite a lot of others to think of!)

You can also extend the collection of expressions to add *conditionals*:

```
if ((2+3)-4) then 4 else ~((2*3)+(3*4))
```

where the value returned is the "then" value if the "if" expression evaluates to 0, and it is the "else" value otherwise.

You could also add *local definitions*, such as the following:

```
let c = ((2+3)-4) in ~((2*c)+(3*4))
```

Or you could add variables, which are set and then used in subsequent expressions.

For all of these extensions, you'll need to think about how to modify all the functions that you have written.

Exercise 3-9: Indexing

A *raw document* is a list of lines (i.e., strings), whereas a *document* is a list of words. Write a function to read a text file into a raw document and then a document.

You want to write an index for a document. This will give a list of words paired with their occurrences, so that the word *Erlang* might have the following entry:

```
{ "Erlang", [1,1,2,4,5,6,6,98,100,102,102] }
```

Write a function that will print this in a *readable form*, such as the following:

```
"Erlang       1-2,4-6,98,100,102"
```

so that duplicates are removed and adjacent numbers are put into a range. You might like to think of doing this via a function that turns the earlier list of occurrences into a list such as this:

```
[{1,2},{4,6},{98,98},{100,100},{102,102}]
```

through a sequence of transformations.

Exercise 3-10: Text Processing

Write a function that will take unstructured text such as this:

```
Write a function that will print this in a readable form,
so that duplicates are removed and adjacent numbers are put into a
range. You might like to think of doing this via a function which turns
the earlier list of occurrences into a list like
[{1,2},{4,6},{98,98},{100,100},{102,102}]
through a sequence of transformations.
```

and transforms it into *filled* text such as this:

```
Write a function that will print this
in a readable form, so that duplicates
are removed and adjacent numbers are put
into a range. You might like to think of
doing this via a function which turns
the earlier list of occurrences into a
list like
[{1,2},{4,6},{98,98},{100,100},{102,102}]
through a sequence of transformations.
```

When you hit a blank line, stop filling. A more fiddly exercise is to *justify* the resulting text:

```
Write  a function  that will  print this
in a   readable form,  so that duplicates
are removed and adjacent numbers are put
into a range. You might like to think of
doing  this via  a function  which turns
the  earlier list of occurrences  into a
list                                 like
[{1,2},{4,6},{98,98},{100,100},{102,102}]
through a sequence of transformations.
```

You don't need to justify the last line.

Concurrent Programming

Concurrency is the ability for different functions to execute in parallel without affecting each other unless explicitly programmed to do so. Each concurrent activity in Erlang is called a *process*. The only way for processes to interact with each other is through *message passing*, where data is sent from one process to another. The philosophy behind Erlang and its concurrency model is best described by Joe Armstrong's tenets:

- The world is concurrent.
- Things in the world don't share data.
- Things communicate with messages.
- Things fail.

The concurrency model and its error-handling mechanisms were built into Erlang from the start. With lightweight processes, it is not unusual to have hundreds of thousands, even millions, of processes running in parallel, often with a small memory footprint. The ability of the runtime system to scale concurrency to these levels directly affects the way programs are developed, differentiating Erlang from other concurrent programming languages.

What if you were to use Erlang to write an instant messaging (IM) server, supporting the transmission of messages between thousands of users in a system such as Google Talk or Facebook? The Erlang design philosophy is to spawn a new process for every event so that the program structure directly reflects the concurrency of multiple users exchanging messages. In an IM system, an event could be a presence update, a message being sent or received, or a login request. Each process will service the event it handles, and terminate when the request has been completed.

You could do the same in C or Java, but you would struggle when scaling the system to hundreds of thousands of concurrent events. An option might be to have a pool of processes handling specific event types or particular users, but certainly not a new process for every event. Erlang gets away with this because it does not use native threads to represent processes. It has its own scheduler in the virtual machine (VM), making the creation of processes very efficient while at the same time minimizing their memory footprint. This efficiency is maintained regardless of the number of concurrent processes in the system. The same argument applies for message passing, where the time to send a message is negligible and constant, regardless of the number of processes. This chapter introduces concurrent programming in Erlang, letting you in on one of the most powerful concurrency models available today.

Creating Processes

So far, we've looked at executing sequential code in a single process. To run concurrent code, you have to create more processes. You do this by spawning a process using the spawn(Module, Function, Arguments) BIF. This BIF creates a new process that evaluates the Function exported from the module Module with the list of Arguments as parameters. The spawn/3 BIF returns a *process identifier*, which from now on we will refer to as a *pid*.

In Figure 4-1, the process we call Pid1 executes the spawn BIF somewhere in its program. This call results in the new process with process identifier Pid2 being created. Process identifier Pid2 is returned as a result of the call to spawn, and will typically be bound to a variable in an expression of the following format:

```
Pid2 = spawn(Module, Function, Arguments).
```

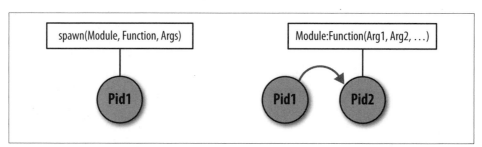

Figure 4-1. Before and after calling spawn

The pid of the new process, Pid2, at this point is known only within the process Pid1, as it is a local variable that has not been shared with anybody. The spawned process starts executing the exported function passed as the second argument to the BIF, and the arity of this function is dictated by the length of the *list* passed as the third argument to the BIF.

 A common error when you start programming Erlang is to forget that the third argument to spawn is a *list* of arguments, so if you want to spawn the function m:f/1 with the argument a, you need to call:

```
spawn(m, f, [a])
```

not:

```
not spawn(m, f, a).
```

Once spawned, a process will continue executing and remain alive until it terminates. If there is no more code to execute, a process is said to terminate *normally*. On the other hand, if a runtime error such as a bad match or a case failure occurs, the process is said to terminate *abnormally*.

Spawning a process will never fail, even if you are spawning a nonexported or even a nonexistent function. As soon as the process is created and spawn/3 returns the pid, the newly created process will terminate with a runtime error:

```
1> spawn(no_module, nonexistent_function, []).
<0.32.0>

=ERROR REPORT==== 29-Feb-2008::21:48:29 ===
Error in process <0.32.0> with exit value:
  {undef,[{no_module,nonexistent_function,[]}]}
```

In the preceding example, note how the error report is formatted. It is different from the ones you saw previously, as the error does not take place in the shell, but in the process with pid <0.32.0>. If the error occurs in a spawned process, it is detected by another part of the Erlang runtime system called the *error logger*, which by default prints an error report in the shell using the format shown earlier. Errors detected by the shell are instead formatted in a more readable form.

The processes() BIF returns a list of all of the processes running in the system. In most cases, you should have no problems using the BIF, but there have been extreme situations in large systems where calling processes() from the shell has been known to result in the runtime system running out of memory!* Don't forget that in industrial applications, you might be dealing with millions of processes running concurrently. In the current implementation of the runtime system, the absolute limit is in the hundreds of millions. Check the Erlang documentation for the latest figures. The default number is much lower, but you can easily change it by starting the Erlang shell with the command erl +P MaxProcceses, where MaxProcesses is an integer.

You can use the shell command i() to find out what the currently executing processes in the runtime system are doing. It will print the process identifier, the function used to spawn the process, the function in which the process is currently executing, as well as other information covered later in this chapter. Look at the example in the following shell printout. Can you spot the process that is running as the error logger?

```
2> processes().
[<0.0.0>,<0.2.0>,<0.4.0>,<0.5.0>,<0.7.0>,<0.8.0>,<0.9.0>,
 <0.10.0>,<0.11.0>,<0.12.0>,<0.13.0>,<0.14.0>,<0.15.0>,
 <0.17.0>,<0.18.0>,<0.19.0>,<0.20.0>,<0.21.0>,<0.22.0>,
 <0.23.0>,<0.24.0>,<0.25.0>,<0.26.0>,<0.30.0>]
3> i().
Pid                 Initial Call                Heap     Reds Msgs
Registered          Current Function            Stack
<0.0.0>             otp_ring0:start/2           987      2684   0
```

---

* Partially because the return values of the operations in the shell are cached.

```
init                init:loop/1                              2
<0.2.0>             erlang:apply/2                        2584    61740    0
erl_prim_loader     erl_prim_loader:loop/3                   5
<0.4.0>             gen_event:init_it/6                    610      219    0
error_logger        gen_event:fetch_msg/5                   11
<0.5.0>             erlang:apply/2                        1597      508    0
...
```

If you are wondering why the processes() BIF returned far more than 20 processes when you created only one that failed right after being spawned, you are not alone. Large parts of the Erlang runtime system are implemented in Erlang, the error_logger and the Erlang shell being two of the many examples. You will come across other system processes as you work your way through the remaining chapters of this book.

Message Passing

Processes communicate with each other using message passing. Messages are sent using the Pid ! Message construct, where Pid is a valid process identifier and Message is a value from *any* Erlang data type (see Figure 4-2).

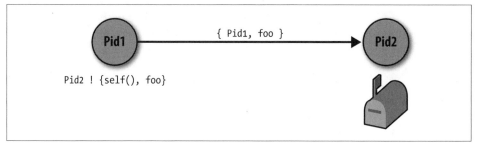

Figure 4-2. Message passing

Each Erlang process has a *mailbox* in which incoming messages are stored. When a message is sent, it is copied from the sending process into the recipient's mailbox for retrieval. Messages are stored in the mailbox in the order in which they are delivered. If two messages are sent from one process to another, the messages are guaranteed to be received in the same order in which they are sent. This guarantee is not extended to messages sent from different processes, however, and in this case the ordering is VM-dependent.

Sending a message will *never fail*; so if you try sending a message to a nonexistent process, it is thrown away without generating an error. Finally, message passing is *asynchronous*: a sending process will not be suspended after sending a message; it will instead immediately continue executing the next expression in its code.

To test sending messages in the shell, let's use the self/0 BIF, which returns the pid of the process in which it is evaluated. The Erlang shell is nothing other than an Erlang process in a read-evaluate-print loop, waiting for you to type in an expression. When you terminate an expression followed by a full stop (.) and press Enter, the shell evaluates what you typed in and prints out a result. Since the shell is an Erlang process, there is nothing stopping us from sending messages to it. To retrieve and display all the messages sent to the shell process, and therefore currently held in the process mailbox, you can use the shell command flush/0, which also has the effect of removing (or *flushing*) those messages from the mailbox:

```
1> Pid = self().
<0.30.0>
2> Pid ! hello.
hello
3> flush().
Shell got hello
ok
4> <0.30.0> ! hello.
* 1: syntax error before: '<'
5> Pid2 = pid(0,30,0).
<0.30.0>
6> Pid2 ! hello2.
hello2
7> flush().
Shell got hello2
ok
```

What is happening in the preceding example? In command 1, the BIF self() returns a pid, which in the shell is bound to the variable Pid and displayed as <0.30.0>. In commands 2 and 3 you see the message being sent to the Pid, and then flushed from the mailbox, using the flush() command in the shell.

You cannot type pids directly in a module or in the shell, as in both cases, they result in a syntax error; this is shown for the shell in command 4. You need either to bind the process identifiers to a variable when BIFs such as self and spawn return them, or generate a pid using the pid/3 shell function, as shown in command 5 and used in command 6. The flush() in command 7 shows that the message indeed went to the shell process.

Pid ! Message is a valid Erlang expression, and as with all valid expressions in Erlang, it has to return a value. The value, in this case, is the *message sent*. So if, for example, you need to send the same message to many processes, you can write either a sequence of message sends, such as Pid1!Msg,Pid2!Msg,Pid3!Msg, or a single expression, such as Pid3!Pid2!Pid1!Message, which is equivalent to writing Pid3!(Pid2!(Pid1!Message)), where Pid1!Message returns the message to send to Pid2, which in turn returns the message to be sent to Pid3.

As we already said, sending messages to nonexistent processes will always succeed. To test this, let's make the shell process crash with an illegal operation. Crashing is the same as an abnormal process termination, something that is considered normal in Erlang, in the sense that Erlang provides mechanisms to deal with it. We will cover abnormal process terminations in more detail in the next chapter, so until then, do not get alarmed. Making the shell crash will automatically result in a new shell process—in this example with pid `<0.38.0>`—being spawned by the runtime system.

With this in mind, we locate the shell pid, make the shell process terminate, and then send a message to it. Based on the semantics of message passing, this will result in the message being thrown away:

```
7> self().
<0.30.0>
8> 1/0.
** exception error: bad argument in an arithmetic expression
     in operator  '/'/2
        called as 1 / 0
9> self().
<0.38.0>
10> pid(0,30,0) ! hello.
hello
11> flush().
ok
```

The reason that message passing and `spawn` always succeed, even if the recipient process does not exist or the spawned process crashes on creation, has to do with *process dependencies*, or rather, their deliberate lack of dependencies. We say that process A *depends on* process B when the fact of B terminating can prevent A from functioning correctly.

Process dependencies are very important and will often influence your design. In massively concurrent systems, you do not want processes to depend on each other unless explicitly specified, and in such cases, you want to have as few dependencies as possible. To give a concrete example of this, imagine an IM server concurrently handling thousands of messages being exchanged by its users. Each message is handled by a process spawned for that particular function. If, due to a bug, one of these processes terminates, you would lose that particular message. Ensuring a lack of dependency between this process and the processes handling all the other messages guarantees that these messages are safely processed and delivered to their recipients regardless of the bug.

Receiving Messages

Messages are retrieved from the process mailbox using the `receive` clause. The `receive` clause is a construct delimited by the reserved words `receive` and `end`, and contains a number of *clauses*. These clauses are similar to `case` clauses, with a pattern in the *head* (to the left of the arrow) and a sequence of expressions in the *body* (to the right).

On executing the `receive` statement, the first (and oldest) message in the mailbox is pattern-matched against each pattern in the `receive` expression in turn:

- If a *successful match* occurs, the message is retrieved from the mailbox, the variables in the pattern are bound to the matching parts of the message, and the body of the clause is executed.

- If *none of the clauses matches*, the subsequent messages in the mailbox are pattern-matched one by one against all of the clauses until either a message matches a clause or all of the messages have failed all of the possible pattern matches.

In the following example, if the message {reset, 151} is sent to the process executing the receive statement, the first clause will pattern-match, resulting in the variable Board being bound to the integer 151. This will result in the function reset(151) being called:

```
receive
  {reset, Board} -> reset(Board);
  _Other         -> {error, unknown_msg}
end
```

Assume now that two new messages—restart and {reset, 151}—are received in that order by the executing process. As soon as the execution flow of the process reaches the receive statement, it will try to match the oldest message in the mailbox, restart. Pattern matching will fail in the first clause, but will match successfully in the second, binding the variable _Other to the atom restart. You can follow this example in Figure 4-3.

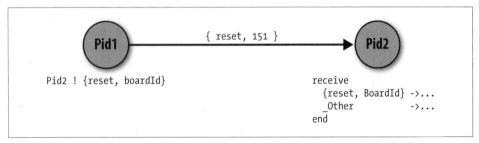

Figure 4-3. Selective receives

A receive statement will return the last evaluated expression in the body of the matched clause. In the example, this will be either the return value of the reset/1 call or the tuple {error, unknown_msg}. Although rarely done, it is possible to write expressions such as Result = receive Msg -> handle(Msg) end, in which a variable (here, Result) is bound to the return value of the receive clause. It is considered a better practice to make the receive clause the body of a separate function, and to bind the return value of the function to the variable.

Take a minute to think about the restart message in the example. Other than being a good practice, nothing is stopping us from sending the message in the format {restart}. It is a common misconception that messages have to be sent in tuples: this is not the case, as *messages can consist of any valid Erlang term*.

Using tuples with only one element is unnecessary, as they consume more memory and are slower to process. It is considered a bad practice to use them not only as messages, but also as terms in your programs. The guideline is that if your tuple has only one element, use the element on its own, without the tuple construct.

When none of the clauses in a case statement match, a runtime error occurs. What happens in receive statements? The syntax and semantics of receive are very similar to those of case, the main difference being that the process is *suspended* in receive statements until a message is matched, whereas in a case statement a runtime error occurs.

The Scheduler

It is probably time to see how *process scheduling* works in Erlang. Go back to the result of the i() shell commands, and you will notice a column headed Reds, an abbreviation for *reductions*. Every command in your program, whether it is a function call, an arithmetical operation, or a BIF, is assigned a number of reduction steps. The VM uses this value to measure the level of activity of a process.

When a process is dispatched, it is assigned a number of reductions[†] it is allowed to execute, a number which is reduced for every operation executed. As soon as the process enters a receive clause where none of the messages matches or its reduction count reaches zero, it is *preempted*. As long as BIFs are not being executed, this strategy results in a fair (but not equal) allocation of execution time among the processes.

Turning to the BIFs, all mathematical operations, for example, have the same number of reductions assigned to them, even if multiplication and division take much longer than addition or subtraction. BIFs such as lists:reverse and lists:member, known to vary in execution time based on their inputs, will interrupt their execution (this is called a *trap*) and bump up the reduction counter. Generally, new BIFs which are added to Erlang are implemented with traps by default. Old BIFs are reimplemented where necessary.

You can increment a process reduction counter using the BIF erlang:bump_reductions(Num), while erlang:yield() can be used to preempt the process altogether. Using yield/0 in the standard symmetric multiprocessing (SMP) emulator will have little, if any, effect. You should not make the behavior of the scheduler influence how you design and program your systems, as this behavior can change

---

[†] The number of reductions will vary between releases. In the R12 release, the number of reductions starts at 2,000 and is reduced by one for every operation. In R13, the number of initial reductions depends on the number of scheduler threads.

without notice in between releases. Knowing how it works, however, might help explain observations when inspecting and profiling your system.

In general, a `receive` statement has the following form:

```
receive
  Pattern1 when Guard1  -> exp11, .., exp1n;
  Pattern2 when Guard2  -> exp21, .., exp2n;
  ...
  Other                 -> expn1, .., expnn
end
```

The keywords used to delimit a `receive` clause are `receive` and `end`. Each pattern consists of any valid Erlang term, including bound and unbound variables as well as optional guards. The expressions are valid Erlang terms or legal expressions that evaluate to terms. The return value of the `receive` clause will be the return value of the last evaluated expression in the body executed, in this case, `exp1n`.

To ensure that the `receive` statement always retrieves the first message in the mailbox you could use an unbound variable (such as `Other` in the first example) or the "don't care" variable if you are not interested in its value.

Lightweight Processes Versus Threads

Erlang processes are lightweight processes whose creation, context switching, and message passing are managed by the VM. There is no relation between OS threads and Erlang processes, making concurrency-related operations not only independent of the underlying operating system, but also very efficient and highly scalable.

Benchmarks comparing Erlang concurrency models to their counterpart in C# or Java are magnitudes better, especially when increasing the number of simultaneously running processes in relation to creation and message passing times. With Erlang, we are dealing with one OS thread per processor (or core), unlike in Java and C#, where each process is represented by an OS thread. As a result, one might argue that comparing these concurrency models is like comparing apples and oranges. Indeed we are, but at the end of the day, if you need to write a massively concurrent system, it does not matter whether you use apples or oranges; what is important is that you are using the right tool for the job. In consequence, this comparison deserves to be made.

Selective and Nonselective Receives

Look at Figure 4-4. What can you deduce from the incoming messages in the `receive` clause about `Pid` being already bound? The variable `Pid` is bound to the value passed to `decode_digit` when it is called, so it is already bound to this value in the pattern part of the `receive` clause. On receiving a message, a successful match will occur only if the first element of the tuple of size 2 sent as a message is exactly equal to (remember the `=:=` construct?) the value stored in the variable `Pid`.

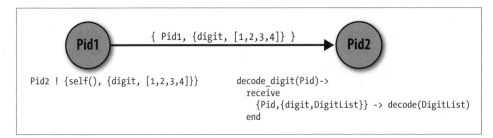

Figure 4-4. Selective receives with bound variables

We call this a *selective receive*, where we retrieve only the messages we are explicitly interested in based on certain criteria, *leaving the remaining messages in the mailbox*. Selective receives often select on process identifiers, but sequence references or other identifiers are also common, as are tagged tuples and guards. Now, contrast the bound variable `Pid` in Figure 4-4 with the unbound variable `DigitList` in Figure 4-5, which will be bound only once a message has been received.

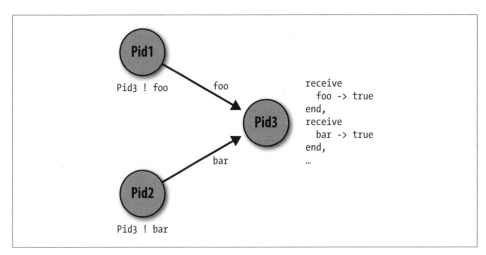

Figure 4-5. Selective reception of multiple messages

In concurrent systems, it is common for *race conditions* to occur. A race condition occurs when the behavior of a system depends on the order in which certain events occur: these events "race" to influence the behavior. Due to the indeterminate nature of concurrency it is not possible to determine the order in which messages sent by different processes will arrive at the receiving process. *This is when selective receive becomes crucial.*

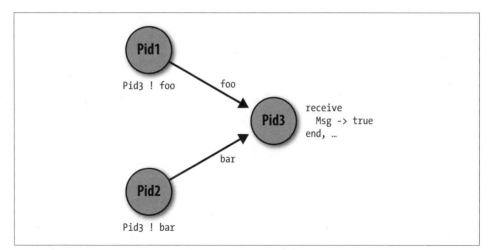

Figure 4-6. Receipt of messages regardless of the sending order

In Figure 4-5, `Pid3` will receive the message `foo` followed by `bar` regardless of the order in which they are delivered. Selective receive of multiple messages is useful when synchronizing processes in a rendezvous, or when data from several sources needs to be collected before it can be processed. Contrast this with a programming language in which it is possible to process messages only in the order in which they arrive: the code for this would have to deal with the possibility of `bar` preceding `foo` and of `foo` preceding `bar`; the code becomes more complex, and much more likely to contain potential flaws.

If the order of the messages is unimportant, you can just bind them to a variable. In Figure 4-6, the first message to arrive at the process `Pid3` will be processed, regardless of the order in which the messages were sent. The variable `Msg` in the `receive` statement will be bound to one of the atoms `foo` or `bar`, depending on which is delivered first.

Figure 4-7 demonstrates how processes share data with each other. The process with `PidA` will send a tagged tuple with its own process identifier, retrieved through a call to the BIF `self()`, to the process with `PidB`. `PidB` will receive it, binding `PidA`'s value to the variable `Pid`. A new tagged tuple is sent to `PidC`, which also pattern-matches the message in its `receive` statement and binds the value of `PidA` to the variable `Pid`. `PidC` now uses the value bound to `Pid` to send the message `foo` back to `PidA`. In this way, processes can share information about each other, allowing communication between processes that initially did not have knowledge of each other.

As processes do not share memory, the only way for them to share data is through message passing. Passing a message results in the data in the message being copied from the heap of the sending process to the heap of the receiving one, so this does not result in the two processes sharing a storage location (which each might read or write) but only in them each having their own copy of the data.

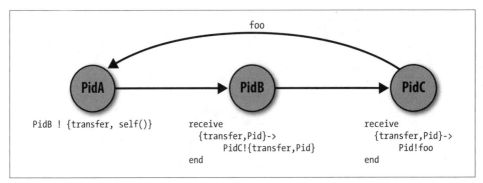

Figure 4-7. Sharing Pid data between processes

An Echo Example

Now that we have covered process creation and message passing, let's use `spawn`, `send`, and `receive`, in a small program. Open your editor and copy the contents of Example 4-1 or download it from the book's website. When doing so, do not forget to export the function you are spawning, in this case `loop/0`. In the example, pay particular notice to the fact that two different processes will be executing and interacting with each other using code defined in the same module.

Example 4-1. The echo process

```erlang
-module(echo).
-export([go/0, loop/0]).

go() ->
    Pid = spawn(echo, loop, []),
    Pid ! {self(), hello},
    receive
      {Pid, Msg} ->
        io:format("~w~n",[Msg])
    end,
    Pid ! stop.


loop() ->
    receive
      {From, Msg} ->
        From ! {self(), Msg},
        loop();
      stop ->
        true
    end.
```

So, what does this program do? Calling the function `go/0` will initiate a process whose first action is to spawn a child process. This child process starts executing in the `loop/0` function and is immediately suspended in the **receive** clause, as its mailbox is empty. The parent, still executing in go/0, uses the Pid for the child process, which is bound as a return value from the spawn BIF, to send the child a message containing a tuple with the parent's process identifier (given as a result of calling self()) and the atom hello.

As soon as the message is sent, the parent is suspended in a receive clause. The child, which is waiting for an incoming message, successfully pattern-matches the {Pid, Msg} tuple where Pid is matched to the process identifier belonging to the parent and Msg is matched to the atom hello. The child process uses the Pid to return the message {self(), hello} back to the parent, where this call to self() returns the pid of the child. See Figure 4-8 for a visual depiction of this process.

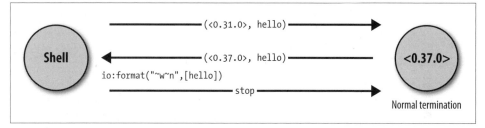

Figure 4-8. Sequence diagram for Example 4-1

At this point, the parent is suspended in the receive clause, and is waiting for a message. Note that it will only pattern-match on the tuple {Pid, Msg}, where the variable Pid is already bound (as a result of the spawn BIF) to the pid of the child process. This is a good (but not entirely secure) way to ensure that the message you receive is, in fact, a message you are expecting, and not just any message consisting of a tuple with two elements sent by another process. The message arrives and is successfully pattern-matched. The atom hello is bound to the Msg variable, which is passed as an argument to the io:format/2 call, printing it out in the shell. As soon as the parent has printed the atom hello in the shell, it sends the atom stop back to the child.

What has the child been doing while the parent was busy receiving the reply and printing it? Remember that processes will terminate if they have no more code to execute, so to avoid terminating, the child called the loop/0 function recursively, suspending it in the receive clause. It receives the stop message sent to it by its parent, returns the atom true as its result, and terminates normally.

Try running the program in the shell and see what happens:

```
1> c(echo).
{ok,echo}
2> echo:go().
hello
stop
```

The atom hello is clearly the result of the io:format/2 call, but where does the atom stop come from? It is the value returned as the result of calling echo:go/0. To further familiarize yourself with concurrency, experiment with the echo example, putting `io:format/2` statements in the `loop/0` process and sending different messages to it. You could also experiment with the `go/0` process, allowing it to send and receive more than one message. When experimenting, you will most likely get the shell to hang in a `receive` clause that will not match. If this happens, you will need to kill the shell and start again.

Registered Processes

It is not always practical to use pids to communicate with processes. To use a pid, a process needs to be notified of it and store its value. It is common to *register* processes that offer specific services with an *alias*, a name that can be used instead of the pid. You register a process with the `register(Alias, Pid)` BIF, where `Alias` is an atom and `Pid` is the process identifier. You do not have to be a parent or a child of the process to call the `register` BIF; you just need to know its process identifier.

Once a process has been registered, any process can send a message to it without having to be aware of its identifier (see Figure 4-9). All the process needs to do is use the `Alias !` `Message` construct. In programs, the alias is usually hardcoded in. Other BIFs which are directly related to process registration include `unregister(Pid)`; `registered()`, which returns a list of registered names; and `whereis(Alias)`, which returns the pid associated with the `Alias`.

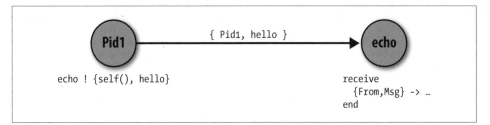

Figure 4-9. Sending a message to a registered process

Look at Example 4-2, which is a variant of Example 4-1. We have removed the `Pid!stop` expression at the end of the `go/0` function, and instead of binding the return value of `spawn/3`, we pass it as the second argument to the `register` BIF. The first argument to `register` is `echo`, the atom we use to name the process. This alias is used to send the message to the newly spawned child.

Example 4-2. The registered echo process

```
-module(echo).
-export([go/0, loop/0]).

go() ->
  register(echo, spawn(echo, loop, [])),
```

```
    echo ! {self(), hello},
    receive
      {_Pid, Msg} ->
        io:format("~w~n",[Msg])
    end.


loop() ->
  receive
    {From, Msg} ->
      From ! {self(), Msg},
      loop();
    stop ->
      true
  end.
```

It is not mandatory, but it is considered a good practice to give your process the same name as the module in which it is defined.

Update your echo module with the changes we just discussed and try out the new BIFs you have just read about in the shell. Test the new implementation of echo, inspecting its state with the i() and regs() shell commands. Note how the shell process sends the stop message to the echo process without knowing its pid, and how whereis/1 returns undefined if the process does not exist:

```
1> c(echo).
{ok,echo}
2> echo:go().
hello
ok
3> whereis(echo).
<0.37.0>
4> echo ! stop.
stop
5> whereis(echo).
undefined
6> regs().

** Registered procs on node nonode@nohost **
Name                    Pid          Initial Call          Reds Msgs
application_controlle   <0.5.0>      erlang:apply/2         4426    0
code_server             <0.20.0>     erlang:apply/2       112203    0
ddll_server             <0.10.0>     erl_ddll:init/1          32    0
erl_prim_loader         <0.2.0>      erlang:apply/2       206631    0
error_logger            <0.4.0>      gen_event:init_it/6     209    0
file_server             <0.19.0>     erlang:apply/2           12    0
file_server_2           <0.18.0>     file_server:init/1    25411    0
global_group            <0.17.0>     global_group:init/1      71    0
global_name_server      <0.12.0>     global:init/1            60    0
inet_db                 <0.15.0>     inet_db:init/1          103    0
init                    <0.0.0>      otp_ring0:start/2      5017    0
kernel_safe_sup         <0.26.0>     supervisor:kernel/1      61    0
kernel_sup              <0.9.0>      supervisor:kernel/1    1377    0
rex                     <0.11.0>     rpc:init/1               44    0
```

```
user                    <0.23.0>       user:server/2        1459    0

** Registered ports on node nonode@nohost **
Name                    Id              Command
ok
```

The shell command `regs()` prints out all the registered processes. It might be an alternative to `i()` when retrieving system information in a system with large quantities of processes. In the preceding example, the `echo` process is not among the processes listed, as we have stopped it. Instead, you are seeing all of the registered *system* processes.

 It is a feature of Erlang memory management that atoms are not garbage collected. Once you've created an atom, it remains in the atom table regardless of whether it is referenced in the code. This can be a potential problem if you decide to register *transient* processes with an alias derived from converting a string to an atom with the `list_to_atom/1` BIF. If you have millions of users logging on to your system every day and you create a registered process for the duration of their sessions, don't be surprised if you end up running out of memory.

You would be much better off storing the mapping of users to pids in a *session table*. It is best to register only processes with a long life span, and if you really must convert a string to use as an alias, use `list_to_existing_atom/1` to ensure that your system does not suffer memory leakages.

Sending messages to nonexistent registered processes causes the calling process to terminate with a `badarg` (see Figure 4-10). This behavior is different from sending a message to a process identifier for a nonexistent process, as registered processes are assumed to provide a service. The absence of a registered process is therefore treated as a bug. If your program might be sending messages to nonexistent registered processes and you do not want the calling process to terminate, wrap a `try ... catch` around the call.

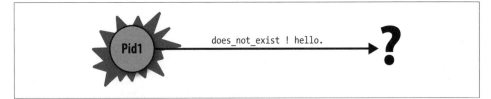

Figure 4-10. Sending messages to non-registered processes

Timeouts

You saw that if a process enters a `receive` statement and none of the messages matches, the process will get suspended. This could be similar to you going to your mailbox at home, discovering there is no mail, and being forced to wait there until the mailman arrives. It might be an option if you are waiting for very urgent mail or have nothing better to do. In most cases, though, all you want to do is check the mailbox, and if nothing has arrived, continue with your household chores. Erlang processes can do just that by using the **receive ... after** construct:

```
receive
  Pattern1 when Guard1 -> exp11, .., exp1n;
  Pattern2 when Guard2 -> exp21, .., exp2n;
  ...
  Other               -> expn1, .., expnn
after
  Timeout -> exp1,  .., expn
end
```

When a process reaches the **receive** statement and no messages pattern-match, it will wait for **Timeout** milliseconds. If after **Timeout** milliseconds no message has arrived, the expressions in the body of the **after** clause are executed. **Timeout** is an integer denoting the time in milliseconds, or the atom **infinity**. Using **infinity** as a timeout value is the same as not including the **after** construct. It is included, as **Timeout** can be a variable set every time the function is called, allowing the **receive ... after** clause to behave as desired in each call (see Figure 4-11).

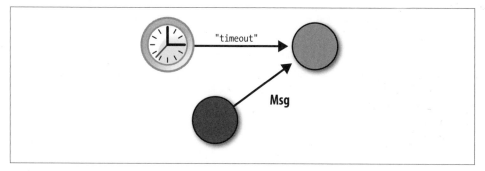

Figure 4-11. Receive timeouts

Assume you have a process registered with the alias **db**, which is acting as a database server. Every time you want to look up an item, you send the database a message and wait for a response. At busy times, however, the request might take too long to be processed, so you return a timeout error by using the **receive ... after** construct. When doing so, however, you will end up receiving the response from the server after the timeout, risking that your replies get out of sync with the sequence of requests sent to the database server. The next time you send the database a request, you will match the oldest message in your **receive** clause. This message will be the response sent back after the timeout, and not the response to the request you just sent. When using **receive ... after**, you need to cater to these cases by flushing your mailbox and ensuring it is empty. In doing so, your code might look something like this:

```
read(Key) ->
  flush(),
  db ! {self(),{read, Key}},
  receive
    {read,R}        -> {ok, R};
    {error, Reason} -> {error, Reason}
  after 1000        -> {error, timeout}
  end.

flush() ->
  receive
    {read, _}  -> flush();
    {error, _} -> flush()
  after 0        -> ok
end.
```

Another use for the `receive ... after` clause is to suspend a process for a period in milliseconds, or to send messages delayed by a certain amount of time. The definition of `sleep/1` in the following code is taken directly from the `timer` library module, while `send_after` will send a message to the calling process after `Time` milliseconds:

```
-module(my_timer).
-export([send_after/2, sleep/1, send/3]).

send_after(Time, Msg) ->
    spawn(my_timer, send, [self(),Time,Msg]).

send(Pid, Time, Msg) ->
  receive
  after
    Time ->
      Pid ! Msg
  end.

sleep(T) ->
  receive
  after
    T -> true
  end.
```

Benchmarking

In this chapter, we have been talking about the low process creation and message passing times in Erlang. To demonstrate them, let's run a benchmark in which the parent spawns a child and sends a message to it. Upon being spawned, the child creates a new process and waits for a message from its parent. Upon receiving the message, it terminates normally. The child's child creates yet another process, resulting in hundreds, thousands, and even millions of processes.

This is a sequential benchmark that will barely take advantage of SMP on a multicore system, because at any one time, only a couple of processes will be executing in parallel:

```
-module(myring).
-export([start/1, start_proc/2]).

start(Num) ->
  start_proc(Num, self()).

start_proc(0, Pid) ->
  Pid ! ok;

start_proc(Num, Pid) ->
  NPid = spawn(?MODULE, start_proc, [Num-1, Pid]),
  NPid ! ok,
  receive ok -> ok end.
```

Let's test the preceding example for 100,000, 1 million, and 10 million processes. To benchmark the program, we use the function call:

```
timer:tc(Module, Function, Arguments)
```

which takes a function and its arguments and executes it. It returns a tuple containing the time in microseconds it took to run the function alongside the return value of the function. Testing the program shows that it takes 0.48 seconds to spawn 100,000 processes, 4.2 seconds to spawn 1 million processes, and about 40 seconds to spawn 10 million processes. Try it out on your computer:

```
1> c(myring).
{ok,myring}
2> timer:tc(myring, start, [100000]).
{484000,ok}
3> timer:tc(myring, start, [1000000]).
{4289360,ok}
4> timer:tc(myring, start, [10000000]).
{40572800,ok}
```

Process Skeletons

There is a common *pattern* to the behavior of processes, regardless of their particular purpose. Processes have to be spawned and their aliases registered. The first action of newly spawned processes is to initialize the process loop data. The loop data is often the result of arguments passed to the spawn BIF and the initialization of the process. It is stored in a variable we refer to as the *process state*. The state is passed to a receive-evaluate function, which receives a message, handles it, updates the state, and passes it back as an argument to a tail-recursive call. If one of the messages it handles is a stop message, the receiving process will clean up after itself and terminate. This is a recurring theme among processes that we usually refer to as a *design pattern*, and it will occur regardless of the task the process has been assigned to perform. Figure 4-12 shows an example skeleton.

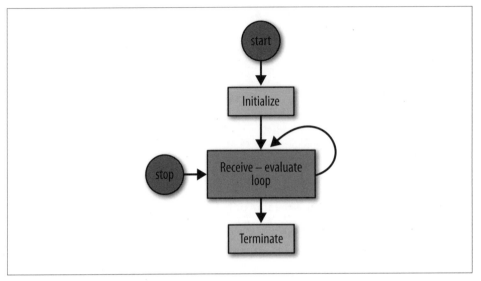

Figure 4-12. A process skeleton

From reoccurring patterns, let's now look at differences among processes:

- The arguments passed to the spawn BIF calls will differ from one process to another.
- You have to decide whether you should register a process, and, if you do register it, what alias should be used.
- In the function that initializes the process state, the actions taken will differ based on the tasks the process will perform.
- The storing of the process state might be generic, but its contents will vary among processes.
- When in the receive-evaluate loop, processes will receive different messages and handle them in different ways.
- And finally, on termination, the cleanup will vary from process to process.

So, even if a *skeleton* of generic actions exists, these actions are complemented by specific ones that are directly related to the specific tasks assigned to the process.

Tail Recursion and Memory Leaks

We mentioned earlier that if processes have no more code to execute, they terminate. Suppose you want to write an echo process that indefinitely continues to send back the message it has received (or that does this until you explicitly send it a message to stop). You would keep the Erlang process alive using a tail-recursive call to the function that contains the receive statement. We often call this function the receive/evaluate loop of the process. Its task is to receive a message, handle it, and then recursively call itself.

This is where the importance of tail recursion in concurrent programming becomes evident. As you do not know how many times the function is going to be called, you must ensure that it executes in *constant memory space* without increasing the recursive call stack every time a message is handled. It is common to have processes handling thousands of messages per second over sustained periods of not only hours, days, or months, but also years! Using tail recursion, where the very last thing the receive/evaluate function does is to call itself, you ensure that this nonstop operation is possible without memory leakages.

What happens when a message doesn't match any of the clauses in a `receive` statement? It remains in the process mailbox indefinitely, causing a memory leakage that over time could also cause the runtime system to run out of memory and crash. Not handling unknown messages should therefore be treated as a bug. Either these messages should not have been sent to this process in the first place, or they should be handled, possibly just by being retrieved from the mailbox and ignored.

The defensive approach of ignoring unknown messages with a "don't care" variable in the `receive` clause, even if convenient, might not always be the best approach. Messages not being handled should probably not have been sent to the process in the first place. And if they were sent on purpose, they were probably not matched because of a bug in one of the `receive` clauses. Throwing these messages away would only make the bug harder to detect. If you do throw unknown messages away, make sure you log their occurrence so that the bugs can be found and corrected.

Concurrency-Related Bottlenecks

Processes are said to act as *bottlenecks* when, over time, they are sent messages at a faster rate than they can handle them, resulting in large mailbox queues. How do processes with many messages in their inbox behave? The answer is badly.

First, the process itself, through a selective `receive`, might match only a specific type of message. If the message is the last one in the mailbox queue, the whole mailbox has to be traversed before this message is successfully matched. This causes a performance penalty that is often visible through high CPU consumption.

Second, processes sending messages to a process with a long message queue are penalized by increasing the number of reductions it costs to send the message. This is an attempt by the runtime system to allow processes with long message queues to catch up by slowing down the processes sending the messages in the first place. The latter bottleneck often manifests itself in a reduction of the overall throughput of the system.

The only way to discover whether there are any bottlenecks is to observe the throughput and message queue buildup when stress-testing the system. Simple remedies to message queue problems can be achieved by optimizing the code and fine-tuning the operating system and VM settings.

Another way to slow down message queue buildups is by suspending the processes generating the messages until they receive an acknowledgment that the message they have sent has been received and handled, effectively creating a synchronous call.

Replacing asynchronous calls with synchronous ones will reduce the maximum throughput of the system when running under heavy load, but never as much as the penalty paid when message queues start building up. Where bottlenecks are known to occur, it is safer to reduce the throughput by introducing synchronous calls, and thus guaranteeing a constant throughput of requests in the system with no degradation of service under heavy loads.

A Case Study on Concurrency-Oriented Programming

When on consulting assignments around the world working with developers coming from a C++ and Java background who have learned Erlang on their own, a common theme we have come across is the use of processes. This theme is irrespective of the experience level of the developers, and of what their system does. Instead of creating a process for every truly concurrent *activity* in the system, they tend to create one for every *task*. Programming concurrent applications in Erlang requires a different strategy for processes, which in turn means reasoning in a different way to what one may be accustomed to. The main difference from other concurrent languages is that with Erlang, processes are so cheap it is best to use a process for each truly concurrent activity in your system, not for every task. This case study is from one of Francesco's first consulting assignments outside of Ericsson, soon after Erlang had been released as open source, and it illustrates clearly what we mean by the difference between a task and an activity.

He worked with a group of engineers who were developing an IM proxy for Jabber. The system received a packet through a socket, decoded it, and took certain actions based on its content. Once the actions were completed, it encoded a reply and sent it to a different socket to be forwarded to the recipient. Only one packet at a time could come through a socket, but many sockets could simultaneously be receiving and handling packets.

As described in Figure 4-13, the original system did not have a process for every truly concurrent *activity*—the processing of a packet from end to end—but instead used a process for every different *task*—decoding, encoding, and so forth. Each open socket in Erlang was associated with a process that was receiving and sending data through this socket. Once the packet was received, it was forwarded to a process that handled the decoding. Once decoded, the decoding process forwarded it to the handler that processed it. The result was sent to an encoding process, which after formatting it, forwarded the reply to the socket process that held the open connection belonging to the recipient.

At its best performance, the system could process five simultaneous messages, with the decoding, handling, and encoding being the bottleneck, regardless of the number of simultaneously connected sockets. There were two other processes, one used for error

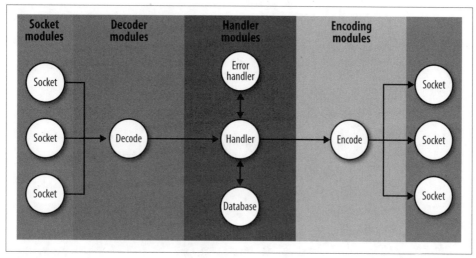

Figure 4-13. A badly designed concurrent system

handling, where all errors were passed, and one managing a database, where data reads, writes, and deletes were executed.

When reviewing the system, we identified what we believed was a truly concurrent activity in the system. It was not the action of decoding, handling, and encoding that was the answer, but the handling of the individual packets themselves. Having a process for every packet and using that process to decode, handle, and encode the packet meant that if thousands of packets were received simultaneously, they would all be processed in parallel. Knowing that a socket can receive only one packet at any one time meant that we could use this socket process to handle the call. Once the packet was received, a function call ensured that it was decoded and handled. The result (possibly an error) was encoded and sent to the socket process managing the connection belonging to the final recipient. The error handler and database processes were not needed, as the consistency of data through the serialization of destructive database operations could have been achieved in the handler process, as could the management of errors.

If you look at Figure 4-14, you will notice that on top of the socket processes, a database process was added to the rewritten program. This was to ensure that data consistency was maintained, as many processes accessing the same data might corrupt it as a result of a race condition. All destructive database operations such as write and delete were serialized through this process. Even if you can execute most of your activities in parallel, it is essential to identify activities that need serializing and place them in a process of their own. By taking care in identifying truly concurrent activites in your Erlang system, and spawning a process for each, you will ensure that you maximize the throughput while reducing the risk of bottlenecks.

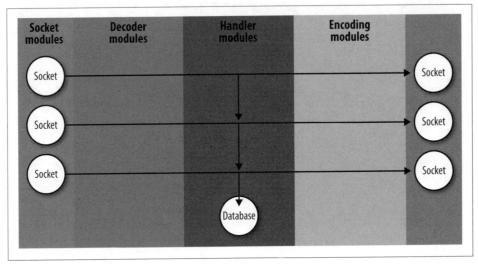

Figure 4-14. A process for each concurrent activity

Race Conditions, Deadlocks, and Process Starvation

Anyone who has programmed concurrent applications before moving to Erlang will have his favorite horror stories on memory corruption, deadlocks, race conditions, and process starvation. Some of these conditions arise as a result of shared memory and the need for semaphores to access them. Others are as a result of priorities. Having a "no shared data" approach, where the only way for processes to share data is by copying data from one process to another, removes the need for locks, and as a result, the vast majority of bugs related to memory corruption deadlocks and race conditions.

Problems in concurrent programs may also arise as a result of synchronous message passing, especially if the communication is across a network. Erlang solves this through asynchronous message passing. And finally, the scheduler, the per-process garbage collection mechanisms, and the massive level of concurrency that can be supported in Erlang systems ensure that all processes get a relatively fair time slice when executing. In most systems, you can expect a majority of the processes to be suspended in a `receive` statement, waiting for an event to trigger a chain of actions.

That being said, Erlang is not completely problem-free. You can avoid these problems, however, through careful and well-thought-out design. Let's start with *race conditions*. If two processes are executing the following code in parallel, what can go wrong?

```
start() ->
    case whereis(db_server) of
        undefined ->
            Pid = spawn(db_server, init, []),
            register(db_server, Pid),
            {ok, Pid};
        Pid when is_pid(Pid) ->
```

```
        {error, already_started}
    end.
```

Assume that the database server process has not been started and two processes simultaneously start executing the start/0 function. The first process calls whereis(db_server), which returns the atom undefined. This pattern-matches the first clause, and as a result, a new database server is spawned. Its process identifier is bound to the variable Pid. If, after spawning the database server, the process runs out of reductions and is preempted, this will allow the second process to start executing. The call whereis(db_server) by the second process also returns undefined, as the first process had not gotten as far as registering it. The second process spawns the database server and might go a little further than the first one, registering it with the name db_server. At this stage, the second process is preempted and the first process continues where it left off. It tries to register the database server it created with the name db_server but fails with a runtime error, as there already is a process with that name. What we would have expected is the tuple {error, already_started} to be returned, instead of a runtime error. Race conditions such as this one in Erlang are rare, but they do happen when you might least expect them. A variant of the preceding example was taken from one of the early Erlang libraries and reported as a bug in 1996.

A second potential problem to keep in mind involves *deadlocks*. A good design of a system based on client/server principles is often enough to guarantee that your application will be deadlock-free. The only rule you have to follow is that if process A sends a message and waits for a response from process B, in effect doing a synchronous call, process B is not allowed, anywhere in its code, to do a synchronous call to process A, as the messages might cross and cause the deadlock. Deadlocks are extremely rare in Erlang as a direct result of the way in which programs are structured. In those rare occasions where they slip through the design phase, they are caught at a very early stage of testing.[‡]

By calling the BIF process_flag(priority, Priority), where Priority can be set to the atom high, normal, or low, the behavior of the scheduler can be changed, giving processes with higher priority a precedence when being dispatched. Not only should you use this feature sparingly; in fact, you should not use it at all! As large parts of the Erlang runtime system are written in Erlang running at a normal priority, you will end up with deadlocks, starvation, and in extreme cases, a scheduler that gives low-priority processes more CPU time than its high-priority counterparts. With SMP, this behavior becomes even more non-deterministic. Endless flame wars and arguments regarding process and priorities have been fought on the Erlang-questions mailing list, deserving a whole chapter on the subject. We will limit ourselves to saying that under no circumstances should you use process priorities. A proper design of your concurrency

---

[‡] In 15 years of working with Erlang on systems with millions of lines of code, Francesco has come across only one deadlock that made it as far as the integration phase.

model will ensure that your system is well balanced and deterministic, with no process starvation, deadlocks, or race conditions. You have been warned!

The Process Manager

The process manager is a debugging tool used to inspect the state of processes in Erlang systems. Whereas the debugger concentrates on tracing the sequential aspects of your program, the process manager deals with the concurrent ones. You can start the process manager by writing **pman:start()** in the shell. A window will open (see Figure 4-15), displaying contents similar to what you saw when experimenting with the i() command. Double-clicking any of the processes will open a trace output window. You can choose your settings by picking options in the File menu.

For each process with an output window, you can trace all the messages that are sent and received. You can trace BIF and function calls, as well as concurrency-related events, such as processes being spawned or terminated. Your can also redirect your trace output from the window to a file. Finally, you can pick the inheritance level of your trace events. A very detailed and well-written user guide comes with the Erlang distribution that we recommend as further reading.

Pid	Current Function	Name	Msgs	Reds	Size
<0.46.0>	io:wait_io_mon_reply/2		0	18	239
<0.45.0>	gstk:worker_init/1		0	16432	612
<0.43.0>	gstk_port_handler:idle/1		0	19645	6767
<0.42.0>	gstk:loop/1		0	136694	6767
<0.41.0>	gs_frontend:loop/1	gs_frontend	0	10131	2588
<0.40.0>	pman_process:pinfo/2		0	50112	2675
<0.38.0>	shell:eval_loop/3		0	1627	1603
<0.34.0>	gen_server:loop/6	kernel_safe_sup	0	65	242
<0.33.0>	gen_server:loop/6		0	50	242
<0.32.0>	shell:get_command1/5		0	4464	17727
<0.31.0>	group:get_line1/4		0	781	255
<0.30.0>	group:server_loop/3	user	0	50	237
<0.29.0>	user_drv:server_loop/5	user_drv	0	522	618

☐ Hide System Processes   ☐ Auto-Hide New   # Hidden: 0

Figure 4-15. The process manager window

At the time of writing, because of its underlying TCL/TK graphics libraries that are no longer supported, the process manager can be unstable when running on Microsoft Windows operating systems.

This chapter introduced the basics of concurrency in Erlang, which is based on message passing between concurrent processes, rather than on shared memory. Message passing is asynchronous, and the selective **receive** facility, under which messages can be handled independently of the order in which they are received, allows modular and concise concurrent programs to be written. In the next chapter, we'll build on this introduction and look at design patterns for process-based systems.

Exercises

Exercise 4-1: An Echo Server

Write the server in Figure 4-16 that will wait in a receive loop until a message is sent to it. Depending on the message, it should either print its contents and loop again, or terminate. You want to hide the fact that you are dealing with a process, and access its services through a functional interface, which you can call from the shell.

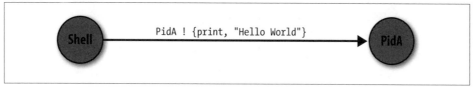

Figure 4-16. An echo server

This functional interface, exported in the echo.erl module, will spawn the process and send messages to it. The function interfaces are shown here:

```
echo:start() ⇒ ok
echo:print(Term) ⇒ ok
echo:stop() ⇒ ok
```

Hint: use the **register/2** built-in function, and test your echo server using the process manager.

Warning: use an internal message protocol to avoid stopping the process when you, for example, call the function echo:print(stop).

Exercise 4-2: The Process Ring

Write a program that will create *N* processes connected in a ring, as shown in Figure 4-17. Once started, these processes will send *M* number of messages around the ring and then terminate gracefully when they receive a quit message. You can start the ring with the call ring:start(M, N, Message).

There are two basic strategies to tackling this exercise. The first one is to have a central process that sets up the ring and initiates sending the message. The second strategy

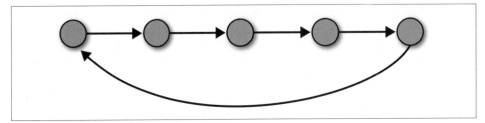

Figure 4-17. The process ring

consists of the new process spawning the next process in the ring. With this strategy, you have to find a method to connect the first process to the second process.

Regardless of the strategy you choose, make sure you have solved this exercise with pen and paper before you start coding. It differs from the ones you have solved before because you will have many processes executing the same function in the same module at the same time. Furthermore, processes will be using this function to interact with each other. When writing your program, make sure your code has many io:format statements in every loop iteration. This will give you a complete overview of what is happening (or not happening) and should help you solve the exercise.

CHAPTER 5

Process Design Patterns

Processes in Erlang systems can act as gateways to databases, handle protocol stacks, or manage the logging of trace messages. Although these processes may handle different requests, there will be similarities in how these requests are handled. We call these similarities *design patterns*. In this chapter, we are going to cover the most common patterns you will come across when working with Erlang processes.

The *client/server* model is commonly used for processes responsible for a *resource* such as a list of rooms, and *services* that can be applied on these resources, such as booking a room or viewing its availability. Requests to this server will allow clients (usually implemented as Erlang processes) to access these resources and services.

Another very common pattern deals with *finite state machines*, also referred to as FSMs. Imagine a process handling events in an instant messaging (IM) session. This process, or finite state machine as we should call it, will be in one of three states. It could be in an offline state, where the session with the remote IM server is being established. It could be in an online state, enabling the user to send and receive messages and status updates. And finally, if the user wants to remain online but not receive any messages or status updates, it could be in a busy state. State changes are triggered through process messages we call *events*. An IM server informing the FSM that the user is logged on successfully would cause a *state transition* from the offline state to the online state. Events received by the FSM do not necessarily have to trigger state transitions. Receiving an instant message or a status update would keep the FSM in an online state while a logout event would cause it to go from an online or busy state to the offline state.

The last pattern we will cover is the *event handler*. Event handler processes will receive messages of a specific type. These could be trace messages generated in your program or stock quotes coming from an external feed. Upon receiving these events, you might want to perform a set of actions such as triggering an SMS (Short Message Service message) or sending an email if certain conditions are met, or simply logging the time stamp and stock price in a file.

Many Erlang processes will fall into one of these three categories. In this chapter, we will look at examples of process design patterns, explaining how they can be used to program client/servers, finite state machines, and event handlers. An experienced Erlang programmer will recognize these patterns in the design phase of the project and use libraries and templates that are part of the OTP framework. For the time being, we will use Erlang without the OTP framework. We will introduce OTP behaviors in Chapter 12.

Client/Server Models

Erlang processes can be used to implement client/server solutions, where both clients and servers are represented as Erlang processes. A server could be a FIFO queue to a printer, window manager, or file server. The resources it handles could be a database, calendar, or finite list of items such as rooms, books, or radio frequencies. Clients use these resources by sending the server requests to print a file, update a window, book a room, or use a frequency. The server receives the request, handles it, and responds with an acknowledgment and a return value if the request was successful, or with an error if the request did not succeed (see Figure 5-1).

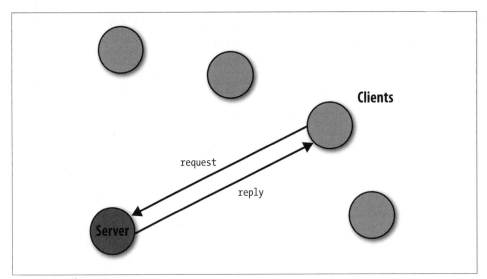

Figure 5-1. The client/server model

When implementing client/server behavior, clients and servers are represented as Erlang processes. Interaction between them takes place through the sending and receiving of messages. Message passing is often hidden in functional interfaces, so instead of calling:

```
printerserver ! {print, File}
```

a client would call:

```
printerserver:print(File)
```

This is a form of *information hiding*, where we do not make the client aware that the server is a process, that it could be registered, and that it might reside on a remote computer. Nor do we expose the message protocol being used between the client and the server, keeping the interface between them safe and simple. All the client needs to do is call a function and expect a return value.

Hiding this information behind a functional interface has to be done with care. The message response times will differ if the process is busy or running on a remote machine. Although this should in most cases not cause any problems, the client needs to be aware of it and be able to cope with a delay in response time. You also need to factor in that things can go wrong behind this function call. There might be a network glitch, the server process might crash, or there might be so many requests that the server response times become unacceptable.

If a client using the service or resource handled by the server expects a reply to the request, the call to the server has to be *synchronous*, as in Figure 5-2. If the client does not need a reply, the call to the server can be *asynchronous*. When you encapsulate synchronous and asynchronous calls in a function call, asynchronous calls commonly return the atom ok, indicating that the request was sent to the server. Synchronous calls will return the value expected by the client. These return values usually follow the format ok, {ok, Result}, or {error, Reason}.

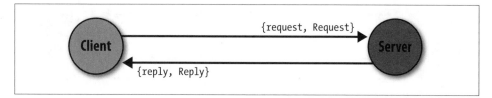

Figure 5-2. Synchronous client/server requests

A Client/Server Example

Enough with the theory! So that you understand what we are talking about, let's walk through a client/server example and test it in the shell. This server is responsible for managing radio frequencies on behalf of its clients, the mobile phones connected to the network. The phone requests a frequency whenever a call needs to be connected, and releases it once the call has terminated (see Figure 5-3).

When a mobile phone has to set up a connection to another subscriber, it calls the frequency:allocate() client function. This call has the effect of generating a synchronous message which is sent to the server. The server handles it and responds with either a message containing an available frequency or an error if all frequencies are being used. The result of the allocate/0 call will therefore be either {ok, Frequency} or {error, no_frequencies}.

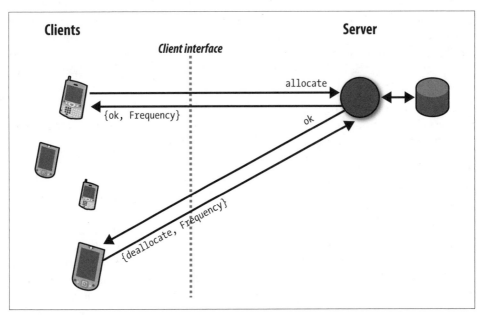

Figure 5-3. A frequency server

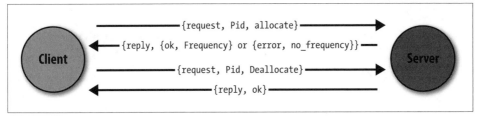

Figure 5-4. Frequency server message sequence diagram

Through a functional interface, we hide the message-passing mechanism, the format of these messages, and the fact that the frequency server is implemented as a registered Erlang process. If we were to move the server to a remote host, we could do so without having to change the client interface.

When the client has completed its phone call and releases the connection, it needs to deallocate the frequency so that other clients can reuse it. It does so by calling the client function `frequency:deallocate(Frequency)`. The call results in a message being sent to the server. The server can then make the frequency available to other clients and responds with the atom `ok`. The atom is sent back to the client and becomes the return value of the `deallocate/1` call. Figure 5-4 shows the message sequence diagram of this example.

The code for the server is in the `frequency` module. Here is the first part:

```
-module(frequency).
-export([start/0, stop/0, allocate/0, deallocate/1]).
-export([init/0]).

%% These are the start functions used to create and
%% initialize the server.

start() ->
  register(frequency, spawn(frequency, init, [])).

init() ->
  Frequencies = {get_frequencies(), []},
  loop(Frequencies).

% Hard Coded
get_frequencies() -> [10,11,12,13,14,15].
```

The **start** function spawns a new process that starts executing the **init** function in the **frequency** module. The **spawn** returns a pid that is passed as the second argument to the **register** BIF. The first argument is the atom **frequency**, which is the alias with which the process is registered. This follows the convention of registering a process with the same name as the module in which it is defined.

 Remember that when spawning a process, you have to export the **init/0** function as it is used by the **spawn/3** BIF. We have put this function in a separate **export** clause to distinguish it from the client functions, which are supposed to be called from other modules. Calling **frequency:init()** from anywhere in your code would be considered a very bad practice, and should not be done.

The newly spawned process starts executing in the **init** function. It creates a tuple consisting of the available frequencies, retrieved through the **get_frequencies/0** call, and a list of the allocated frequencies—initially given by the empty list—as the server has just been started. The tuple, which forms what we call the *state* or *loop data*, is bound to the **Frequencies** variable and passed as an argument to the receive-evaluate function, which in this example we've called **loop/1**.

In the **init/0** function, we use the variable **Frequencies** for readability reasons, but nothing is stopping us from creating the tuple directly in the **loop/1** call, as in the call **loop({get_frequencies(), []})**.

Here is how the *client functions* are implemented:

```
%%  The client Functions

stop()           -> call(stop).
allocate()       -> call(allocate).
deallocate(Freq) -> call({deallocate, Freq}).

%% We hide all message passing and the message
%% protocol in a functional interface.
```

```
call(Message) ->
  frequency ! {request, self(), Message},
  receive
    {reply, Reply} -> Reply
  end.
```

Client and supervisor[*] processes can interact with the frequency server using what we refer to as *client functions*. These exported functions include start, stop, allocate, and deallocate. They call the call/1 function, passing the message to be sent to the server as an argument. This function will encapsulate the message protocol between the server and its clients, sending a message of the format {request, Pid, Message}. The atom request is a tag in the tuple, Pid is the process identifier of the calling process (returned by calling the self() BIF in the calling process), and Message is the argument originally passed to the call/1 function.

When the message has been sent to the process, the client is suspended in the receive clause waiting for a response of the format {reply, Reply}, where the atom reply is a tag and the variable Reply is the actual response. The server response is pattern-matched, and the contents of the variable Reply become the return value of the client functions.

Pay special attention to how message passing and the message protocol have been abstracted to a format independent of the action relating to the message itself; this is what we referred to earlier as *information hiding*, allowing the details of the protocol and the message structure to be modified without affecting any of the client code.

Now that we have covered the code to start and interact with the frequency server, let's take a look at its receive-evaluate loop:

```
%% The Main Loop

loop(Frequencies) ->
  receive
    {request, Pid, allocate} ->
      {NewFrequencies, Reply} = allocate(Frequencies, Pid),
      reply(Pid, Reply),
      loop(NewFrequencies);
    {request, Pid , {deallocate, Freq}} ->
      NewFrequencies = deallocate(Frequencies, Freq),
      reply(Pid, ok),
      loop(NewFrequencies);
    {request, Pid, stop} ->
      reply(Pid, ok)
  end.

reply(Pid, Reply) ->
  Pid ! {reply, Reply}.
```

---

[*] We will cover supervisors in the next chapter.

The receive clause will accept three kinds of requests originating from the client functions, namely allocate, deallocate, and stop. These requests follow the same format defined in the call/1 function, that is, {request, Pid, Message}. The Message is pattern-matched in the expression and used to determine which clause is executed. This, in turn, determines the internal functions that are called. These internal functions will return the new loop data, which in our example consists of the new lists of available and allocated frequencies, and where needed, a reply to send back to the client. The client pid, sent as part of the request, is used to identify the calling process and is used in the reply/2 call.

Assume a client wants to *initiate* a call. To do so, it would request a frequency by calling the frequency:allocate() function. This function sends a message of the format {request, Pid, allocate} to the frequency server, pattern matching in the first clause of the receive statement. This message will result in the server function allocate(Frequencies, Pid) being called, where Frequencies is the loop data containing a tuple of allocated and available frequencies. The allocate function (defined shortly) will check whether there are any available frequencies:

- If so, it will return the updated loop data, where the newly allocated frequency has been moved from the available list and stored together with the pid in the list of allocated ones. The reply sent to the client is of the format {ok, Frequency}.

- If no frequencies are available, the loop data is returned unchanged and the {error, no_frequency} message is returned as a reply.

The Reply is sent to the reply(Pid, Message) call, which formats it to the internal client/server message format and sends it back to the client. The function then calls loop/1 recursively, passing the new loop data as an argument.

Deallocation works in a similar way. The client function results in the message {request, Pid, deallocate} being sent and matched in the second clause of the receive statement. This makes a call to deallocate(Frequencies, Frequency) and the deallocate function moves the Frequency from the allocated list to the deallocated one, returning the updated loop data. The atom ok is sent back to the client, and the loop/1 function is called recursively with the updated loop data.

If the stop request is received, ok is returned to the calling process and the server terminates, as there is no more code to execute. In the previous two clauses, loop/1 was called in the final expression of the case clause, but not in this case.

We complete this system by implementing the allocation and deallocation functions:

```
%% The Internal Help Functions used to allocate and
%% deallocate frequencies.

allocate({[], Allocated}, _Pid) ->
  {{[], Allocated}, {error, no_frequency}};
allocate({[Freq|Free], Allocated}, Pid) ->
  {{Free, [{Freq, Pid}|Allocated]}, {ok, Freq}}.
```

```
deallocate({Free, Allocated}, Freq) ->
  NewAllocated=lists:keydelete(Freq, 1, Allocated),
  {[Freq|Free], NewAllocated}.
```

The allocate/2 and deallocate/2 functions are local to the frequency module, and are what we refer to as *internal help* functions:

- If there are no available frequencies, allocate/2 will pattern-match in the first clause, as the first element of the tuple containing the list of available frequencies is empty. This clause returns the {error, no_frequency} tuple alongside the unchanged loop data.

- If there is at least one available frequency, the second clause will match successfully. The frequency is removed from the list of available ones, paired up with the client pid, and moved to the list of allocated frequencies.

The updated frequency data is returned by the allocate function. Finally, deallocate will remove the newly freed frequency from the list of allocated ones using the lists:keydelete/3 library function and concatenate it to the list of available frequencies.

This frequency allocator example has used all of the key sequential and concurrent programming concepts we have covered so far. They include pattern matching, recursion, library functions, process spawning, and message passing. Spend some time making sure you understand them. You should test the example using the debugger and the process manager, following the message passing protocols between the client and server and the sequential aspects of the loop function. You can see an example of the frequency allocator in action now:

```
1> c(frequency).
{ok,frequency}
2> frequency:start().
true
3> frequency:allocate().
{ok,10}
4> frequency:allocate().
{ok,11}
5> frequency:allocate().
{ok,12}
6> frequency:allocate().
{ok,13}
7> frequency:allocate().
{ok,14}
8> frequency:allocate().
{ok,15}
9> frequency:allocate().
{error,no_frequency}
10> frequency:deallocate(11).
ok
11> frequency:allocate().
{ok,11}
12> frequency:stop().
ok
```

A Process Pattern Example

Now let's look at similarities between the client-server example we just described and the process skeleton we introduced in Chapter 4. Picture an application, either a web browser or a word processor, which handles many simultaneously open windows centrally controlled by a window manager. As we aim to have a process for each truly concurrent activity, spawning a process for every window is the way to go. These processes would probably not be registered, as many windows of the same type could be running concurrently.

After being spawned, each process would call the `initialize` function, which draws and displays the window and its contents. The return value of the `initialize` function contains references to the widgets displayed in the window. These references are stored in the state variable and are used whenever the window needs updating. The state variable is passed as an argument to a tail-recursive function that implements the receive-evaluate loop.

In this loop function, the process waits for events originating in or relating to the window it is managing. It could be a user typing in a form or choosing a menu entry, or an external process pushing data that needs to be displayed. Every event relating to this window is translated to an Erlang message and sent to the process. The process, upon receiving the message, calls the `handle` function, passing the message and state as arguments. If the event were the result of a few keystrokes typed in a form, the `handle` function might want to display them. If the user picked an entry in one of the menus, the `handle` function would take appropriate actions in executing that menu choice. Or, if the event was caused by an external process pushing data, possibly an image from a webcam or an alert message, the appropriate widget would be updated. The receipt of these events in Erlang would be seen as a generic pattern in all processes. What would be considered specific and change from process to process is how these events are handled.

Finally, what if the process receives a `stop` message? This message might have originated from a user picking the Exit menu entry or clicking the Destroy button, or from the window manager broadcasting a notification that the application is being shut down. Regardless of the reason, a `stop` message is sent to the process. Upon receiving it, the process calls a `terminate` function, which destroys all of the widgets, ensuring that they are no longer displayed. After the window has been shut down, the process terminates because there is no more code to execute.

Look at the following process skeleton. Could you not fit all of the specific code into the `initialize/1`, `handle_msg/2`, and `terminate/1` functions?

```
-module(server).
-export([start/2, stop/1, call/2]).
-export([init/1]).

start(Name, Data) ->
  Pid = spawn(generic_handler, init,[Data])
```

```
      register(Name, Pid), ok.

  stop(Name) ->
    Name ! {stop, self()},
    receive {reply, Reply} -> Reply end.

  call(Name, Msg) ->
    Name ! {request, self(), Msg},
    receive {reply, Reply} -> Reply end.

  reply(To, Msg) ->
    To ! {reply, Msg}.

  init(Data) ->
    loop(initialize(Data)).

  loop(State) ->
    receive
      {request, From, Msg} ->
        {Reply,NewState} = handle_msg(Msg, State),
        reply(From, Reply),
        loop(NewState);
      {stop, From}  ->
        reply(From, terminate(State))
    end.

  initialize(...)    -> ...
  handle_msg(...,...) -> ...
  terminate(...)      -> ...
```

Using the generic code in the preceding skeleton, let's go through the GUI example one last time:

- The `initialize` function draws the window and displays it, returning a reference to the widget that gets bound to the `state` variable.

- Every time an event arrives in the form of an Erlang message, the event is taken care of in the `handle_msg` function. The call takes the message and the state as arguments and returns an updated `State` variable. This variable is passed to the recursive `loop` call, ensuring that the process is kept alive. Any reply is also sent back to the process where the request originated.

- If the `stop` message is received, `terminate` is called, destroying the window and all the widgets associated with it. The `loop` function is not called, allowing the process to terminate normally.

Finite State Machines

Erlang processes can be used to implement finite state machines. A finite state machine, or FSM for short, is a model that consists of a finite number of states and events. You can think of an FSM as a model of the world which will contain abstractions from the details of the real system. At any one time, the FSM is in a specific state. Depending on the incoming event and the current state of the FSM, a set of actions and a transition to a new state will occur (see Figure 5-5).

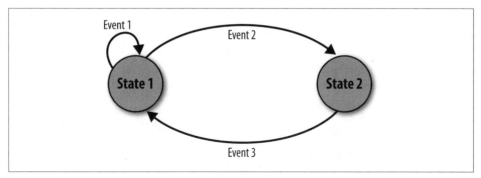

Figure 5-5. A finite state machine

In Erlang, each state is represented as a tail-recursive function, and each event is represented as an incoming message. When a message is received and matched in a `receive` clause, a set of actions are executed. These actions are followed by a state transition achieved by calling the function corresponding to the new state.

An FSM Example

As an example, think of modeling a fixed-line phone as a finite state machine (see Figure 5-6). The phone can be in the `idle` state when it is plugged in and waiting either for an incoming phone call or for a user to take it off the hook. If you receive an incoming call from your aunt,[†] the phone will start ringing. Once it has started ringing, the state will change from `idle` to `ringing` and will wait for one of two events. You can pretend to be asleep, hopefully resulting in your aunt giving up on you and putting the phone on her end back on the hook. This will result in the FSM going back to the `idle` state (and you going back to sleep).

If instead of ignoring it, you take your phone off the hook, it would stop ringing and the FSM would move to the `connected` state, leaving you to talk to your heart's content. When you are done with the call and hang up, the state reverts to `idle`.

If the phone is in the `idle` state and you take it off the hook, a dial tone is started. Once the dial tone has started, the FSM changes to the `dial` state and you enter your aunt's phone number. Either you can hang up and your FSM goes back to the `idle` state, or your aunt picks up and you go to the `connected` state.

State machines are very common in all sorts of processing applications. In telecom systems, they are used not only to handle the state of equipment, as in the preceding example, but also in complex protocol stacks. The fact that Erlang handles them

---

† Or any other relative of your choice who tends to call you very early on a Saturday morning.

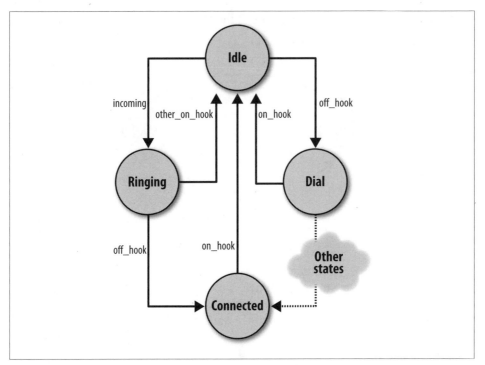

Figure 5-6. Fixed-line phone finite state machine

graciously is not a surprise. When prototyping with the early versions of Erlang between 1987 and 1991, it was the Plain Old Telephony System (POTS) finite state machines described in this section that the development team used to test their ideas of what Erlang should look like.

With a tail-recursive function for every state, actions implemented as function calls, and events represented as messages, this is what the code for the `idle` state would look like:

```erlang
idle() ->
  receive
    {Number, incoming} ->
      start_ringing(),
      ringing(Number);
    off_hook ->
      start_tone(),
      dial()
  end.

ringing(Number) ->
  receive
    {Number, other_on_hook} ->
      stop_ringing(),
      idle();
    {Number, off_hook} ->
```

```
        stop_ringing(),
        connected(Number)
    end.

start_ringing() -> ...
start_tone()    -> ...
stop_ringing()  -> ...
```

We leave the coding of the functions for the other states as an exercise.

A Mutex Semaphore

Let's look at another example of a finite state machine, this time implementing a mutex semaphore. A *semaphore* is a process that serializes access to a particular resource, guaranteeing mutual exclusion. Mutex semaphores might not be the first thing that comes to mind when working with Erlang, as they are commonly used in languages with shared memory. However, they can be used as a general mechanism for managing any resource, not just memory.

Assume that only one process at a time is allowed to use the file server, thus guaranteeing that no two processes are simultaneously reading or writing to the same file. Before making any calls to the file server, the process wanting to access the file calls the `mutex:wait()` function, putting a lock on the server. When the process has finished handling the files, it calls the function `mutex:signal()`, removing the lock (see Figure 5-7).

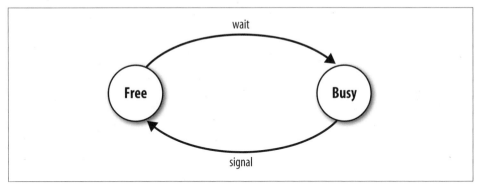

Figure 5-7. The mutex semaphore state diagram

If a process called `PidB` tries to call `mutex:wait()` when the semaphore is busy with `PidA`, `PidB` is suspended in its `receive` clause until `PidA` calls `signal/0`. The semaphore becomes available, and the process whose wait message is first in the message queue, `PidB` in our case, will be allowed to access the file server. The message sequence diagram in Figure 5-8 demonstrates this.

Look at the following code to get a feel for how to use tail-recursive functions to denote the states, and messages to denote events. And before reading on, try to figure out what the `terminate` function should do to clean up when the mutex is terminated.

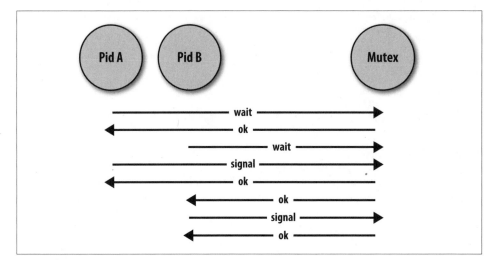

Figure 5-8. The mutex message sequence diagram

```erlang
-module(mutex).
-export([start/0, stop/0]).
-export([wait/0, signal/0]).
-export([init/0]).

start() ->
  register(mutex, spawn(?MODULE, init, [])).

stop() ->
  mutex ! stop.

wait() ->
  mutex ! {wait, self()},
  receive ok -> ok end.

signal() ->
  mutex ! {signal, self()}, ok.

init() ->
  free().

free() ->
  receive
    {wait, Pid} ->
      Pid ! ok,
      busy(Pid);
    stop ->
      terminate()
  end.

busy(Pid) ->
  receive
    {signal, Pid} ->
      free()
```

```
      end.

    terminate() ->
      receive
        {wait, Pid} ->
          exit(Pid, kill),
          terminate()
      after
        0 -> ok
      end.
```

The `stop/0` function sends a stop message that is handled only in the `free` state. Prior to terminating the mutex process, all processes that are waiting for or holding the semaphore are allowed to complete their tasks. However, any process that attempts to wait for the semaphore after `stop/0` is called will be killed unconditionally in the `terminate/0` function.

Event Managers and Handlers

Try to picture a process that receives trace events generated in your system. You might want to do many things with these trace events, but you might not necessarily want to do all of them at the same time. You probably want to log all the trace events to file. If you are in front of the console, you might want to print them to standard I/O. You might be interested in statistics to determine how often certain errors occur, or if the event requires some action to be taken, you might want to send an SMS or SNMP[‡] trap.

At any one time, you will want to execute some, if not all, of these actions, and toggle between them. But if you walk away from your desk, you might want to turn the logging to the console off while maintaining the gathering of statistics and logging to file.

An *event manager* does what we just described. It is a process that receives a specific type of event and executes a set of actions determined by the type of event. These actions can be added and removed dynamically throughout the lifetime of the process, and are not necessarily defined or known when the code implementing the process is first written. They are collected in modules we call the *event handlers*.

Large systems usually have an event manager for every type of event. Event types commonly include alarms, equipment state changes, errors, and trace events, just to mention a few. When they are received, one or more actions are applied to each event.

The most common form of event manager found in almost all industrial-grade systems handles alarms (see Figure 5-9). Alarms are raised when a problem occurs and are cleared when it goes away. They might require automated or manual intervention, but this is not always the case. An alarm would be raised if the data link between two devices

---

[‡] SNMP stands for Simple Network Management Protocol. It is a standard used for controlling and monitoring systems over IP-based networks.

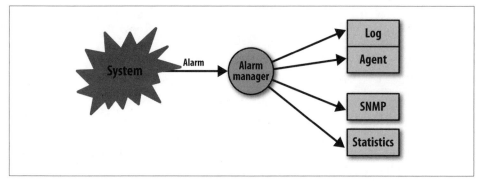

Figure 5-9. An alarm manager implemented as an event handler

is lost and be cleared if it recovers. Other examples include a cabinet door being opened, a fan breaking, or a TCP/IP connection being lost.

The alarm handler will often log these alarms, collect statistics, and filter and forward them to agents. Agents might receive the events and try to resolve the issues themselves. If a communication link is down, for example, an agent would automatically try to reconfigure the system to use the standby link, requesting human intervention only if the standby link goes down as well.

A Generic Event Manager Example

Here is an example of an event manager that allows you to add and remove handlers during runtime. The code is completely generic and independent of the individual handlers. Handlers can be implemented in separate modules and have to export a number of functions, referred to as *callback functions*. These functions can be called by the event manager. We will cover them in a minute. Let's first look at how we've implemented the event manager, starting with its client functions:

start(Name, HandlerList)
> Will start a generic event manager, registering it with the alias Name. HandlerList is a list of tuples of the form {Handler, Data}, where Handler is the name of the handler callback module and Data is the argument passed to the handler's init callback function. HandlerList can be empty at startup, as handlers can be subsequently added using the add_handler/2 call.

stop(Name)
> Will terminate all the handlers and stop the event manager process. It will return a list of items of the form {Handler, Data}, where Data is the return value of the terminate callback function of the individual handlers.

add_handler(Name, Handler, Data)
> Will add the handler defined in the callback module Handler, passing Data as an argument to the handler's init callback function.

delete_handler(Name, Handler)
: Will remove the handler defined in the callback module Handler. The handler's terminate callback function will be called, and its return value will be the return value of this call. This call returns the tuple {error, instance} if Handler does not exist.

get_data(Name, Handler)
: Will return the contents of the state variable of the Handler. This call returns the tuple {error, instance} if Handler does not exist.

send_event(Name, Event)
: Will forward the contents of Event to all the handlers.

Here is the code for the generic event manager module:

```
-module(event_manager).
-export([start/2, stop/1]).
-export([add_handler/3, delete_handler/2, get_data/2, send_event/2]).
-export([init/1]).

start(Name, HandlerList) ->
  register(Name, spawn(event_manager, init, [HandlerList])), ok.

init(HandlerList) ->
  loop(initialize(HandlerList)).

initialize([]) -> [];
initialize([{Handler, InitData}|Rest]) ->
  [{Handler, Handler:init(InitData)}|initialize(Rest)].
```

Here is an explanation of what the code is doing:

- The start(Name, HandlerList) function spawns the event manager process and registers it with the alias Name.
- The newly spawned process starts executing in the init/1 function with a Handler List tuple list of the format {Handler, Data} as an argument.
- We traverse the list in the initialize/1 function calling Handler:init(Data) for every entry.
- The result of this call is stored in a list of the format {Handler, State}, where State is the return value of the init function.
- This list is passed as an argument to the event manager's loop/1 function.

When stopping the event manager process, we send a stop message received in the loop/1 function. If you are looking for loop/1, you will find it with the generic code at the end of this module. Receiving the stop message results in terminate/1 traversing the list of handlers and calling Handler:terminate(Data) for every entry. The return value of these calls, a list of the format {Handler, Value}, is sent back to the process that originally called stop/1 and becomes the return value of this function:

```
stop(Name) ->
  Name ! {stop, self()},
  receive {reply, Reply} -> Reply end.

terminate([]) -> [];
terminate([{Handler, Data}|Rest]) ->
  [{Handler, Handler:terminate(Data)}|terminate(Rest)].
```

Now we'll look at the client functions used to add, remove, and inspect the event handlers, as well as forwarding them the events. Through the call/2 function, they send the request to the event manager process which handles them in handle_msg/2. Pay particular attention to the send_event/2 call, which traverses the list of handlers, calling the callback function Handler:handle_event(Event, Data). The return value of this call replaces the old Data and is used by the handler the next time one of its callbacks is invoked:

```
add_handler(Name, Handler, InitData) ->
  call(Name, {add_handler, Handler, InitData}).

delete_handler(Name, Handler) ->
  call(Name, {delete_handler, Handler}).

get_data(Name, Handler) ->
  call(Name, {get_data, Handler}).

send_event(Name, Event) ->
  call(Name, {send_event, Event}).

handle_msg({add_handler, Handler, InitData}, LoopData) ->
  {ok, [{Handler, Handler:init(InitData)}|LoopData]};

handle_msg({delete_handler, Handler}, LoopData) ->
  case lists:keysearch(Handler, 1, LoopData) of
    false ->
      {{error, instance}, LoopData};
    {value, {Handler, Data}} ->
      Reply = {data, Handler:terminate(Data)},
      NewLoopData = lists:keydelete(Handler, 1, LoopData),
      {Reply, NewLoopData}
  end;

handle_msg({get_data, Handler}, LoopData) ->
  case lists:keysearch(Handler, 1, LoopData) of
    false                    -> {{error, instance}, LoopData};
    {value, {Handler, Data}} -> {{data, Data}, LoopData}
  end;

handle_msg({send_event, Event}, LoopData) ->
  {ok, event(Event, LoopData)}.

event(_Event, []) -> [];
event(Event, [{Handler, Data}|Rest]) ->
  [{Handler, Handler:handle_event(Event, Data)}|event(Event, Rest)].
```

The following code, together with the **start** and **stop** functions we already covered, is a direct rip off from the process pattern example. By now, you should have spotted the recurring theme—*processes that handle very different tasks do so in similar ways, following a pattern*:

```erlang
call(Name, Msg) ->
  Name ! {request, self(), Msg},
  receive {reply, Reply} -> Reply end.

reply(To, Msg) ->
  To ! {reply, Msg}.

loop(State) ->
  receive
    {request, From, Msg} ->
      {Reply, NewState} = handle_msg(Msg, State),
      reply(From, Reply),
      loop(NewState);
    {stop, From}   ->
      reply(From, terminate(State))
  end.
```

Event Handlers

In our event manager implementation, our event handlers have to export the following three callback functions:

init(InitData)
> Initializes the handler and returns a value that is used the next time a callback function belonging to the handler is invoked.

terminate(Data)
> Allows the handler to clean up. If we have opened files or sockets in the init/1 callback, they would be closed here. The return value of terminate/1 is passed back to the functions that originally instigated the removal of the handler. In our event manager example, they are the delete_handler/2 and stop/1 calls.

handle_event(Event, Data)
> Is called when an event is forwarded to the event manager through the send_event/2 call. Its return value will be used the next time a callback function for this handler is invoked.

Using these callback functions, let's write two handlers—one that pretty-prints the events to the shell, and one that logs the events to file.

The io_handler event handler filters out events of the format {raise_alarm, Id, Type} and {clear_alarm, Id, Type}. All other events are ignored. In the init/1 function, we set a counter which is incremented every time an event is handled.

The handle_event/2 callback uses this counter every time an alarm event is received, displaying it together with information on the alarm:

```
-module(io_handler).
-export([init/1, terminate/1, handle_event/2]).

init(Count) -> Count.

terminate(Count) -> {count, Count}.

handle_event({raise_alarm, Id, Alarm}, Count) ->
  print(alarm, Id, Alarm, Count),
  Count+1;
handle_event({clear_alarm, Id, Alarm}, Count) ->
  print(clear, Id, Alarm, Count),
  Count+1;
handle_event(Event, Count) ->
  Count.

print(Type, Id, Alarm, Count) ->
  Date = fmt(date()), Time = fmt(time()),
  io:format("#~w,~s,~s,~w,~w,~p~n",
            [Count, Date, Time, Type, Id, Alarm]).

fmt({AInt,BInt,CInt}) ->
  AStr = pad(integer_to_list(AInt)),
  BStr = pad(integer_to_list(BInt)),
  CStr = pad(integer_to_list(CInt)),
  [AStr,$:,BStr,$:,CStr].

pad([M1])  -> [$0,M1];
pad(Other) -> Other.
```

The second handler that we implement logs all the events of the format {EventType, Id, Description} in a comma-separated file, ignoring everything else that is not a tuple of size 3.

We open the file in the init/1 function, write to it in handle_event/2, and close it in the terminate function. As this file will probably be read and manipulated by other programs, we will provide more detail in the information we write to it and spend less effort with its formatting. Instead of time() and date(), we use the now() BIF which gives us a timestamp with a much higher level of accuracy. It returns a tuple containing the mega seconds, seconds, and microseconds that have elapsed since January 1, 1970. When the log_handler is deleted from the event manager, the terminate/2 call will close the file:

```
-module(log_handler).

-export([init/1, terminate/1, handle_event/2]).

init(File) ->
  {ok, Fd} = file:open(File, write),
  Fd.

terminate(Fd) -> file:close(Fd).

handle_event({Action, Id, Event}, Fd) ->
```

```
        {MegaSec, Sec, MicroSec} = now(),
        Args = io:format(Fd, "~w,~w,~w,~w,~w,~p~n",
                         [MegaSec, Sec, MicroSec, Action, Id, Event]),
        Fd;
  handle_event(_, Fd) ->
      Fd.
```

Try out the event manager and the two handlers we've implemented in the shell. We
start the event manager with the log_handler, after which we add and delete the
io_handler. In between, we generate a few alarms and test the other client functions
we've implemented in the event manager work:

```
1> event_manager:start(alarm, [{log_handler, "AlarmLog"}]).
ok
2> event_manager:send_event(alarm, {raise_alarm, 10, cabinet_open}).
ok
3> event_manager:add_handler(alarm, io_handler, 1).
ok
4> event_manager:send_event(alarm, {clear_alarm, 10, cabinet_open}).
#1,2009:03:16,08:33:14,clear,10,cabinet_open
ok
5> event_manager:send_event(alarm, {event, 156, link_up}).
ok
6> event_manager:get_data(alarm, io_handler).
{data,2}
7> event_manager:delete_handler(alarm, stats_handler).
{error,instance}
8> event_manager:stop(alarm).
[{io_handler,{count,2}},{log_handler,ok}]
```

Exercises

Exercise 5-1: A Database Server

Write a database server that stores a database in its loop data. You should register the
server and access its services through a functional interface. Exported functions in the
my_db.erl module should include:

```
my_db:start()          ⇒ ok.
my_db:stop()           ⇒ ok.
my_db:write(Key, Element) ⇒ ok.
my_db:delete(Key)      ⇒ ok.
my_db:read(Key)        ⇒ {ok, Element} | {error, instance}.
my_db:match(Element)   ⇒ [Key1, ..., KeyN].
```

Hint: use the db.erl module as a backend and use the server skeleton from the echo
server from Exercise 4-1 in Chapter 4. Example:

```
1> my_db:start().
ok
2> my_db:write(foo, bar).
ok
3> my_db:read(baz).
```

```
{error, instance}
4> my_db:read(foo).
{ok, bar}
5> my_db:match(bar).
[foo]
```

Exercise 5-2: Changing the Frequency Server

Using the frequency server example in this chapter, change the code to ensure that only the client who allocated a frequency is allowed to deallocate it. Make sure that deallocating a frequency that has not been allocated does not make the server crash.

Hint: use the self() BIF in the allocate and deallocate functions called by the client.

Extend the frequency server so that it can be stopped only if no frequencies are allocated.

Finally, test your changes to see whether they still allow individual clients to allocate more than one frequency at a time. This was previously possible by calling allocate_frequency/0 more than once. Limit the number of frequencies a client can allocate to three.

Exercise 5-3: Swapping Handlers

What happens if you want to close and open a new file in the log_handler? You would have to call event_manager:delete_handler/2 immediately followed by event_manager:add_handler/2. The risk with this is that in between these two calls, you might miss an event. Therefore, implement the following function:

```
event_manager:swap_handlers(Name, OldHandler, NewHandler)
```

which swaps the handlers atomically, ensuring that no events are lost. To ensure that the state of the handlers is maintained, pass the return value of OldHandler:terminate/1 to the NewHandler:init/1 call.

Exercise 5-4: Event Statistics

Write a stats_handler module that takes the first and second elements of the event tuple {Type, Id, Description} in our example and keep a count of how many times the combination of {Type, Description} occurs. Users should be able to retrieve these statistics by using the client function event_manager:get_data/2.

Exercise 5-5: Phone FSM

Complete the coding of the phone FSM example, and then instrument it with logging using an event handler process. This should record enough information to enable billing for the use of the phone.

CHAPTER 6

Process Error Handling

Whatever the programming language, building distributed, fault-tolerant, and scalable systems with requirements for high availability is not for the faint of heart. Erlang's reputation for handling the fault-tolerant and high-availability aspects of these systems has its foundations in the simple but powerful constructs built into the language's concurrency model. These constructs allow processes to monitor each other's behavior and to recover from software faults. They give Erlang a competitive advantage over other programming languages, as they facilitate development of the complex architecture that provides the required fault tolerance through isolating errors and ensuring nonstop operation. Attempts to develop similar frameworks in other languages have either failed or hit a major complexity barrier due to the lack of the very constructs described in this chapter.

Process Links and Exit Signals

You might have heard of the "let it crash and let someone else deal with it" and "crash early" approaches. That's the Erlang way! If something goes wrong, let your process terminate as soon as possible and let another process deal with the problem. The link/1 BIF will have been used by this other process to allow it to monitor and detect abnormal terminations and handle them generically.

The link/1 BIF takes a pid as an argument and creates a bidirectional link between the calling process and the process denoted by the pid. The spawn_link/3 BIF will yield the same result as calling spawn/3 followed by link/1, except that it will do so *atomically* (i.e., in a single step, so either both calls succeed or neither one does). In diagrams of Erlang processes, you denote processes linked to each other with a line, as shown in Figure 6-1.

As links are bidirectional, it does not matter whether process A linked to process B or B to A; the result will be the same. If a linked process terminates abnormally, an exit signal will be sent to all the processes to which the failing process is linked. The process receiving the signal will exit, propagating a new exit signal to the processes that it is linked to (this collection is also know as its *link set*).

Figure 6-1. Linked processes

The exit signal is a tuple of the format {'EXIT', Pid, Reason}, containing the atom 'EXIT', the Pid of the terminating process, and the Reason for its termination. The process on the receiving end will terminate with the same reason and propagate a new exit signal with its own pid to all the processes in its link set, as shown in Figure 6-2.

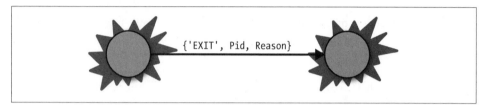

Figure 6-2. Exit signals

When process A fails, its exit signal propagates to process B. Process B terminates with the same reason as A, and its exit signal propagates to process C (see Figure 6-3). If you have a group of mutually dependent processes in a system, it is a good design practice to link them together to ensure that if one terminates, they will all terminate.

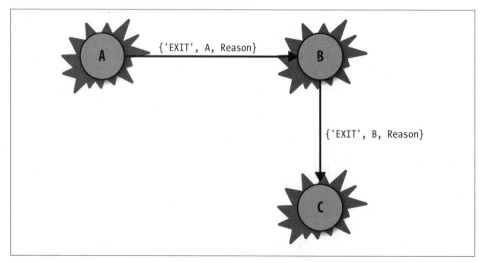

Figure 6-3. Propagation of exit signals

Type the following example in an editor, or download it from the book's website. It is a simple program that spawns a process that links to its parent. When sent an integer N by the process in the message {request, Pid, N}, the process adds one to N and returns the result to the Pid. If it takes more than one second to compute the result, or if the process crashes, the request/1 function returns the atom timeout. There are no checks to the arguments passed to request/1, so sending anything but an integer will cause a runtime error that terminates the process. As the process is linked to its parent, the exit signal will propagate to the parent and terminate it with the same reason:

```erlang
-module(add_one).
-export([start/0, request/1, loop/0]).

start() ->
  register(add_one, spawn_link(add_one, loop, [])).

request(Int) ->
  add_one ! {request, self(), Int},
  receive
    {result, Result} ->  Result
    after 1000       ->  timeout
  end.

loop() ->
  receive
    {request, Pid, Msg} ->
      Pid ! {result, Msg + 1}
  end,
  loop().
```

Test this program by sending it a noninteger, and see how the shell reacts. In the example that follows, we send the atom one, causing the process to crash. As the add_one process is linked to the shell, the propagation of the exit signal will cause the shell process to terminate as well. The error report shown in the shell comes from the add_one process, while the exception exit printout comes from the shell itself. Note how we get a different pid when calling the self() BIF before and after the crash, indicating that the shell process has been restarted:

```erlang
1> self().
<0.29.0>
2> add_one:start().
true
3> add_one:request(1).
2
4> add_one:request(one).

=ERROR REPORT==== 21-Jul-2008::16:29:38 ===
Error in process <0.37.0> with exit value: {badarith,[{add_one,loop,0}]}

** exception exit: badarith
     in function  add_one:loop/0
5> self().
<0.40.0>
```

So far, so good, but you are now probably asking yourself how a process can handle abnormal terminations and recovery strategies if the only thing it can do when it receives an exit signal is to terminate itself. The answer is by *trapping exits*.

Trapping Exits

Processes can *trap exit signals* by setting the process flag `trap_exit`, and by executing the function call `process_flag(trap_exit, true)`. The call is usually made in the initialization function, allowing exit signals to be converted to messages of the format `{'EXIT', Pid, Reason}`. If a process is trapping exits, these messages are saved in the process mailbox in exactly the same way as other messages. You can retrieve these messages using the `receive` construct, pattern matching on them like any other message.

If an exit signal is trapped, it does not propagate further. All processes in its link set, other than the one that terminated, are not affected. You usually denote processes that are trapping exits with a double circle, as shown in Figure 6-4.

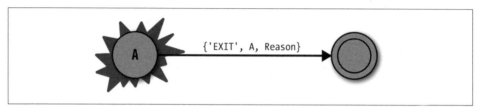

Figure 6-4. Trapping exits

Let's take a specific example, shown in Figure 6-5. Process B, marked with a double circle, is trapping exits. If a runtime error occurs in process A, it will terminate and send out an exit signal of the format `{'EXIT', A, Reason}`, where A is the pid of the process that has failed and Reason is the reason for its termination. The atom `'EXIT'` is used to tag the tuple and facilitate pattern matching. This message is stored in process B's mailbox without affecting C. Unless B explicitly informs C that A has terminated, C will never know.

Let's revisit the `add_one` example, letting the shell trap exits. The result, instead of a crash, should be an `'EXIT'` message sent to the shell. You can retrieve this signal using the `flush/0` command, as it will not pattern-match against any of the `receive` clauses in the `request` function:

```
1> process_flag(trap_exit, true).
false
2> add_one:start().
true
3> add_one:request(one).

=ERROR REPORT==== 21-Jul-2008::16:44:32 ===
Error in process <0.37.0> with exit value: {badarith,[{add_one,loop,0}]}
```

```
timeout
4> flush().
Shell got {'EXIT',<0.37.0>,{badarith,[{add_one,loop,0}]}}
ok
```

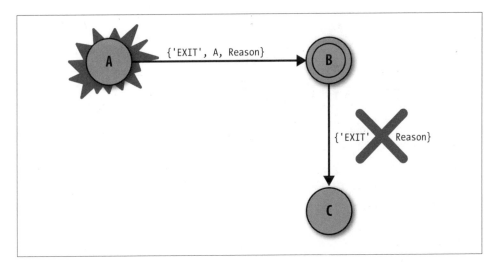

Figure 6-5. Propagation of exit signals

It is also possible to explicitly catch the exit message sent to the shell. In the following
variant of the earlier program, the `request` function has an additional pattern match
over `{'EXIT',_,_}` to trap the exit message from the `loop()` process:

```
-module(add_two).
-export([start/0, request/1, loop/0]).

start() ->
  process_flag(trap_exit, true),
  Pid = spawn_link(add_two, loop, []),
  register(add_two, Pid),
  {ok, Pid}.

request(Int) ->
  add_two ! {request, self(), Int},
  receive
    {result, Result}      -> Result;
    {'EXIT', _Pid, Reason} -> {error, Reason}
    after 1000            -> timeout
  end.

loop() ->
  receive
    {request, Pid, Msg} ->
        Pid ! {result, Msg + 2}
  end,
  loop().
```

Note how we call process_flag(trap_exit, true) in the start/0 function. Run the program and you should observe the following output:

```
1> c(add_two).
{ok, add_two}
2> add_two:start().
{ok, <0.119.0>}
3> add_two:request(6).
8
4> add_two:request(six).
{error,{badarith,[{add_two,loop,0}]}}

=ERROR REPORT==== 24-Aug-2008::18:59:30 ===
Error in process <0.36.0> with exit value: {badarith,[{add_two,loop,0}]}
```

The response to command 4 in the shell comes from matching the {'EXIT',_,_} raised when loop() fails on the atom six.

If you want to stop trapping exits, use process_flag(trap_exit, false). It is considered a bad practice to toggle the trap_exit flag, as it makes your program hard to debug and maintain. The trap_exit flag is set to false by default when a process is spawned.

The monitor BIFs

Links are bidirectional. When the need arose to monitor processes unidirectionally, the erlang:monitor/2 BIF was added to Erlang. When you call the following:

```
erlang:monitor(process, Proc)
```

a monitor is created from the calling process to the process denoted by Proc, where Proc can be either a process identifier or a registered name. When the process with the pid terminates, the message {'DOWN',Reference,process,Pid,Reason} is sent to the monitoring process. This message includes a *reference* to the monitor. References, which we cover in more detail in Chapter 9, are (essentially) unique values that you can use to identify entities such as responses to particular requests; you can compare references for equality and use them in pattern-matching definitions.

If you try to link to a nonexistent process, the calling process terminates with a runtime error. The monitor BIF behaves differently. If Pid has already terminated (or never existed) when monitor is called, the 'DOWN' message is immediately sent with the Reason set to noproc. Repeated calls to erlang:monitor(process,Pid) will return different references, creating *multiple* independent monitors. They will all send their 'DOWN' message when Pid terminates.

Monitors are removed by calling erlang:demonitor(Reference). The 'DOWN' message could have been sent right before the call to demonitor, so the process using the monitor should not forget to flush its mailbox. To be on the safe side, you can use erlang:demonitor(Reference, [flush]), which will turn off the monitor while removing any 'DOWN' message from the Reference provided.

In the following example, we spawn a process that crashes immediately, as the module it is supposed to execute does not exist. We start monitoring it and immediately receive the 'DOWN' message. When retrieving the message, we pattern-match on both the Reference and the Pid, returning the reason for termination. Notice that the reason is noproc, the error stating that process Pid does not exist: compare this with the runtime error we get when trying to link to a nonexistent process:

```
1> Pid = spawn(crash, no_function, []).

=ERROR REPORT==== 21-Jul-2008::15:32:02 ===
Error in process <0.32.0> with exit value: {undef,[{crash,no_function,[]}]}

<0.32.0>
2> Reference = erlang:monitor(process, Pid).
#Ref<0.0.0.31>
3> receive
       {'DOWN',Reference,process,Pid,Reason} -> Reason
   end.
noproc
4> link(Pid).
** exception error: no such process or port
     in function  link/1
         called as link(<0.32.0>)
```

When would you pick monitoring over linking? Links are established permanently, as in supervision trees, or when you want the propagation path of the exit signal to be bidirectional. Monitors are ideal for monitoring the client call to a behavior process, where you do not want to affect the state of the process you are calling and don't want it to receive an exit signal if the client terminates as a result of a link to another process.

The exit BIFs

The BIF call exit(Reason) causes the calling process to terminate, with the reason for termination being the argument passed to the BIF. The terminating process will generate an exit signal that is sent to all the processes to which it is linked. If exit/1 is called within the scope of a try...catch construct (see Figure 6-6), it can be caught, as it can within a catch itself.

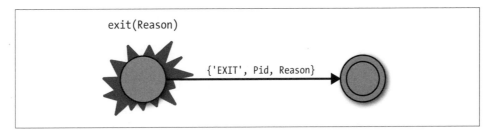

Figure 6-6. The exit/1 BIF

If you want to send an exit signal to *a particular process*, you call the `exit` BIF using `exit(Pid, Reason)`, as shown in Figure 6-7. The result is almost the same as the message sent by a process to its link set on termination, except that the `Pid` in the `{'EXIT',Pid,Reason}` signal is the process identifier of the receiving process itself, rather than the process that has terminated. If the receiving process is trapping exits, the signal is converted to an exit message and stored in the process mailbox. If the process is not trapping exits and the reason is not `normal`, it will terminate and propagate an exit signal. Exit signals sent to a process cannot be caught with a `catch` and will result in the process terminating. We discuss propagation of errors in more detail in the section "Propagation Semantics" on page 148.

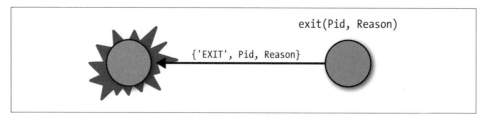

Figure 6-7. The exit/2 BIF

BIFs and Terminology

Before trying some examples and looking at these constructs in more depth, let's review the terminology and the most important BIFs dealing with termination:

- A *link* is a bidirectional propagation path for exit signals set up between processes.
- An *exit signal* is a signal transmitted by a process upon termination containing the reason for the termination.
- *Error trapping* is the ability of a process to handle exit signals as though they were messages.

The BIFs that are related to concurrent error handling include:

link(Pid)
    Sets a bidirectional link between the calling process and `Pid`.

unlink(Pid)
    Removes a link to `Pid`.

spawn_link(Mod, Fun, Args)
    Atomically spawns and sets a link between the calling process and the spawned process.

spawn_monitor(Mod, Fun, Args)
    Atomically spawns and sets a monitor between the calling process and the spawned process.

```
process_flag(trap_exit, Flag)
```
Sets the current process to convert exit signals to exit messages. The Flag contains the atoms true (turning on exit trapping) and false (turning it off).

```
erlang:monitor(process,Pid)
```
Creates a unidirectional monitor toward Pid. It returns a reference to the calling process; this reference can be used to identify the terminated process in a pattern match.

```
erlang:demonitor(Reference)
```
Clears the monitor so that monitoring no longer takes place. Don't forget to flush messages that might have arrived prior to calling the demonitor BIF.

```
erlang:demonitor(Reference,[flush])
```
Is the same as demonitor/1, but removes the {_,Reference,_,_,_} message if it was sent as a result of a race condition.

```
exit(Reason)
```
Causes the calling process to terminate with reason Reason.

```
exit(Pid,Reason)
```
Sends an exit signal to Pid.

Even in code written using monitor and link, race conditions can occur. Look at the following code fragments. The first statement spawns a child, binds the process identifier to the variable Pid, and allows the process executing the code to link to it:

```
link(Pid = spawn(Module, Function, Args))
```

The second statement spawns a child, links to it, and binds the process identifier to the variable Pid:

```
Pid = spawn_link(Module, Function, Args)
```

At first sight, both examples appear to do the same thing, except that spawn_link/3 does it *atomically*. By *atomic operation*, we mean an operation that has to be completed before the process can be suspended. When dealing with concurrency, what is apparently a small detail, such as whether an operation is atomic, can make all the difference in how your program behaves:

- If you use spawn_link/3, the process is spawned and linked to the parent. This operation cannot be suspended in between the spawn and the link, as all BIFs are atomic. The earliest the process can be suspended is after executing the BIF.

- If you instead execute spawn and link as two separate operations, the parent process calling spawn might get suspended right after spawning the process and binding the variable Pid, but before calling link/1. The new process starts executing, encounters a runtime error, and terminates. The parent process is preempted, and the first thing it does is to link to a nonexistent process. This will result in a runtime error instead of an exit signal being received.

This problem is similar to the example of race conditions that we looked at in Chapter 4, where the outcome may vary depending on the order of events, itself a consequence of where the processes are suspended and which core they are running on. A rule of thumb is to always use `spawn_link` unless you are toggling between linking and unlinking to the process, or if it is not a child process you are linking to. Before going on to the next section, ask yourself how to solve the preceding problem using the `monitor` BIF.

Propagation Semantics

Now that we have covered the most important terminology and BIFs, let's look at the details of the propagation semantics associated with the exit signals as summarized in Table 6-1. When a process terminates it sends an exit signal to the processes in its link set. These exit signals can be *normal* or *nonnormal*. Normal exit signals are generated either when the process terminates because there is no more code to execute, or by calling the exit BIFs with the reason `normal`.

A process that is not trapping exit signals terminates if it receives a nonnormal exit signal. Exit signals with reason `normal` are ignored. A process that is trapping exit signals converts all incoming exit signals, normal and nonnormal, to conventional messages that are stored in the mailbox and handled in a `receive` statement.

If `Reason` in any of the exit BIFs is `kill`, the process is terminated unconditionally, regardless of the `trap_exit` flag.[*] An exit signal is propagated to processes in its linked set with reason `killed`. This will ensure that processes trapping exits are not terminated if one of the peers in its link set is killed unconditionally.

Table 6-1. Propagation semantics

Reason	Trapping exits `trap_exit=true`	Not trapping exits `trap_exit=false`
normal	Receives `{'EXIT', Pid, normal}`	Nothing happens
kill	Terminates with reason `killed`	Terminates with reason `killed`
Other	Receives `{'EXIT', Pid, Other}`	Terminates with reason `Other`

Robust Systems

In Erlang, you build robust systems by *layering*. Using processes, you create a tree in which the leaves consist of the *application layer* that handles the operational tasks while the interior nodes *monitor* the leaves and other nodes below them, as shown in Figure 6-8. Processes at any level will trap errors occurring at a level immediately below them. A process whose only task is to supervise children—in our case the nodes of the tree—is called a *supervisor*. A leaf process performing operational tasks is called a

---

[*] You can catch the call `exit(kill)`. A process will be unconditionally killed only when `exit(Pid,kill)` is used.

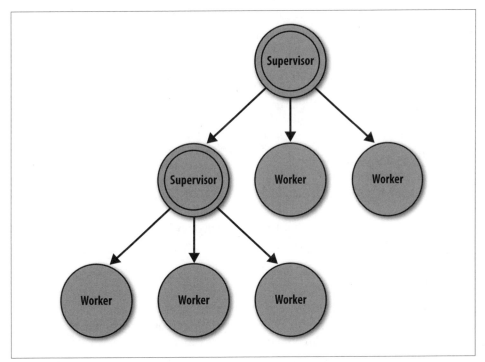

Figure 6-8. Fault tolerance by layering

worker. When we refer to child processes, we mean both supervisors and workers belonging to a particular supervisor.

In well-designed systems, application programmers will not have to worry about error-handling code. If a worker crashes, the exit signal is sent to its supervisor, which isolates it from the higher levels of the system. Based on a set of preconfigured parameters and the reason for termination, the supervisor will decide whether the worker should be restarted.

Supervisors aren't the only processes that might want to monitor other processes, however. If a process has a dependency on another process that is not necessarily its child, it will want to link itself to it. Upon abnormal termination, both processes can take appropriate action.

In large Erlang systems, you should never allow processes that are not part of a supervision tree; all processes should be linked either to a supervisor or to another worker. As Erlang programs will run for years without having to be restarted, millions, if not billions, of processes will be created throughout the system's lifetime. You need to have full control of these processes, and, if necessary, be able to take down supervision trees. You never know how bugs manifest themselves; the last thing you want is to miss an abnormal process termination as a result of it not being linked to a supervisor. Another danger is hanging processes, possibly as (but not limited to) a result of errors or time-outs, causing a memory leakage that might take months to detect.

 Imagine an upgrade where you have to kill all processes dealing with a specific type of call. If these processes are part of a supervision tree, all you need to do is terminate the top-level supervisor, upgrade the code, and restart it. We get the shivers just thinking of the task of having to go into the shell and manually find and terminate all the processes which are not linked to their parent or supervisor. If you do not know what processes are running in your system, the only practical way to do it is to restart the shell, something which goes against the whole principle of high availability.

If you are serious about your fault tolerance and high availability, make sure all of your processes are linked to a supervision tree.

Monitoring Clients

Remember the section "A Client/Server Example" on page 119 in Chapter 5? The server is unreliable! If the client crashes before it sends the frequency release message, the server will not deallocate the frequency and allow other clients to reuse it.

Let's rewrite the server, making it reliable by monitoring the clients. When a client is allocated a frequency, the server links to it. If a client terminates before deallocating a frequency, the server will receive an exit signal and deallocate it automatically. If the client does not terminate, and deallocates the frequency using the client function, the server removes the link. Here is the code from Chapter 5; all of the new code is highlighted:

```erlang
-module(frequency).
-export([start/0, stop/0, allocate/0, deallocate/1]).
-export([init/0]).

%% These are the start functions used to create and
%% initialize the server.

start() ->
  register(frequency, spawn(frequency, init, [])).

init() ->
process_flag(trap_exit, true),
  Frequencies = {get_frequencies(), []},
  loop(Frequencies).

% Hard Coded
get_frequencies() -> [10,11,12,13,14,15].

%%  The client Functions
```

```erlang
stop()           -> call(stop).
allocate()       -> call(allocate).
deallocate(Freq) -> call({deallocate, Freq}).

%% We hide all message passing and the message
%% protocol in a functional interface.
call(Message) ->
  frequency ! {request, self(), Message},
  receive
    {reply, Reply} -> Reply
  end.

reply(Pid, Message) ->
  Pid ! {reply, Message}.

loop(Frequencies) ->
  receive
    {request, Pid, allocate} ->
      {NewFrequencies, Reply} = allocate(Frequencies, Pid),
      reply(Pid, Reply),
      loop(NewFrequencies);
    {request, Pid , {deallocate, Freq}} ->
      NewFrequencies=deallocate(Frequencies, Freq),
      reply(Pid, ok),
      loop(NewFrequencies);
    {'EXIT', Pid, _Reason} ->
      NewFrequencies = exited(Frequencies, Pid),
      loop(NewFrequencies);
    {request, Pid, stop} ->
      reply(Pid, ok)
  end.

allocate({[], Allocated}, _Pid) ->
  {{[], Allocated}, {error, no_frequencies}};
allocate({[Freq|Frequencies], Allocated}, Pid) ->
  link(Pid),
  {{Frequencies,[{Freq,Pid}|Allocated]},{ok,Freq}}.

deallocate({Free, Allocated}, Freq) ->
  {value,{Freq,Pid}} = lists:keysearch(Freq,1,Allocated),
  unlink(Pid),
  NewAllocated=lists:keydelete(Freq,1,Allocated),
  {[Freq|Free],  NewAllocated}.

exited({Free, Allocated}, Pid) ->
  case lists:keysearch(Pid,2,Allocated) of
    {value,{Freq,Pid}} ->
       NewAllocated = lists:keydelete(Freq,1,Allocated),
      {[Freq|Free],NewAllocated};
    false ->
      {Free,Allocated}
  end.
```

Note how in the exited/2 function we ensure that the pair consisting of the client Pid and the frequency is a member of the list containing the allocated frequencies. This is to avoid a potential race condition where the client correctly deallocates the frequency but terminates before the server is able to handle the deallocate message and unlink itself from the client. As a result, the server will receive the exit signal from the client, even if it has already deallocated the frequency.

```
1> frequency:start().
true
2> frequency:allocate().
{ok,10}
3> exit(self(), kill).
** exception exit: killed
4> frequency:allocate().
{ok,10}
```

In this example, we used a bidirectional link instead of the unidirectional monitor. This design decision is based on the fact that if our frequency server terminates abnormally, we want all of the clients that have been allocated frequencies to terminate as well.

A Supervisor Example

Supervisors are processes whose only task is to start children and monitor them. How are they implemented in practice? Children can be started either in the initialization phase of the supervisor, or dynamically, once the supervisor has been started. Supervisors will trap exits and link to their children when spawning them. If a child process terminates, the supervisor will receive the exit signal. The supervisor can then use the Pid of the child in the exit signal to identify the process and restart it.

Supervisors should manage process terminations and restarts in a uniform fashion, making decisions on what actions to take. These actions might include doing nothing, restarting the process, restarting the whole subtree, or terminating, making its supervisor resolve the problem.

Supervisors should behave in a similar manner, irrespective of what the system does. Together with clients/servers, finite state machines, and event handlers, they are considered a *process design pattern*:

- The *generic* part of the supervisor starts the children, monitors them, and restarts them in case of a termination.
- The *specific* part of the supervisor consists of the children, including when and how they are started and restarted.

In the following example, the supervisor we have implemented takes a child list of tuples of the form {Module, Function, Arguments}. This list describes the children that the supervisor has to supervise by giving the functions that have to be called to start the child processes: an example is given by {add_two, start, []}, introduced at the beginning of the chapter. In doing this, we assume that the child process is started using the spawn_link/3 BIFs, and that the function, if successful, returns the tuple {ok, Pid}; you can verify that this is the case for add_two:start/0.

The supervisor, which is also started with the `spawn_link/3` BIF as it needs to be linked to its parent, starts executing in the `init/1` function. It starts trapping exits, and by calling the `start_children/1` function, it spawns all of the children. If the `apply/3` call creating the child was successful and the function returned `{ok, Pid}`, the entry `{Pid, {Module, Function, Arguments}}` is added to the list of spawned children that is passed to the `loop/1` function:

```erlang
-module(my_supervisor).
-export([start_link/2, stop/1]).
-export([init/1]).

start_link(Name, ChildSpecList) ->
  register(Name, spawn_link(my_supervisor, init, [ChildSpecList])), ok.

init(ChildSpecList) ->
  process_flag(trap_exit, true),
  loop(start_children(ChildSpecList)).

start_children([]) -> [];
start_children([{M, F, A} | ChildSpecList]) ->
  case (catch apply(M,F,A)) of
    {ok, Pid} ->
      [{Pid, {M,F,A}}|start_children(ChildSpecList)];
    _ ->
      start_children(ChildSpecList)
  end.
```

The `loop` of the supervisor waits in a `receive` clause for `EXIT` and `stop` messages. If a child terminates, the supervisor receives the `EXIT` signal and restarts the terminated child, replacing its entry in the list of children stored in the `ChildList` variable:

```erlang
restart_child(Pid, ChildList) ->
  {value, {Pid, {M,F,A}}} = lists:keysearch(Pid, 1, ChildList),
  {ok, NewPid} = apply(M,F,A),
  [{NewPid, {M,F,A}}|lists:keydelete(Pid,1,ChildList)].

loop(ChildList) ->
  receive
    {'EXIT', Pid, _Reason} ->
      NewChildList = restart_child(Pid, ChildList),
      loop(NewChildList);
    {stop, From}  ->
      From ! {reply, terminate(ChildList)}
  end.
```

We stop the supervisor by calling the synchronous client function `stop/0`. Upon receiving the `stop` message, the supervisor runs through the `ChildList`, terminating the children one by one. Having terminated all the children, the atom `ok` is returned to the process that initiated the `stop` call:

```erlang
stop(Name) ->
  Name ! {stop, self()},
  receive {reply, Reply} -> Reply end.
```

```
terminate([{Pid, _} | ChildList]) ->
  exit(Pid, kill),
  terminate(ChildList);
terminate(_ChildList) -> ok.
```

In our example, the supervisor and the children are linked to each other. Can you think of a reason why you should not use the `monitor` BIF? The reason for this design choice is similar to the one in our frequency server example. Should the supervisor terminate, we want it to bring down all of its children, no matter how horrid that may sound!

Let's run the example in the shell and make sure it works:

```
1> my_supervisor:start_link(my_supervisor, [{add_two, start, []}]).
ok
2> whereis(add_two).
<0.125.0>
3> exit(whereis(add_two), kill).
true
4> add_two:request(100).
102
5> whereis(add_two).
<0.128.0>
```

This supervisor example is relatively simple. We extend it in the exercises that follow.

Exercises

Exercise 6-1: The Linked Ping Pong Server

Modify processes A and B from Exercise 4-1 in Chapter 4 by linking the processes to each other. When the `stop` function is called, instead of sending the `stop` message, make the first process terminate abnormally. This should result in the `EXIT` signal propagating to the other process, causing it to terminate as well.

Exercise 6-2: A Reliable Mutex Semaphore

Suppose that the mutex semaphore from the section "Finite State Machines" on page 126 in Chapter 5 is unreliable. What happens if a process that currently holds the semaphore terminates prior to releasing it? Or what happens if a process waiting to execute is terminated due to an exit signal? By trapping exits and linking to the process that currently holds the semaphore, make your mutex semaphore reliable.

In your first version of this exercise, use `try...catch` when calling `link(Pid)`. You have to wrap it in a `catch` just in case the process denoted by `Pid` has terminated before you handle its request.

In a second version of the exercise, use `erlang:monitor(type, Item)`. Compare and contrast the two solutions. Which one of them do you prefer?

Exercise 6-3: A Supervisor Process

The supervisor we provided in the example is very basic. We want to expand its features and allow it to handle more generic functionality. In an iterative test-and-develop cycle, add the following features one at a time:

- If a child terminates both normally and abnormally, the supervisor will receive the exit signal and restart the child. We want to extend the {Module, Function, Argument} child tuple to include a Type parameter, which can be set to permanent or transient. If the child is transient, it is not restarted if it terminated normally. Restart it only upon abnormal termination.

- What happens if the supervisor tries to spawn a child whose module is not available? The process will crash, and its EXIT signal is sent to the supervisor that immediately restarts it. Our supervisor does not handle the infinite restart case. To avoid this case, use a counter that will restart a child a maximum of five times per minute, removing the child from the child list and printing an error message when this threshold is reached.

- Your supervisor should be able to start children even once the supervisor has started. Add a unique identifier in the child list, and implement the function start_child(Module, Function, Argument), which returns the unique Id and the child Pid. Don't forget to implement the stop_child(Id) call, which stops the child. Why do we choose to identify the child through its Id instead of the Pid when stopping it?

Base your supervisor on the example in this chapter. You can download the code from the website for the book as a starting point.

To test your supervisor, start the mutex semaphore and database server processes, as shown in Figure 6-9.

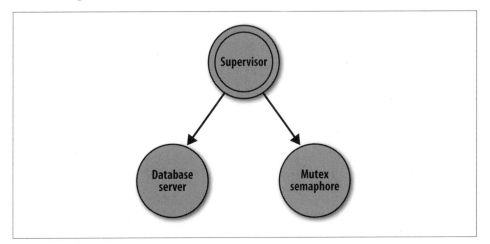

Figure 6-9. The supervision tree

- You will have to change the `start` function to ensure that the processes link themselves to their parent and return `{ok,Pid}`.

- Kill your processes using `exit(whereis(ProcName), kill)`.

- See whether they have been restarted by calling `whereis(ProcName)` and ensure that you are getting different process IDs every time.

- If the process is not registered, kill it by calling `exit(Pid, kill)`. You will get `Pid` from the return value of the `start_child` function. (You can then start many processes of the same type.)

- Once killed, check whether the process has been restarted by calling the `i()` help function in the shell.

CHAPTER 7

Records and Macros

As soon as your first Erlang product reaches the market and is deployed around the world, you start working on feature enhancements for the second release. Imagine 15,000 lines of code, which incidentally happens to be the size of the code base of the first Erlang product Ericsson shipped, the Mobility Server. In your code base, you have tuples that contain data relating to the existing features and constants that have been hardcoded. When you add new features, you need to add fields to these tuples. The problem is that the fields need to be updated not only in the code base where you are adding these features, but also in the remaining 15,000 lines of code where you aren't adding them. Missing one tuple will cause a runtime error. Assuming your constants also need to be updated, you need to change the hardcoded values everywhere they are used. And even more costly than implementing these software changes is the fact that the entire code base needs to be retested to ensure that no new bugs have been introduced or fields and constant updates have been omitted.

One of the most common constructions in computing is to bring together a number of pieces of data as a single item. Erlang tuples provide the basic mechanism for collecting data, but they do have some disadvantages, particularly when a larger number of data items are collected as a single object. In the first part of this chapter, you will learn about records, which overcome most of these disadvantages and which also make code evolution easier to achieve. The key to this is the fact that records provide data abstraction by which the actual representation of the data is hidden from the programs that access it.

Macros allow you to write abbreviations that are expanded by the Erlang preprocessor. Macros can be used to make programs more readable, to extend the language, and to write debugging code. We conclude the chapter by describing the include directive, by which header files containing macro and record definitions are used in Erlang projects.

Although neither is essential for writing Erlang programs, both are useful in making programs easier to read, modify, and debug, facilitating code enhancements and support of deployed products. It is no coincidence that records and macros, the two constructs described in this chapter, were added to the language soon after Ericsson's

Mobility Server went into production and developers started to support it while working on enhancing its feature set.

Records

To understand the advantages of records, we will first introduce a small example dealing with information about people. Suppose you want to store basic information about a person, including his name, age, and telephone number. You could do this using three-element tuples of the form {Name,Age,Phone}:

```
-module(tuples1).
-export([test/1, test/2]).

birthday({Name,Age,Phone}) ->
   {Name,Age+1,Phone}.

joe() ->
   {"Joe", 21, "999-999"}.

showPerson({Name,Age,Phone}) ->
   io:format("name: ~p  age: ~p  phone: ~p~n", [Name,Age,Phone]).

test1() ->
   showPerson(joe()).

test2() ->
   showPerson(birthday(joe())).
```

At every point in the program where the person representation is used, it must be presented as a complete tuple: {Name,Age,Phone}. Although not apparently a problem for a three-element tuple, adding new fields means you would have to update the tuple everywhere, even in the code base where the new fields are not used. Missing an update will result in a badmatch runtime error when pattern-matching the tuple. Furthermore, tuples do not scale well when dealing with sizes of 30 or even 10 elements, as the potential for misunderstanding or error is much greater.

Introducing Records

A *record* is a data structure with a fixed number of *fields* that are accessed by *name*, similar to a C structure or a Pascal record. This differs from *tuples*, where *fields* are accessed by *position*. In the case of the person example, you would define a record type as follows:

```
-record(person, {name,age,phone}).
```

This introduces the record type person, where each record instance contains three fields named name, age, and phone. Field names are defined as atoms. Here is an example of a record instance of this type:

```
#person{name="Joe",
        age=21,
        phone="999-999"}
```

In the preceding code, `#person` is the *constructor* for person records. It just so happens in this example that we listed the fields in the same order as in the definition, but this is not necessary. The following expression gives the same value:

```
#person{phone="999-999",
        name="Joe",
        age=21}
```

In both examples, we defined all the fields, but it is possible to give default values for the fields in the record definition, as in the following:

```
-record(person, {name,age=0,phone=""}).
```

Now a person record like this one:

```
#person{name="Fred"}
```

will have age zero and an empty phone number; in the absence of a default value being specified, the "default default" is the atom `undefined`.

The general definition of a record *name* with fields named *field1* to *fieldn* will take the following form:

```
-record{name, {field1 [ = default1 ],
               field2 [ = default2 ],
               ...
               fieldn [ = defaultn ] }
```

where the parts enclosed in square brackets are *optional* declarations of default field values. The same field name can be used in more than one record type; indeed, two records might share the same list of names. The name of the record can be used in only one definition, however, as this is used to identify the record.

Working with Records

Suppose you are given a record value. How can you access the fields, and how can you describe a modified record? Given the following example:

```
Person = #person{name="Fred"}
```

you access the fields of the record like this: `Person#person.name`, `Person#person.age`, and so on. What will be the values of these? The general form for this field access will be:

```
RecordExp#name.fieldName
```

where the *name* and *fieldName* cannot be variables and *RecordExp* is an expression denoting a record. Typically, this will be a variable, but it might also be the result of a function application or a field access for another record type.

Suppose you want to modify a single field of a record. You can write this directly, as in the following:

```
NewPerson = Person#person{age=37}
```

In such a case, the record syntax is a real advantage. You have mentioned only the field whose value is modified; those that are unchanged from Person to NewPerson need not figure in the definition. In fact, the record mechanism allows for any selection of the fields to be updated, as in:

```
NewPerson = Person#person{phone="999-999",age=37}
```

The general case will be:

```
RecordExp#name{..., fieldNamei=valuei, ... }
```

where the field updates can occur in any order, but each field name can occur, at most, only once.

Functions and Pattern Matching over Records

Using pattern matching over records it is possible to extract field values and to affect the control flow of computation. Suppose you want to define the birthday function, which increases the age of the person by one. You could define the function using field selection and update like this:

```
birthday(P) ->
    P#person{age = P#person.age + 1}.
```

But it is clearer to use pattern matching:

```
birthday(#person{age=Age} = P) ->
    P#person{age=Age+1}.
```

The preceding code makes it clear that the function is applied to a person record, as well as extracting the age field into the variable Age. It is also possible to match against field values so that you increase only Joe's age, keeping everyone else the same age:

```
joesBirthday(#person{age=Age,name="Joe"} = P) ->
    P#person{age=Age+1};
joesBirthday(P) -> P.
```

Revisiting the example from the beginning of the section, you can give the definitions using records:

```
-module(records1).
-export([birthday/1, joe/0, showPerson/1]).

-record(person, {name,age=0,phone}).

birthday(#person{age=Age} = P) ->
  P#person{age=Age+1}.

joe() ->
  #person{name="Joe",
```

```
                          age=21,
                          phone="999-999"}.

        showPerson(#person{age=Age,phone=Phone,name=Name}) ->
           io:format("name: ~p   age: ~p   phone: ~p~n", [Name,Age,Phone]).
```

Although the notation used here is a little more verbose, this is more than compensated for by the clarity of the code, which makes clear our intention to work with records of people, as well as concentrating on the relevant details: it is clear from the definition of birthday that it operates on the age field and leaves the others unchanged. Finally, the code is more easily modified if the composition of the record is changed or extended; the first exercise at the end of this chapter gives you a chance to verify this for yourself.

Record fields can contain any valid Erlang data types. As records are valid data types, fields can contain other records, resulting in nested records. For example, the content of the name field in a person record could itself be a record:

```
        -record(name, {first, surname}).

        P = #person{name = #name{first = "Robert",
                                 surname = "Virding"}}
        First = (P#person.name)#name.first.
```

Furthermore, field selection of a nested field can be given by a single expression, as in the definition of First earlier.

Records in the Shell

Records in Erlang are a compile-time feature, and they don't have their own types in the virtual machine (VM). Because of this, the shell deals with them differently than it does other constructions.

Using the command rr(*moduleName*) in the shell, all record definitions in the module *moduleName* are loaded. You can otherwise define records directly in the shell itself using the command rd(name, {field1, field2, ... }), which defines the record *name* with fields field1, field2, and so on. This can be useful in testing and debugging, or if you do not have access to the module in which you've defined the record. Finally, the command rl() lists all the record definitions currently visible in the shell. Try them out in the shell:

```
        1> c("/Users/Francesco/records1", [{outdir, "/Users/Francesco/"}]).
        {ok,records1}
        2> rr(records1).
        [person]
        3> Person = #person{name="Mike",age=30}.
        #person{name = "Mike",age = 30,phone = undefined}
        4> Person#person.age + 1.
        31
        5> NewPerson = Person#person{phone=5697}.
        #person{name = "Mike",age = 30,phone = 5697}
        6> rd(name, {first, surname}).
```

```
     name
7> NewPerson = Person#person{name=#name{first="Mike",surname="Williams"}}.
#person{name = #name{first = "Mike",surname = "Williams"},
        age = 30,phone = undefined}
8> FirstName = (NewPerson#person.name)#name.first.
"Mike"
9> rl().
-record(name,{first,surname}).
-record(person,{name,age = 0,phone}).
ok
10> Person = Person#person{name=#name{first="Chris",surname="Williams"}}.
** exception error: no match of right hand side value
                    #person{name = #name{first = "Mike",surname = "Williams"},
                            age = 30,phone = undefined}
```

In the preceding example, we load the `person` record definition from the `records1` module, create an instance of it, and extract the `age` field. In command 6, we create a new record of type `name`, with the fields `first` and `surname`. We bind the `name` field of the record stored in the variable `Person` to a new record instance we create in one operation. Finally, in command 8, we extract the first name by looking up the `name` field in the record of type `person` stored in the variable `NewPerson`, all in one operation.

Look at what happens in command 10. This is a very common error made by beginners and seasoned programmers, that is, forgetting that Erlang variables are single assignment and that the = operator is nondestructive. In command 10, you might think you are changing the value of the `name` field to a new name, but you are in fact pattern-matching a record you've just created on the right side with the contents of the bound variable `Person` on the left. The pattern matching fails, as the record name contains the fields `"Mike"` and `"Williams"` on the left and the fields `"Chris"` and `"Williams"` on the right.

Finally, the shell commands `rf(RecordName)` and `rf()` forget one or all of the record definitions currently visible in the shell.

Record Implementation

We are now about to let you in on a poorly kept secret. We would rather not tell you, but when testing with records from the shell, using debugging tools to troubleshoot your code, or printing out internal data structures, you are bound to come across this. The Erlang compiler implements records before programs are run. Records are translated into tuples, and functions over records translate to functions and BIFs over the corresponding tuples. You can see this from this shell interaction:

```
11> records1:joe().
#person{name = "Joe",age = 21,phone = "999-999"}
12> records1:joe() == {person,"Joe",21,"999-999"}.
true
13> Tuple = {name,"Francesco","Cesarini"}.
#name{first = "Francesco",surname = "Cesarini"}
```

```
14> Tuple#name.first.
"Francesco"
```

From the preceding code, you can deduce that person is a 4-tuple, the first element being the atom person "tagging" the tuple and the remaining elements being the tuple fields in the order in which they are listed in the declaration of the record. The name record is a 3-tuple, where the first element is the atom name, the second is the first name field, and the third is the surname field.

Note how the shell by default assumes that a tuple is a record. This will unfortunately be the same in your programs, so whatever you do, *never, ever use the tuple representations of records in your programs*. If you do, the authors of this book will disown you and deny any involvement in helping you learn Erlang. We mean it!

> Why should you never use the tuple representation of records? Using the representation breaks data abstraction, so any modification to your record type will not be reflected in the code using the tuples. If you add a field to the record, the size of the tuple created by the compiler will change, resulting in a badmatch error when trying to pattern-match the record to your tuple (where you obviously forgot to add the new element). Swapping the field order in the record will not affect your code if you are using records, as you access the fields by name. If in some places, however, you use a tuple and forget to swap all occurrences, your program may fail, or worse, may behave in an unexpected and unintended way. Finally, even though this should be the least of your worries, the internal record representation might change in future releases of Erlang, making your code nonbackward-compatible.

To view the code produced as a source code transformation on records, compile your module and include the 'E' option. This results in a file with the E suffix. As an example, let's compile the records1 module using compile:file(records1, ['E']) or the shell command c(records1, ['E']), producing a file called *records1.E*. No beam file containing the object code is produced. Note the slightly different syntax to what you have read so far, and pay particular attention to the record operations and tests which have been mapped to tuples, as well as the module_info functions which have been added. We will not go into the details of the various commands, as they are implementation-dependent and outside the scope of this book. They are, however, still interesting to see:

```
-file("/Users/Francesco/records1.erl", 1).

birthday({person,_,Age,_} = P) ->
    begin
        Rec0 = Age + 1,
        Rec1 = P,
        case Rec1 of
            {person,_,_,_} ->
                setelement(3, Rec1, Rec0);
            _ ->
                erlang:error({badrecord,person})
```

```
        end
    end.

joe() ->
    {person,"Joe",21,"999-999"}.

showPerson({person,Name,Age,Phone}) ->
    io:format("name: ~p  age: ~p  phone: ~p~n", [Name,Age,Phone
]).

module_info() ->
    erlang:get_module_info(records1).

module_info(X) ->
    erlang:get_module_info(records1, X).
```

Record BIFs

The BIF record_info will give information about a record type and its representation.
The function call record_info(fields, *recType*) will return the list of field names in the
recType, and the function call record_info(size, *recType*) will return the size of the
representing tuple, namely the number of fields plus one. The position of a field in
the representing tuple is given by *#recType.fieldName*, where both *recType* and *field-Name* are atoms:

```
15> #person.name.
2
16> record_info(size, person).
4
17> record_info(fields, person).
[name,age,phone]
18> RecType = person.
person
19> record_info(fields, RecType).
* 1: illegal record info
20> RecType#name.
* 1: syntax error before: '.'
```

Note how command 19 failed. If you type the same code in a module as part of a
function and compile it, the compilation will also fail. The reason is simple. The
record_info/2 BIF and the #RecordType.Field operations must contain literal atoms;
they may not contain variables. This is because they are handled by the compiler and
converted to their respective values before the code is run and the variables are bound.

A BIF that you can use in guards is is_record(*Term*, *RecordTag*). The BIF will verify that
Term is a tuple, that its first element is *RecordTag*, and that the size of the tuple is correct.
This BIF returns the atom true or false.

Macros

Macros allow you to write abbreviations of Erlang constructs that the Erlang Preprocessor (EPP) expands at compile time. You can use macros to make programs more readable and to implement features outside the language itself. With conditional macros, it becomes possible to write programs that can be customized in different ways, switching between debugging and production modes or among different architectures.

Simple Macros

The simplest macro can be used to define a constant, as in:

```
-define(TIMEOUT, 1000).
```

The macro is used by putting a ? in front of the macro name, as in:

```
receive
    after ?TIMEOUT -> ok
end
```

After macro expansion in epp, the preceding code will give the following Erlang program:

```
receive
    after 1000 -> ok
end
```

The general form of a simple macro definition is:

```
-define(Name,Replacement).
```

where it is customary—but not required—to CAPITALIZE the Name. In the earlier example, the Replacement was the literal 1000; it can, in fact, be any sequence of Erlang tokens—that is, a sequence of "words" such as variables, atoms, symbols, or punctuation. The result need not be a complete Erlang expression or a *top-level form* (i.e., a function definition or compiler directive). It is *not* possible to build new tokens through macro expansion. As an example, consider the following:

```
-define(FUNC,X).
-define(TION,+X).

double(X) -> ?FUNC?TION.
```

Here, you can see that the replacement for TION is not an expression, but on expansion a legitimate function (or *top-level form*) definition is produced. Note that when appending macros, a space delimiting their results is added to the result by default:

```
double(X) -> X + X.
```

Parameterized Macros

Macros can take parameters which are indicated by variable names. The general form for parameterized macros is:

```
-define(Name(Var1,Var2,...,VarN), Replacement).
```

where, as for normal Erlang variables, the variables Var1, Var2, ..., VarN need to begin with a capital letter. Here is an example:

```
-define(Multiple(X,Y),X rem Y == 0).

tstFun(Z,W) when ?Multiple(Z,W) -> true;
tstFun(Z,W)                     -> false.
```

The macro definition is used here to make a guard expression more readable; a macro rather than a function needs to be used, as the syntax for guards precludes function calls in guards. After macro expansion, the call is "inlined" thus:

```
tstFun(Z,W) when Z rem W == 0 -> true;
tstFun(Z,W)                   -> false.
```

Another example of parameterized macros could be for diagnostic printouts. It is not uncommon to come across code where two macros have been defined, but one is commented out:

```
%-define(DBG(Str, Args), ok).
-define(DBG(Str, Args), io:format(Str, Args)).

birthday(#person{age=Age} = P) ->
    ?DBG("in records1:birthday(~p)~n", [P]),
    P#person{age=Age+1}.
```

When developing the system, you have all of the debug printouts on in the code. When you want to turn them off, all you need to do is comment the second definition of DBG and uncomment the first one before recompiling the code.

Debugging and Macros

One of the major uses of macros in Erlang is to allow code to be instrumented in various ways. The advantage of the macro approach is that in using conditional macros (which we will describe in this section), it is possible to generate different versions of code, such as a debugging version and a production version.

The first aspect of this is the ability to get hold of the argument to a macro as a *string*, made up of the tokens comprising the argument. You do this by prefixing the variable with ??, as in ??Call:

```
-define(VALUE(Call),io:format("~p = ~p~n",[??Call,Call])).
test1() -> ?VALUE(length([1,2,3])).
```

The first use of the `Call` parameter is as `??Call`, which will be expanded to the text of the parameter as a string; the second call will be expanded to a call to `length` so that in the shell, you would see the following:

```
36> macros1: test1().
"length ( [ 1 , 2 , 3 ] )" = 3
```

Second, there is a set of predefined macros that are commonly used in debugging code:

`?MODULE`
> This expands to the name of the module in which it is used.

`?MODULE_STRING`
> This expands to a string consisting of the name of the module in which it is used.

`?FILE`
> This expands to the name of the file in which it is used.

`?LINE`
> This expands to the line number of the position at which it is used.

`?MACHINE`
> This expands to the name of the VM that is being used; currently, the only possible value for this is `BEAM`.

Finally, it is possible to define conditional macros, which will be expanded in different ways according to different flags passed to the compiler. Conditional macros are a more elegant and effective way to get the same effect as the earlier `?DBG` example, where given two macros, the user comments one out. The following directives make this possible:

`-undef(Flag).`
> This will unset the `Flag`.

`-ifdef(Flag).`
> If `Flag` is set, the statements that follow are executed.

`-ifndef(Flag).`
> If `Flag` is *not* set, the statements that follow are executed.

`-else.`
> This provides an alternative catch-all case: if this case is reached, the statements that follow are executed.

`-endif.`
> This terminates the conditional construct.

Here is an example of their use:

```
-ifdef(debug).
    -define(DBG(Str, Args), io:format(Str, Args)).
-else.
    -define(DBG(Str, Args), ok).
-endif.
```

In the code this is used as follows:

```
?DBG("~p:call(~p) called~n",[?MODULE, Request])
```

To turn on system debugging, you need to set the debug flag. You can do this in the shell using the following command:

```
c(Module,[{d,debug}]).
```

Or, you can do it programmatically, using `compile:file/2` with similar flags. You can unset the flag by using `c(Module,[{u,debug}])`.

Conditional macro definitions such as these need to be properly nested, and cannot occur within function definitions.

To *debug* macro definitions, it is possible to get the compiler to dump a file of the results of applying `epp` to a file. You do this in a shell with `c(Module,['P'])` and in a program with `compile:file/2`; these commands dump the result in the file *Module.P*. The `'P'` flag differs from the `'E'` flag in that code transformations necessary for record operations are not done by `'P'`.

Include Files

It is customary to put record and macro definitions into an *include file* so that they can be shared across multiple modules throughout a project, and not simply in a single module. To make the definitions available to more than one module, you place them in a separate *file* and include them in a module using the −include directive, usually placed after the `module` and `export` directives:

```
-include("File.hrl").
```

In the preceding directive, the quotes `"..."` around the filename are *mandatory*. Include files customarily have the suffix *.hrl*, but this is not enforced.

The compiler has a list of paths to search for include files, the first of which is the current directory followed by the directory containing the source code being compiled. You can include other paths in the path list by compiling your code using the `i` option: `c(Module, [{i, Dir}])`. Several directories can be specified, where the directory specified last is searched first.

Exercises

Exercise 7-1: Extending Records

Extend the `person` record type to include a field for the address of the person. Which of the existing functions over `person` need to be modified, and which can be left unchanged?

Exercise 7-2: Record Guards

Using the `record` BIF `record(P, person)`, it is possible to check whether the variable `P` contains a `person` record. Explain how you would use this to modify the function `foobar`, defined as follows:

```
foobar(P) when P#person.name == "Joe" -> ...
```

so that it will not fail if applied to a nonrecord.

Exercise 7-3: The db.erl Exercise Revisited

Revisit the database example *db.erl* that you wrote in Exercise 3-4 in Chapter 3. Rewrite it using records instead of tuples. As a record, you could use the following definition:

```
-record{data, {key, data}).
```

You should remember to place this definition in an include file. Test your results using the database server developed in Exercise 5-1 in Chapter 5.

Exercise 7-4: Records and Shapes

Define a record type to represent circles; define another to represent rectangles. You should assume the following:

- A circle has a radius.
- A rectangle has a length and a width.

Give functions that work over these types to give the *perimeter* and *area* of these geometric figures. Once this is completed, add the code for triangles to your type definitions and functions, where you can assume that the triangle is described by the lengths of its three sides.

Exercise 7-5: Binary Tree Records

Define a record type to represent *binary trees* with numerical values held at internal nodes and at the leaves. Figure 7-1 shows an example.

Define functions over the record type to do the following:

- Sum the values contained in the tree.
- Find the maximum value contained in the tree (if any).

A tree is ordered if, for all nodes, the values in the left subtree below the node are smaller than or equal to the value held at the node, and this value is less than or equal to all the values in the right subtree below the node. Figure 7-2 shows an example:

- Define a function to check whether a binary tree is ordered.
- Define a function to insert a value in an ordered tree so that the order is preserved.

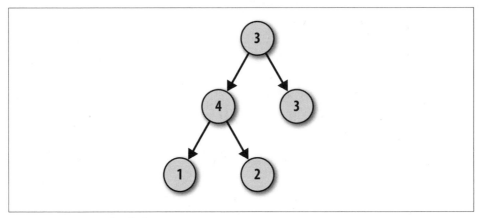

Figure 7-1. An example of a binary tree

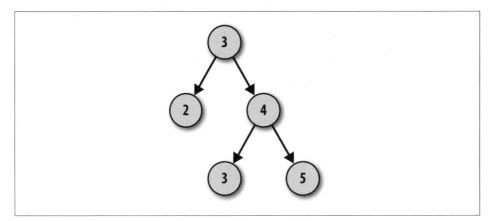

Figure 7-2. An ordered binary tree

Exercise 7-6: Parameterized Macros

Define a parameterized macro SHOW_EVAL that will simply return the result of an expression when the show mode is switched off, but which will also print the expression and its value when the show flag is on. You should ensure that the expression is evaluated only once whichever case holds.

Exercise 7-7: Counting Calls

How can you use the Erlang macro facility to count the number of calls to a particular function in a particular module?

Exercise 7-8: Enumerated Types

An enumerated type consists of a finite number of elements, such as the days of the week or months of the year. How can you use macros to help the implementation of enumerated types in Erlang?

Exercise 7-9: Debugging the db.erl Exercise

Extend the database example in Exercise 7-3 so that it includes optional debugging code reporting on the actions requested of the database as they are executed.

Software Upgrade

You receive a bug report that in one of your instant messaging (IM) servers, the euro symbol is reaching its final destination in a garbled form. You find the error in the library module that maps special characters, correct the bug, recompile the code, and test it. When it's validated, you transfer the patch to the live servers and load it in the Erlang runtime system. The next time a euro symbol is received, the patched module will be used and the euro symbol will be mapped correctly.

You achieve all of this without strange or complex workarounds, without having to restart your system, and most importantly, without affecting any of the other IM-related events currently being handled. If all of this sounds simple, well, it is. And not only is it simple, but it is also really, really cool! Originally inspired by the Smalltalk language, the software upgrade capability is a feature very rarely found in modern programming.

The ability to load new and updated modules during runtime allows systems to run without interruption, not only when errors are being fixed but also when new functionality is added. It also reduces the turnaround time of bugs and facilitates testing, as in most cases, systems do not have to be restarted for patches to be validated and deployed. The software upgrade mechanism relies on a set of simple but powerful constructs on which more powerful tools are built. These upgrade tools are used in pretty much every Erlang-based system where downtime has to be reduced to a minimum.

Upgrading Modules

You have probably come across the loading of new module versions in the runtime system when trying the examples and working on the exercises in this book without realizing what was going on. Think of the incremental approach to software development that we use in Erlang; we'll go through a short case study to show update in practice next, before we cover the details of how it works and other code-handling features of Erlang in the rest of the chapter.

We start by writing a database module similar to the one described in Exercise 3-4 in Chapter 3. We will use the key value dictionary store defined in the dict module to create a bit of variety, and to have the excuse to introduce a new library. Have a look at its manual page. We want to create a module that exports two functions: create/0, which returns an empty database; and write/3, which inserts a Key and Element pair into the database.

The code is described in the db module. Remember the –vsn(1.0) attribute we discussed in Chapter 2? Even if it is not mandatory, it will help us keep track of which module version we have loaded in the runtime system at any point:

```erlang
-module(db).
-export([new/0,write/3,read/2, delete/2,destroy/1]).
-vsn(1.0).

new()                  -> dict:new().

write(Key, Data, Db) -> dict:store(Key, Data, Db).

read(Key, Db) ->
  case dict:fetch(Key, Db) of
    error      -> {error, instance};
    {ok, Data} -> {ok, Data}
  end.

delete(Key, Db) -> dict:erase(Key, Db).

destroy(_Db)     -> ok.
```

Let's now compile and test the code we wrote by adding two elements to the database and looking up one that has not been inserted. The data structure returned by the dict module might appear strange at first. It is visible here because we are testing the module from the shell and binding the value to a series of variables we pass to the dict functions. In a normal implementation, the Db variable would be passed around in the receive-eval loop data, and it would not be visible:

```erlang
1> c(db).
{ok,db}
2> Db = db:new().
{dict,0,16,16,8,80,48,
      {[],[],[],[],[],[],[],[],[],[],[],[],[],[],[],[]},
      {{[],[],[],[],[],[],[],[],[],[],[],[],[],[],[],[]}}}
3> Db1 = db:write(francesco, san_francisco, Db).
{dict,1,16,16,8,80,48,
      {[],[],[],[],[],[],[],[],[],[],[],[],[],[],[],[]},
      {{[],[],[],[],[],[],[],[],[],[],[],[],[],[],[],
        [[francesco|san_francisco]]}}}
4> Db2 = db:write(alison, london, Db1).
{dict,2,16,16,8,80,48,
      {[],[],[],[],[],[],[],[],[],[],[],[],[],[],[],[]},
      {{[],[],[],[],[],[],[],[],[],[],[],[],[],[],[],
        [[alison|london]],
        [[francesco|san_francisco]]}}}
```

```
5> db:read(francesco, Db2).
** exception error: no case clause matching san_francisco
    in function  db:read/2
6> dict:fetch(francesco, Db2).
san_francisco
```

Hold it! Something went wrong. When calling read/2, instead of returning {ok, san_francisco}, we got a case clause error. Looking at our implementation and the manual page for the dict module, we quickly realize that we used dict:fetch/2 instead of dict:find/2. A call directly to dict:fetch/2 confirms that the function returns Data (not {ok, Data}) if the entry is in the dictionary, and raises an exception otherwise. The dict:find/2 function, on the other hand, returns the tuple {ok, Data} if it finds the entry, or the atom error otherwise.

Let's fix the bug, replacing the read function with the following code, and while doing so, bump up the version of this module to 1.1:

```
...
-vsn(1.1).
...
read(Key, Db) ->
  case dict:find(Key, Db) of
    error      -> {error, instance};
    {ok, Data} -> {ok, Data}
  end.
...
```

Before doing anything with the shell, let's use the module_info/0 function to get the version of the code, which is 1.0. We'll save the corrections to the db module, compile it in the same shell where we ran the previous iteration, and continue testing the module with the same entries we previously inserted in the database. The call to read/2 now works, and when we do a call to module_info/1, we get the new module version in the attribute list:

```
7> db:module_info().
[{exports,[{new,0},
           {write,3},
           {read,2},
           {destroy,1},
           {delete,2},
           {module_info,0},
           {module_info,1}]},
 {imports,[]},
 {attributes,[{vsn,[1.0]}]},
 {compile,[{options,[{outdir,"/Users/Francesco/"}]},
           {version,"4.5.2"},
           {time,{2008,8,11,3,9,42}},
           {source,"/Users/Francesco/db.erl"}]}]
8> c(db).
{ok,db}
9> db:read(francesco, Db2).
{ok,san_francisco}
10> db:read(martin, Db2).
```

```
{error,instance}
11> db:module_info(attributes).
[{vsn,[1.1]}]
```

In our example, we fixed a bug, but we instead could have added new functionality, or possibly done both. So, although you might not have realized it, when developing and testing our code, we were actually using the software upgrade functionality. When doing so, the data stored by the process (in our example, the data in the Db variables of the shell) was not affected by the upgrade and was still available after loading the new module.

Behind the Scenes

So, how does the software upgrade feature work behind the scenes? At any one time, two versions of a module may be loaded in the runtime system. We call them the *old* and the *current* versions. Before we explain the details of software upgrade, we'll quickly go over the ways in which functions in one module can refer to functions in another.

- You might not remember, but function calls of the following format:

    `Module:Function(Arg1, .., ArgN)`

    where the `module` name is prefixed to an exported `Function` name, are usually referred to as *fully qualified function calls*. This is one of the ways in which a function in one module (A, say) may refer to a function defined in another (let's call it B).

- The other mechanism by which module A can refer to functions in module B is for A to import some of the functions of B:

    `-import(B, [f/1]).`

    Within A it is then possible to refer *directly* to f, as well as to B:f.

Within a module, it is possible to refer to *another function in the same module* either directly or by a fully qualified name. We'll come back to this distinction shortly.

Now we'll explain the software upgrade process, first for intermodule calls, and then for intramodule calls. Each running process that refers to functions in module A will be linked to a version of module A. When a new version of module A is loaded, that becomes the *current* version, and the previous current version becomes the *old* version.

If a process p, defined in module A, calls a function from module B, either directly or by a fully qualified call (see Figure 8-1, 1), and a new version (version 2) of module B is loaded, the process will remain linked to the same version of B, which has now become the old version of B (see Figure 8-1, 2). At the next call of a function from B, either directly or in fully qualified form, the link will be switched to the new version, version 2 (see Figure 8-1, 3). This will apply to all functions from B, and not simply the function whose call initiated the switch.

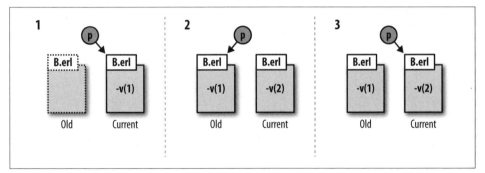

Figure 8-1. Upgrading module B

The case of a running process whose defining module is upgraded is more complicated, and in particular, it depends on the way in which functions within the module are called, either directly or through fully qualified calls:

- If the function call is not fully qualified, the process will continue running the old version of the module.

- When a new version is loaded, however, global calls to the old module are no longer possible, whereas local calls can be made only by processes in a recursive loop.

To show this in more detail, let's look at an example module and the effect of software upgrade on it:

```
1   -module(modtest2).
2
3   -export([main/0,loop/0,a/1,do/1]).
4
5   main() ->
6     register(foo,spawn(modtest2,loop,[])).
7
8   loop() ->
9     receive
10      {Sender, N} ->
11        Sender ! a(N)
12    end,
13    loop().
14
15  do(M) ->
16    foo ! {self(),M},
17    receive Y ->
18      Y
19    end.
20
21   a(N) -> N+2.
```

The main program spawns a named process, `foo`, which runs the `loop()` function. The effect of this loop is to serve values of the function `a/1`: values are sent to the `loop` process by the `do/1` function. Here is an example of the program:

```
1> c(modtest2).
{ok,modtest2}
2> modtest2:main().
true
3> modtest2:do(99).
101
```

Suppose you now upgrade the definition of a/1 in line 21 to the following:

```
a(N) -> N.
```

and recompile; the effect is as follows:

```
4> c(modtest2).
{ok,modtest2}
5> modtest2:do(99).
101
```

So, it is evident that no change has occurred. If, on the other hand, you modify the call a(N) to a fully qualified call, as in the following:

```
loop() ->
  receive
    {Sender, N} ->
      Sender ! modtest2:a(N)
  end,
  loop().
```

the effect of the same software upgrade will be evident after recompilation:

```
6> c(modtest2).
{ok,modtest2}
7> modtest2:do(99).
99
```

As a final example, it is possible to upgrade a running loop, when the recursive call is fully qualified:

```
loop() ->
  receive
    {Sender, N} ->
      Sender ! a(N)
  end,
  modtest2:loop().
```

If you insert a print statement:

```
loop() ->
  receive
    {Sender, N} ->
      Sender ! a(N)
  end,
  io:put_chars("boo!~n"),
  modtest2:loop().
```

you can see the effect of the change in the following interaction:

```
1> c(modtest2).
{ok,modtest2}
```

```
2> modtest2:main().
true
3> modtest2:do(99).
101
4> c(modtest2).
{ok,modtest2}
5> modtest2:do(99).
101
6> modtest2:do(99).
boo!
101
```

In our earlier database example, we were always running the latest version of the db module, because all calls from the shell to the library were fully qualified.

As only two versions of a module may exist in the runtime system at any one time, when a third version is loaded the oldest version is purged (removed) and the current version becomes the old version, as shown in Figure 8-2. Any process running the oldest version of the purged module is terminated. Any process running what has now become the old version will continue doing so until it executes a fully qualified function call.

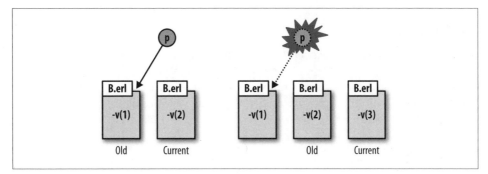

Figure 8-2. Linkage to an old version

Loading Code

Code is loaded in the Erlang runtime system in a variety of ways. The first is by *calling a function in a module that has not yet been loaded.* The code server, a process which is part of the Erlang kernel, will search for the compiled (*.beam*) file of that module, and if it finds it, it will load it; note that it is not compiled automatically if the beam file is missing. The process that made the call to that module is now able to call the function.

Another way to load a module is by *compiling* it. You can use the c(Module) shell command, the function compile:file(Module), or one of its many derivatives, all documented in the compile library module. In our example, we loaded the latest version of the db module in the shell every time it was compiled.

Finally, you can *explicitly load* a module using the `code:load_file(Module)` call. This call is useful because it can appear in programs as well as being used from the shell. From the shell, however, you can use the equivalent shell command `l(Module)`.

All of these calls result in the oldest version of the module (if any) being purged, the current version becoming the old one, and the newly loaded one becoming the current one. Note that the terms *old* and *current* refer not to the compilation time or to a higher revision in the `vsn` attribute, but to the order in which you loaded the modules in the runtime system.

There are also a number of ways to see whether a module is already loaded:

- Try typing in part of the module name in the Erlang shell and pressing the Tab key; if the name *autocompletes*, the module is loaded.
- Just pressing the Tab key will list *all* the modules currently loaded.
- Another way to find out whether a module is loaded is to call the function `code:is_loaded(Module)`, which returns the location of the beam file if the `Module` is already loaded; otherwise, it returns the atom `false`.

The Code Server

We briefly mentioned the *code server* in the preceding section. Here, we'll look at it in more detail. The code server's primary task is to manage the dynamic loading and purging of modules. A set of library functions accessible in the `code.erl` module provide the programmer with the flexibility needed to manage and configure the server to handle the system's code base.

Loading modules

The dynamic loading of code in Erlang will be triggered as a result of a call to a module that is not loaded, or when it is explicitly requested to do so through the `code:load_file/1` call. The code server will search the directories in the *code search path* for a compiled version of the module, and when the code server finds it, it will load it in the virtual machine.

The code search path consists of a list of directories. These directories are sequentially searched for the compiled version of the module you want to load. Try viewing the default search paths of the runtime system by using the `code:get_path()` call. Default directories will include the current working directory together with all of the paths to the default library applications which come as part of the Erlang/OTP distribution. You can find all of these library applications in the *$ERLANGROOT/lib* directory. To find out the Erlang root directory of your installation, use the `code:root_dir()` call.

In your code search path, you might have several versions of the same module. When you load a new version, the directories are scanned sequentially, resulting in the first occurrence of the module being picked up. It is thus not uncommon to create a *patches* directory that appears first in the code search path. Any patches in this directory will be picked up first, overriding the original versions of the module they were trying to patch.

Sticky Directories

Have you ever tried to create a module called `lists.erl`? If so, when you tried to compile it, you were almost certainly greeted with an error telling you that the runtime system can't load a module that resides in the *sticky* directory.

If this happens with a module other than `lists`, you must have picked the name of a module residing in the kernel, the Standard Library (*stdlib*), or the Systems Architecture Support Library (*sasl*) application directories. These three directories are marked as sticky directories by default, so any attempt to override a module defined in them fails.

You can create your own sticky directories by calling the function `code:stick_dir(Dir)`. If you want to override one of the modules in a sticky directory, use the `code:unstick_dir(Dir)` call or start Erlang using the –nostick flag.

Manipulating the code search path

You can add directories to the code search path by using `code:add_patha(Dir)` to add a directory to the beginning of the list, and `code:add_pathz(Dir)` to append one to the end. In the following example, note how the current working directory becomes the second element. The code module also provides functions to delete, replace, and override directories:

```
3> code:add_patha("/tmp").
true
4> code:get_path().
["/tmp",".","/usr/local/lib/erlang/lib/kernel-2.12.3/ebin",
 "/usr/local/lib/erlang/lib/stdlib-1.15.3/ebin",
 "/usr/local/lib/erlang/lib/xmerl-1.1.9/ebin",
 "/usr/local/lib/erlang/lib/wx-0.97.0718/ebin",
 "/usr/local/lib/erlang/lib/webtool-0.8.3.2/ebin",
 "/usr/local/lib/erlang/lib/typer-0.1.3/ebin",
 ..............]
```

In the example, we show just some of the library module directories. Try the command on your machine, and `cd` into any of the listed directories. When inspecting the contents, you should find all of the beam files relating to that particular application.

The explanation in the preceding text of how the code server looks for modules will hopefully explain why you had to change the current working directory of the Erlang shell to the directory that contained your beam files. From the `code:get_path/0` example, unless you have modified the code search path structure, you can see that the current working directory (.) is the first directory the code server searches when trying to load a module. Finally, you can also add directories when starting the Erlang shell by passing the `erl -pa Path` or `erl -pz Path` directive to the `erl` command to add the directory at the beginning (a) or end (z) of the path.

Shell Modes: Interactive and Embedded

The Erlang shell is started by default in what we call *interactive* mode. This means that at startup, only the modules the runtime system needs are loaded. Other code is dynamically loaded when a fully qualified function is made to a function in that module.

In *embedded* mode, all the modules listed in a binary boot file are loaded at startup. After startup, calls to modules that have not been loaded result in a runtime error. Embedded mode enforces strict revision control, as it requires that all modules are available at startup. Lastly, it does not impact the soft real-time aspects of the system by stopping, searching, and loading a module during a time-critical operation, as might be the case for interactive mode. You can choose your mode by starting Erlang with the `erl -mode Mode` directive, where `Mode` is either `embedded` or `interactive`.

Purging Modules

The code server can get rid of or *purge* old versions of a module by calling `code:purge(Module)`, but if any processes were running that code, they will first be terminated, after which the old version of the code is deleted. The function returns `true` if any processes were terminated as a result of the call or `false` otherwise.

If you do not want to terminate any processes running the old version of the code, use `code:soft_purge(Module)`. The call will remove the old version of the module only if there are no processes running the code. If any processes are still running the code, it will return `false` and will do nothing else. If it was successful in deleting the old module, it will return `true`.

The OTP framework, which we cover in Chapter 12, provides a supervisor mechanism that is designed to deal with process termination in an organized way; an obvious application of this sort of supervision is for processes that terminate after a software upgrade.

Upgrading Processes

Now that we have looked at software upgrade in more detail, let's go through a practical example where the format of loop data needs to be changed in a running loop.

We implement a `db_server` module that provides a process storing the database in the dictionary format used by version 1.1 of the `db` module. Alongside the exported client functions, pay particular attention to the `upgrade/1` function. We will tell you more about it in just a second.

```
-module(db_server).
-export([start/0, stop/0, upgrade/1]).
```

```
-export([write/2, read/1, delete/1]).
-export([init/0, loop/1]).
-vsn(1.0).

start() ->
  register(db_server, spawn(db_server, init, [])).

stop()->
  db_server ! stop.

upgrade(Data) ->
  db_server ! {upgrade, Data}.

write(Key, Data) ->
  db_server ! {write, Key, Data}.

read(Key) ->
  db_server ! {read, self(), Key},
  receive Reply -> Reply end.

delete(Key) ->
  db_server ! {delete, Key}.

init() ->
  loop(db:new()).

loop(Db) ->
  receive
    {write, Key, Data} ->
       loop(db:write(Key, Data, Db));
    {read, Pid, Key} ->
       Pid ! db:read(Key, Db),
       loop(Db);
    {delete, Key} ->
       loop(db:delete(Key, Db));
    {upgrade, Data} ->
      NewDb = db:convert(Data, Db),
      db_server:loop(NewDb);
    stop ->
      db:destroy(Db)
  end.
```

The upgrade function takes a variable as an argument and forwards it on to the db_server process. This variable is passed to the db:convert/2 function, which returns the database in a possibly updated format. The convert/2 function was not included in version 1.1 of the db module, as all we did was fix a bug that did not require us to change the internal format of the data. Read through the db_server code and make sure you understand it. If anything is unclear, copy it and test it from the shell, and read the manual pages for the dict library.

Let's now create a new db module, this time basing it on general balanced trees, using the gb_trees library module. When implementing it, we include a convert/2 function. Given the data structure returned by the dict module, this function extracts all of the elements from the dictionary and inserts them in a binary tree, returning a data structure that the gb_trees module can now use:

```erlang
-module(db).
-export([new/0, destroy/1, write/3, delete/2, read/2, convert/2]).
-vsn(1.2).

new() -> gb_trees:empty().

write(Key, Data, Db) -> gb_trees:insert(Key, Data, Db).

read(Key, Db) ->
  case gb_trees:lookup(Key, Db) of
    none          -> {error, instance};
    {value, Data} -> {ok, Data}
  end.

destroy(_Db) -> ok.

delete(Key, Db) -> gb_trees:delete(Key, Db).

convert(dict,Dict) ->
  dict(dict:fetch_keys(Dict), Dict, new());
convert(_, Data) ->
  Data.

dict([Key|Tail], Dict, GbTree) ->
  Data = dict:fetch(Key, Dict),
  NewGbTree = gb_trees:insert(Key, Data, GbTree),
  dict(Tail, Dict, NewGbTree);
dict([], _, GbTree) -> GbTree.
```

We can now perform an upgrade from version 1.1 to version 1.2 of the db module, changing the internal format of the db_server loop data. To do so, we need a *patches* directory at the top of the code search path. We start the Erlang runtime system using the –pa patches flag (or add the directory dynamically using code:add_patha/1). Next, we place the compiled 1.2 version of the db module in the *patches* directory. Finally, we load the new db module, (soft) purge the old one, and call the upgrade client function. Example 8-1 shows the interaction in full.

Example 8-1. Software upgrade in action

```erlang
1> cd("/Users/Francesco/database/").
/Users/Francesco/database
ok
2> make:all([load]).
Recompile: db
Recompile: db_server
up_to_date
3> db:module_info().
[{exports,[{new,0},
           {write,3},
           {read,2},
           {destroy,1},
```

```
                    {delete,2},
                    {module_info,0},
                    {module_info,1}]},
    {imports,[]},
    {attributes,[{vsn,[1.1]}]},
    {compile,[{options,[]},
                    {version,"4.5.2"},
                    {time,{2008,8,11,16,34,48}},
                    {source,"/Users/Francesco/database/db.erl"}]}]
4> db_server:start().
true
5> db_server:write(francesco, san_francisco).
{write,francesco,san_francisco}
6> db_server:write(alison, london).
{write,alison,london}
7> db_server:read(alison).
{ok,london}
8> db_server:read(martin).
{error,instance}
9> code:add_patha("/Users/Francesco/patches").
true
10> code:load_file(db).
{module,db}
11> code:soft_purge(db).
true
12> db_server:upgrade(dict).
{upgrade,dict}
13> db:module_info().
[{exports,[{new,0},
                    {write,3},
                    {read,2},
                    {destroy,1},
                    {delete,2},
                    {convert,2},
                    {module_info,0},
                    {module_info,1}]},
    {imports,[]},
    {attributes,[{vsn,[1.2]}]},
    {compile,[{options,[{outdir,"/Users/Francesco/patches/"}]},
                    {version,"4.5.2"},
                    {time,{2008,8,11,16,30,33}},
                    {source,"/Users/Francesco/patches/db.erl"}]}]
14> db_server:write(martin, cairo).
{write,martin,cairo}
15> db_server:read(francesco).
{ok,san_francisco}
16> db_server:read(martin).
{ok,cairo}
```

The server is still up and running, with the same key-element data, but stored in a
different format and using an upgraded version of the db module. It is as simple as that,
and, as you can see, extremely powerful.

Other important issues to keep in mind when doing a software upgrade include non-backward-compatible modules, functions for downgrading in case an upgrade fails, and synchronizing upgrades in distributed environments. Although the basics of upgrading your code in runtime are very simple, if your system is complex and upgrades are major, your routines and procedures might not be as simple. Make sure you test the upgrade steps thoroughly and cover all potential scenarios. The SASL application, part of the OTP middleware, has complex tools for handling software upgrades built on the principles we just covered.

The .erlang File

It is time to introduce the *.erlang* file. This file is placed in the user's home directory or in the Erlang root directory. It should contain valid Erlang expressions, all of which are read and executed at startup. It is useful for setting paths to the modules and tools of the development environment you are using. For example:

```
code:add_patha("/home/cesarini/patches").
code:add_patha("/home/cesarini/erlang/buildtools-1.0/ebin").
```

You also could use the *.erlang* file for configuration or startup purposes. We will provide more examples in Chapter 11.

Exercise

Exercise 8-1: Software Upgrade During Runtime

Take this chapter's `db.erl` module, based on general balanced trees, and add an extra function called `code_upgrade/1`. This function will take a database of the format used in the lists version of the database from Chapter 3, Exercise 3-4. It should create an ETS table and store in it all the elements passed to it. Its return value should be the table that is created.

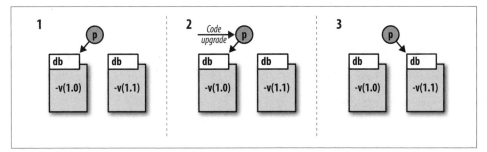

Figure 8-3. Upgrading the database server

Interface:

```
db:code_upgrade([RecordList]) -> gb_tree().
```

1. Test the function in the shell, and place the beam and source code files in a sub-directory called *patches*. In part 2 of this exercise, continue working in your current working directory where you have stored the db.erl module that uses lists. Be careful with the different compiled versions of the two modules so that you do not mix them up.

2. Add the client function code_upgrade() to the my_db.erl server module. This function should send a message to the server which will load the new db module, call the db:code_upgrade/1 function with the old database, and then continue looping with the new ETS database representation, as shown in Figure 8-3. After the code upgrade, all the data that was there prior to the operation should still be available, and the client should be able to insert, delete, and do queries on elements as though nothing has happened.

3. Test your program by starting it using the lists version of the db module. Insert a few elements in the database and switch over to the tree version of the db module. To switch over to the new version, you must first load it into the system. (In larger systems, there will be tools to handle upgrades. As this database server is of more modest proportions, you can load it manually.) Once the ETS version of the db.erl module has been loaded, call my_db:code_upgrade(). Read the elements, write some new ones, and delete the old ones. Make sure the server is stable.

The following interaction gives an example of this in action:

```
1> my_db:start().
ok
2> my_db:write(bob, handyman).
ok
3> my_db:write(chris, projectleader).
ok
4> my_db:read(bob).
{ok,handyman}
5> code:add_patha("/home/cesarini/erlang/patches/").
true
6> my_db:code_upgrade().
```

```
ok
7> my_db:read(bob).
{ok,handyman}
8> my_db:delete(bob).
ok
9> my_db:write(bob, manager).
ok
10> code:soft_purge(db).
true
```

CHAPTER 9

More Data Types and High-Level Constructs

At this stage of the book, we have covered all the fundamentals: sequential programming, concurrency, fault tolerance, and error recovery. The Erlang language—and in particular, its many libraries—offer more to help the programmer be as effective as possible. The various language features covered in this chapter, many of them derived from functional programming languages, are tools that will improve the productivity of a working Erlang programmer.

Functional Programming for Real

Erlang is a *functional programming language*—among other things—but what does that really mean? Prominent in Erlang are function definitions, but that's also the case in Pascal, C, and many other languages. What makes a true functional programming language is the fact that functions can be handled just like any other sort of data. Functional data types in Erlang are called *funs*. They can be passed as arguments to other functions and they can be stored in data structures like tuples and records or sent as messages to other processes. Most of all, they can be *the results of other functions* so that the functions passed around as data can be created dynamically within the program, and are not just pointers to statically defined functions. This, in turn, lets you write concise, abstract, reusable functions that are parameterized over particular behaviors, "wrapped up" as function arguments. As a result, your code becomes not only more compact, but also easier to write, understand, and maintain.

List comprehension is another powerful construct whose roots lie in functional programming. List comprehension constructs allow you to generate lists, merge them together, and filter the results based on a set of predicates. The result is a list of elements derived from the generators for which the predicate evaluated to true. Just like funs, list comprehensions result in compact, powerful code, enhancing programmer productivity.

A *binary* is another Erlang data type which, even if not directly related to functional programming, has in Erlang been influenced by it. A binary is nothing other than a sequence of ones and zeros; an untyped chunk of content stored in memory. All socket and port communication is binary-based, as is all file-related I/O. The power of using binaries in Erlang is the ability to pattern-match on a bit level, efficiently extracting relevant bits and bytes with a minimal amount of effort and code. This makes Erlang perfectly suited to handling protocol stacks and IP-related traffic, encoding and decoding message frames sent and received as a result of these protocols with a minimal amount of code.

Finally, the *reference* data type, elements of which are commonly known as *refs*, provides you with unique tags across processes in distributed environments. In particular, we use reference data values for comparisons, many of which are related to message passing.

Funs and Higher-Order Functions

To understand what funs are all about, it is best to start with an example. Type the following assignment clause in an Erlang shell, binding the variable Bump to a fun:

```
Bump = fun(Int) -> Int + 1 end.
```

The fun takes a variable as an argument, binds it to the variable Int, and "bumps up" its numerical value by one. You call the fun by following it with its arguments in parentheses, just as though you were calling a function. You can use its name if it has been assigned to a variable:

```
1> Bump = fun(Int) -> Int + 1 end.
#Fun<erl_eval.6.13229925>
2> Bump(10).
11
```

Or, you can call it directly:

```
3> (fun(Int) -> Int + 1 end)(9).
10
```

A fun is a function, but instead of uniquely identifying it with a module, function name, and arity, you identify it using the variable it is bound to, or its definition. In the following sections, we will explain why funs are so relevant and useful.

Functions As Arguments

One of the most common operations on lists is to visit every element and transform it in some way. For instance, the following code compares functions to double all elements of a numeric list and to reverse every list in a list of lists:

```
doubleAll([]) ->                       revAll([]) ->
    [];                                    [];

doubleAll([X|Xs]) ->                   revAll([X|Xs]) ->
    [X*2 | doubleAll(Xs)].                 [reverse(X) | revAll(Xs)].
```

Can you see a common pattern between the two functions? All that differs in the two examples is the transformation affecting the element X, italicized in the example; this can be captured by a function, giving the map function, whose first argument, F, is the function to be applied to each element in the list:

```
map(F,[]) ->
    [];
map(F,[X|Xs]) ->
    [F(X) | map(F,Xs)].
```

Another common list operation is to filter out elements with a particular property— for example, even numbers or lists that are palindromes (lists that are the same when they are reversed):

```
3> hof1:evens([1,2,3,4]).
[2,4]
4> hof1:palins([[2,2],[1,2,3],[1,2,1]]).
[[2,2],[1,2,1]]
```

The following code shows these two example functions in action:

```
evens([]) ->                           palins([]) ->
    [];                                    [];
evens([X|Xs]) ->                       palins([X|Xs]) ->
    case X rem 2 == 0 of                   case palin(X) of
        true ->                                true ->
            [X| evens(Xs)];                        [X| palins(Xs)];
        _ ->                                   _ ->
            evens(Xs)                              palins(Xs)
    end.                                   end.
```

In the preceding code, palin/1 is defined by:

```
palin(X) -> X == reverse(X).
```

in the module hof1.

The filter function encapsulates this "filtering out" behavior in a single definition, where the function P embodies the property to be tested: P(X) will return true if X has the property and false if not:

```
filter(P,[]) ->
    [];
filter(P,[X|Xs]) ->
    case P(X) of
        true ->
            [X| filter(P,Xs)];
        _ ->
            filter(P,Xs)
    end.
```

Funs such as P(X) that return `true` or `false` are called *predicates*. So far, so good, but how can you actually *use* the functions `map` and `filter` to give you the behavior you want? To apply them, you need to be able to write down the funs that you pass as their arguments.

Writing Down Functions: fun Expressions

Most likely, you're used to writing down functions when you write function definitions: you have to give the *arguments* to the function—perhaps using *pattern matching* to distinguish among various cases—and in the last expression executed, you return the *result*. A fun expression does the same thing, but without giving the function a name. Let's start with a series of examples.

A function to double its argument is given by:

```
fun(X) -> X*2 end
```

A function to add two numbers is:

```
fun(X,Y) -> X+Y end
```

A function to give the head of a list (and `null` if it is empty) is given by:

```
fun([])    -> null;
   ([X|_]) -> X
end
```

As you can see, these are very similar to function definitions, except the expression is enclosed in a single `fun ... end`. Sometimes you see the `end` followed by a period, or full stop. That is because the full stop signals the end of a definition, or the end of a line of input to the shell. It is *not part of the function expression itself*. Multiple clauses use syntax similar to `case`: the keyword `fun` does not appear before each different pattern-matching case. Remember, though, that arguments are enclosed in parentheses, as in (...), even if there is only one variable as the argument.

Using fun expressions you can now use `map` and `filter` to define `doubleAll` and `palins`:

```
doubleAll(Xs) ->
  map( fun(X) -> X*2 end , Xs).

palins(Xs) ->
  filter( fun(X) -> X == reverse(X) end , Xs).
```

In each case, you can see the fun expression exactly "wraps up" the particular behavior to be mapped or the property to be filtered.

Fun expressions can also encapsulate *side effects*, so the following:

```
fun(X) -> io:format("Element: ~p~n",[X]) end
```

is the function that will print information about its argument on standard output, and this:

```
fun(X) -> Pid ! X end
```

is the function that sends its argument as a message to `Pid`. Each of these can themselves be used as arguments to `foreach`, which performs an action for each element of a list:

```
foreach(F,[]) ->
  ok;
foreach(F,[X|Xs]) ->
  F(X),
  foreach(F,Xs).
```

Here is `foreach` in action:

```
5> hof1:foreach(fun(X) -> io:format("Element: ~p~n",[X]) end, [2,3,4]).
Element: 2
Element: 3
Element: 4
ok
6> hof1:foreach(fun(X) -> self() ! X end, [2,3,4]).
ok
7> flush().
Shell got 2
Shell got 3
Shell got 4
ok
```

Note that `foreach` is defined not only in the `hof1` example, but also in the `lists` module. A function that takes a fun as an argument is called a *higher-order function*. The `lists` module has a collection of them, which we will review shortly.

Functions As Results

You have seen how functions can be arguments to other functions, and how to describe "anonymous" functions using fun expressions. Now, we'll look at how you can combine them in writing and use functions that have functions as their results. Let's look at an example to start:

```
times(X) ->
  fun (Y) -> X*Y end.
```

This is a function that takes one argument: `X`,

```
times(X) -> ...
```

and whose result is this expression:

```
fun (Y) -> X*Y end.
```

When the preceding function is applied to `Y`, it multiplies it by `X`. Let's see it in action:

```
8> Times = hof1:times(3).
#Fun<hof1.3.75238199>
9> hof1:times(3)(2).
* 1: syntax error before: '('
10> (hof1:times(3))(2).
6
11> Times(2).
6
```

Evaluating `hof1:times(3)` gives a result that says, somewhat cryptically, this value is a function. So, we can try to apply that to 2, using the usual function application in command 9. In the first attempt, we get a syntax error, but bracketing things properly in command 10, we see that `times(3)` applied to 2 gives the result 6, as we would expect.

Using this, we can give yet another definition of `doubleAll`!

```
double(Xs) ->
  map(times(2),Xs).
```

The first argument to `map` here is `times(2)`, a function which is the *result* of applying `times` to 2.

Another example is given by the following function:

```
sendTo(Pid) ->
  fun (X) ->
    Pid ! X
  end.
```

which, given a `Pid`, returns the function that sends its argument, X, as a message to that `Pid`.

 At the start of the chapter, we said that functions in Erlang were just like any other data value. There's just one exception to this: one thing you can't do with a function created dynamically is to *spawn* it, because it doesn't belong to any particular module.

Using Already Defined Functions

Erlang also provides ways to use functions that are already defined as arguments to other functions, or as values of variables. Within a module M, a local function F with arity n can be denoted by `fun F/n`. Outside the module, it will be denoted by `fun M:F/n`, as in the following example:

```
12> hof1:filter(fun hof1:palin/1,[[2,2],[2,3]]).
[[2,2]]
13> Pal = fun hof1:palin/1.
#Fun<hof1.palin.1>
14> hof1:filter(Pal,[[2,2],[2,3]]).
[[2,2]]
```

where `palin/1` is defined by:

```
palin(X) -> X == reverse(X).
```

in the module `hof1`.

Something such as fun foo/2 is only syntactic sugar[*] for the notation fun(A1, A2) -> foo(A1,A2) end. However, the "fully qualified" notation fun M:F/N has a special significance for software update, as we explained in Chapter 8. When this fun is called, it will always use the definition in the current version of the module, whereas calls to other funs will go to the module in which the fun was defined, which could be the old version of the module.

Remember that when passing funs to other Erlang nodes, if local and global function calls are made in the fun body, the modules in which those called functions are defined must be in the code search path of the remote Erlang node.

 In Erlang legacy code, you might come across the notation {Module, Function}, where the tuple denotes a fun. The arity of the fun is determined at runtime, based on the number of arguments it is called with. This usage is now deprecated and should no longer be used.

Functions and Variables

All variables that are introduced in fun expressions are considered to be new variables, so they *shadow* any variables that already exist with the same name. This means the shadowed variables are not accessible within the fun expression, but will still be accessible on exit from the fun. All other variables are accessible within the fun expression. Let's look at an example to make this clear:

```
foo() ->                         bar() ->
  X = 2,                           X = 3,
  Bump = fun(X) -> X+1 end.        Add = fun(Y) -> X+Y end,
  Bump(10).                        Add(10)

15> funs:foo()                   1> funs:bar()
11                               13
```

In foo, the local variable X in Bump shadows the earlier X, and the result of foo() shows that within Bump(10), X has the value 10 as passed in by the parameter. In the example of bar, it is evident that the original definition of X is accessible within the body of Add.

Predefined, Higher-Order Functions

The lists module contains a collection of higher-order functions—that is, functions that take funs as arguments and apply them to a list. These higher-order functions allow you to hide the recursive pattern in the function call to the lists module while isolating all side effects and operations on the list elements in a fun. The following list contains

---

[*] The phrase *syntactic sugar* was invented by Peter Landin, and is used for shorthand notations that make a language "sweeter to use," but don't add to its expressiveness.

the most common recursive patterns defined in the higher-order functions of the lists module:

all(Predicate, List)
> Returns true if the Predicate—which is a fun that returns a Boolean value when it is applied to one of the list elements—returns true for all the elements in the list, and returns false otherwise.

any(Predicate, List)
> Returns true if the Predicate returns true for any element in the list, and returns false otherwise.

dropwhile(Predicate, List)
> Drops the head of the list and recurses on the tail for as long as Predicate returns true.

filter(Predicate, List)
> Removes all elements in the list for which Predicate is false, returning the list of remaining elements.

foldl(Fun, Accumulator, List)
> Takes a two-argument function, Fun. These arguments will be an Element from the List and an Accumulator value. The fun returns the new Accumulator, which is also the return value of the call once the list has been traversed. Unlike its sister function, lists:foldr/3, lists:foldl/3 traverses the list from left to right in a tail-recursive way:

```
foldl(F, Accu, [Hd|Tail]) ->
  foldl(F, F(Hd, Accu), Tail);
foldl(F, Accu, []) when is_function(F, 2) -> Accu.

foldr(F, Accu, [Hd|Tail]) ->
  F(Hd, foldr(F, Accu, Tail));
foldr(F, Accu, []) when is_function(F, 2) -> Accu.
```

map(Fun, List)
> Takes a Fun, which it applies to every element in the list. It returns the list containing the results of the fun applications.

partition(Predicate, List)
> Takes a list and divides it into two lists, one containing the elements for which the Predicate returned true, and the other containing the elements for which the Predicate returned false.

If you open the manual page for the lists module, you can review the higher-order functions we just listed alongside those that we have not covered in this chapter. And most important, before reading on, you can try them out in the shell to be sure you understand what is going on:

```
16> Bump = fun(X) -> X+1 end.
#Fun<erl_eval.19.120858100>
17> lists:map(Bump, [1,2,3,4,5]).
```

```
[2,3,4,5,6]
18> Positive = fun(X) -> if X < 0 -> false;
                          X >= 0 -> true
                 end
        end.
#Fun<erl_eval.19.120858100>
19> lists:filter(Positive, [-2,-1,0,1,2]).
[0,1,2]
20> lists:all(Positive, [0,1,2,3,4]).
true
21> lists:all(Positive, [-1,0,1]).
false
22> Sum = fun(Element, Accumulator) -> Element + Accumulator end.
#Fun<erl_eval.12.113037538>
23> lists:foldl(Sum, 0, [1,2,3,4,5,6,7]).
28
```

Lazy Evaluation and Lists

From our coverage of the recursive definition of lists in Chapter 2, you might recall that
a proper or well-formed list is either the empty list or a list where the head is an element
and the tail is itself a proper list. The Erlang evaluation mechanism means that when
a list is passed to a function, the list is evaluated fully before the function body is
executed. Other functional languages, notably Haskell, embody demand-driven (or
lazy) evaluation. Under lazy evaluation, arguments are evaluated only when necessary
in the function body; for data structures such as lists, only the parts that are needed
will be evaluated.

It is possible to build a representation of lazy lists in Erlang. To do this, you use lists in
which the tail is a fun returning a new head and a recursive fun. This avoids you having
to generate a large list that is then traversed. Instead, you reduce the memory imprint
and generate subsequent values only when you need them.

As an example, the following list contains the infinite sequence of whole numbers,
starting at zero. The head is the first sequence number and the tail is a construct that
will recursively generate the next sequence number together with the new tail:

```
next(Seq) ->
    fun() -> [Seq|next(Seq+1)] end.
```

Running the call in the shell or a recursive function that dispatches the sequence numbers would give you something like this:

```
24> SeqFun0 = sequence:next(0).
#Fun<sequence.0.31154838>
25> [Seq1|SeqFun1] = SeqFun0().
[0|#Fun<sequence.0.31154838>]
26> [Seq2|SeqFun2] = SeqFun1().
[1|#Fun<sequence.0.31154838>]
```

It is possible to build a library of functions that perform similar roles over lazy lists to
functions such as map/2 and foldl/3. We leave this as an exercise for the reader.

List Comprehensions

Typical of operations on lists are mapping—applying a function to every element of a list—and filtering—selecting those elements of a list that have a particular property. Often these are used together, and the *list comprehension* notation gives you a powerful and compact way to write down lists constructed in this way. We'll begin with some simple examples, and then look in more detail at a simple database case study.

A First Example

The list comprehension in the following code has the generator X <- [1,2,3], which means X *runs through* the values 1, 2, and 3 in turn. Going into the output for each of these is X+1, the expression before the symbol ||, 1 leading to 2, and so on. This gives the result [2,3,4].

```
1> [X+1 || X <- [1,2,3]].
[2,3,4]
```

In the next code snippet, there is a filter, X rem 2 == 0, which selects only those Xs that pass this test. Here, that X is even:

```
2> [X || X <- [1,2,3], X rem 2 == 0].
[2]
```

The following code combines the two: the even elements of [1,2,3] are selected, and 1 is added to each of them.

```
3> [X+1 || X <- [1,2,3], X rem 2 == 0].
[3]
```

These are the basics of list comprehension. Next, we'll discuss what a list comprehension is like in general, before looking at a larger example.

General List Comprehensions

In general, a list comprehension has three component parts:

```
[ Expression || Generators, Guards, Generators, ... ]
```

Generators
A generator has the form Pattern <- List, where Pattern is a pattern that is matched with elements from the List expression. You can read the symbol <- as "comes from"; it's also like the mathematical symbol ∈, meaning "is an element of."

Guards
Guards are just like guards in function definitions, giving a true or false result. The variables in the guards are those which appear in generators to the left of the guard (and any other variables defined at the outer level).

Expression
The expression specifies what the elements of the result will look like.

Pattern matching performs two tasks, just as it does in a function definition: it allows a complex element—such as a tuple—to have its component parts matched, and it selects only those elements which match the pattern, as shown in the following example:

```
1> Database = [ {francesco, harryPotter}, {simon, jamesBond},
                {marcus, jamesBond}, {francesco, daVinciCode} ].
...
2> [Person || {Person,_} <- Database].
[francesco,simon,marcus,francesco]
3> [Book || {Person,Book} <- Database, Person == francesco].
[harryPotter,daVinciCode]
4> [Book || {francesco,Book} <- Database].
[harryPotter,daVinciCode]
5> [Person || {Person,daVinciCode} <- Database].
[francesco]
6> [Book || {Person,Book} <- Database, Person /= marcus].
[harryPotter,jamesBond,daVinciCode]
7> [Person || {Person,Book} <- Database, Person /= marcus].
[francesco,simon,francesco]
8> [{Book, [Person || {Person,B} <- Database, Book==B ]} || {_,Book} <- Database].
[{harryPotter,[francesco]},
 {jamesBond,[simon,marcus]},
 {jamesBond,[simon,marcus]},
 {daVinciCode,[francesco]}]
9> [{Book,[ Person || {Person,Book} <- Database ]} || {_,Book} <- Database].
[{harryPotter,[francesco,simon,marcus,francesco]},
 {jamesBond,[francesco,simon,marcus,francesco]},
 {jamesBond,[francesco,simon,marcus,francesco]},
 {daVinciCode,[francesco,simon,marcus,francesco]}]
```

In command 4, we have the generator {francesco,Book} <- Database; the pattern will match only those pairs whose first element is francesco. The result of the list comprehension is therefore all the Books borrowed and never returned by francesco. The examples at commands 5, 6, and 7 show how generators and guards are used to give more complex results.

In command 8, you can see a list comprehension appear as part of the result expression. The main generator runs through all the books in the database, and for each of these it generates the pair [Person || {Person,B} <- Database, Book==B], which is the list of all people borrowing the Book.

Command 9 is included for contrast and shows that *each variable appearing in a generator is a new variable*. So, the Book variable appearing in the inner generator is new, and won't pattern-match the value of the outer Book variable as you might expect. This is evident in the difference in results between commands 8 and 9.

Multiple Generators

A list comprehension will, in general, have more than one generator, which you can intersperse with guards. In the next example, we used `lists:seq(N,M)`, which is the list of integers from N to M (e.g., `lists:seq(1,4)` is `[1,2,3,4]`):

```
10> [ {X,Y} || X <- lists:seq(1,3), Y <- lists:seq(X,3) ].
[{1,1},{1,2},{1,3},{2,2},{2,3},{3,3}]
11> [ {X,Y} || X <- lists:seq(1,4), X rem 2 == 0, Y <- lists:seq(X,4) ].
[{2,2},{2,3},{2,4},{4,4}]
12> [ {X,Y} || X <- lists:seq(1,4), X rem 2 == 0, Y <- lists:seq(X,4), X+Y>4 ].
[{2,3},{2,4},{4,4}]
```

The generators are *nested* so that in command 10, a choice is made for X—1, say—and then Y will run through the list `[1,2,3]`, which is the value of the generator list when X is 1; next, X is given the value and Y ranges through `[2,3]`, and so on.

In command 11, a guard follows the generator for X, allowing only even values for X. In command 12, the final guard shows how multiple variables can take part in a guard.

Standard Functions

You can use list comprehensions to define some of the standard list processing functions found in the `lists` module. Take a look at the implementation of the `lists` module itself, and you will find that in some cases it is using list comprehensions:

```
map(F,Xs)    -> [ F(X) || X <-Xs ].
filter(P,Xs) -> [ X || X <-Xs, P(X) ].
append(Xss)  -> [ X || Xs <- Xss, X <- Xs ].
```

A longer example is the function to find all the permutations of a list:

```
perms([]) ->
  [[]];
perms([X|Xs]) ->
  [ insert(X,As,Bs) || Ps <- perms(Xs),
                    {As,Bs} <- splits(Ps) ].

splits([]) ->
  [{[],[]}];
splits([X|Xs] = Ys) ->
  [ {[],Ys} | [ { [X|As] , Bs } || {As,Bs} <- splits(Xs) ] ].

insert(X,As,Bs) ->
  lists:append([As,[X],Bs]).
```

The algorithm here builds all the permutations of `[X|Xs]` by finding all permutations of Xs, and all the ways in which X can be inserted into that permutation, as given by the `splits` function. An alternative algorithm works by generating an element Y of the list, and all the permutations P of the list with Y removed, returning `[Y|P]` (coding this is left as an exercise for the reader).

We'll conclude this discussion with a three-line implementation of a quicksort using list comprehensions. Given a list, we break it up into a head and a tail. We use the head as a *pivot*, generating two new lists. The first list contains all elements smaller than or equal to the pivot, and the second list contains all elements larger than the pivot. We then recursively call quicksort on the newly generated lists, appending them together with the pivot in between:

```
qsort([]) -> [];
qsort([X|Xs]) ->
    qsort([Y || Y<-Xs, Y =< X]) ++ [X] ++ qsort([Y || Y<-Xs, Y > X]).
```

Funs and List Comprehensions Are Born

In the early days, long before Erlang was released as open source, Joe Armstrong waved me (Francesco) into his office and enthusiastically started showing me some new features that had just been added to the language. He described how you could abstract recursive patterns—encapsulating all of the side effects into a fun—and use a higher-order function to traverse the list. He went on to demonstrate the power of list comprehensions, using the same quicksort example included in this book. I didn't think much about the fact that, when I was leaving his office, he asked me to use funs and list comprehensions in my programs, but not to tell anyone about them!

A few months later, when the 5.4 release of Erlang made it into production, an email was sent to the internal Ericsson Erlang mailing list. Someone had accidentally discovered the ++ construct and was wondering whether any other "undocumented features" had found their way into the release. The product manager for Erlang also started wondering, as a request to add funs and list comprehensions to the language had been put on hold as a result of other priorities a few months earlier. He went through the parser used by the compiler and quickly caught on. His diplomatic response stated that any undocumented features in Erlang, ++ included, were not officially supported and were not guaranteed to be part of any future releases.

Based on the reactions on the mailing list, however, it was obvious that I was not the only person Joe had been speaking to. Major projects had (without telling anyone) started to use these "undocumented features," creating a critical mass that ensured that list comprehensions and funs became a permanent (and documented) part of Erlang.

Binaries and Serialization

Sometimes large amounts of structured data have to be transferred between computers or stored. How should protocols ensure that data is generated and transmitted as quickly and efficiently as possible? The answer is to make use of every possible bit of storage, packing the bits in each word with as much information as possible. Erlang *binaries* give a pattern-matching notation for manipulating binary data structures, making this kind of lower-level programming easier, robust, and more space efficient than using tuples or lists. In addition to binaries, in the following subsections we will also look at efficient ways that high-level, nonflat data structures can be serialized and deserialized, using trees as an example.

Binaries

A binary is a reference to a chunk of raw, untyped memory. It was originally used by the Erlang runtime system for loading code over the network, but was quickly applied in a more generic setting of socket-based communication. Binaries are effective in moving large amounts of data, with BIFs provided for coding, decoding, and binary manipulation to this extent:

```
1> Bin1 = term_to_binary({test,12,true,[1,2,3]}).
<<131,104,4,100,0,4,116,101,115,116,97,12,100,0,4,116,114,
   117,101,107,0,3,1,2,3>>
2> Term1 = binary_to_term(Bin1).
{test,12,true,[1,2,3]}
3> Bin2 = term_to_binary({cat,dog}).
<<131,104,2,100,0,3,99,97,116,100,0,3,100,111,103>>
4> Bin3 = list_to_binary([Bin1, Bin2]).
<<131,104,4,100,0,4,116,101,115,116,97,12,100,0,4,116,114,
   117,101,107,0,3,1,2,3,131,104,2,100,...>>
5> Term2 = binary_to_term(Bin3).
{test,12,true,[1,2,3]}
6> {Bin4,Bin5} = split_binary(Bin3,25).
{<<131,104,4,100,0,4,116,101,115,116,97,12,100,0,4,116,
   114,117,101,107,0,3,1,2,3>>,
 <<131,104,2,100,0,3,99,97,116,100,0,3,100,111,103>>}
7> Term4 = binary_to_term(Bin5).
{cat,dog}
8> is_binary(Term4).
false
9> is_binary(Bin4).
true
```

In the preceding code, the inverse BIFs `term_to_binary/1` and `binary_to_term/1` code and decode terms as binaries, that is, sequences of bytes. The function `is_binary/1` is a guard that tests for binaries. When dealing with octet streams, namely lists of integers, you can use `list_to_binary/1` and `binary_to_list/1`.

> It is best to code a sequence of elements in a single binary as a single list, using `term_to_binary`. Coding the elements separately, joining the results using the function `list_to_binary`, and then decoding will give unexpected results. For example, in command 5 in the preceding code, the application `binary_to_term(Bin3)` will decode only the first element in the segment. To decode them all, it is necessary to split them first, and then to decode them separately (as in commands 6 and 7).

The BIFs discussed here work at the level of bytes, and can be seen as a little cumbersome in comparison with the bit syntax which can be used to manipulate bit strings of any length using a pattern-matching style (including joining and splitting such strings). We turn to this next.

The Bit Syntax

The *bit syntax* described in this section allows binaries to be seen as a number of segments, which are sequences of bits but which need not be bytes (nor be aligned on byte boundaries). We use the term *bitstring* to mean an arbitrary-length sequence of bits, and the term *binary* to mean a string whose length is divisible by eight so that it can be seen as a sequence of bytes.

You can construct `Bins` using the following bit syntax:

```
Bin = <<E1, E2, ...,En>>
```

You can pattern-match them like this:

```
<<E1, E2, ...,En>> = Bin
```

Here is the bit syntax in action:

```
1> Bin1 = <<1,2,3>>.
<<1,2,3>>
2> binary_to_list(Bin1).
[1,2,3]
3> <<E,F>> = Bin1.
** exception error: no match of right hand side value <<1,2,3>>
4> <<E,F,G>> = Bin1.
<<1,2,3>>
5> E.
1
6> [B|Bs] = binary_to_list(Bin1).
[1,2,3]
7> Bin2 = list_to_binary(Bs).
<<2,3>>
```

In the preceding code, the BIFs `binary_to_list/1` and `list_to_binary/2` provide translations to and from binaries for ease of list-style manipulation of binaries.

The real strength of binaries lies in the fact that each expression in `Bin` can be *qualified by a size and/or type qualification*:

```
Expr:Size/Type
```

These qualifications allow fine control of formats of numbers, both integers and floats, and mean that a bit program can read like the high-level specification of a protocol rather than its low-level (and opaque) implementation; we will see this in action for TCP segment manipulation shortly.

We now look at sizes and types in detail.

Sizes

The size specified is in bits. The default size of an integer is 8, and of a float 64 (eight bytes).

Types

The type is a list of type specifiers, separated by hyphens. Type specifiers can be any of the following:

A type
> The valid types are `integer`, `float`, `binary`, `byte`, `bits`, and `bitstring`.

A sign
> The valid signs are `signed` and `unsigned` (the default). In the `signed` case, the first bit determines the sign: 0 is positive, 1 is negative.

An endian value
> Endian values are CPU-dependent. Endian values can be `big` (the default), `little`, or `native`. In the `big`-endian case, the first byte is the least significant one; in `little` endian, the first byte is the most significant. You will have to take endian values only if you are transferring binaries across different CPU architectures. If you want your endian value to be determined at runtime, use `native`.

A unit specification, such as `unit:Val`
> The number of bits used by the entry will be `Val*N`, where `N` is the size of the value. The default unit for `bits` and `bitstring` is `1`; for `binary` and `bytes` it is `8`.

The following code snippet shows these types in action:

```
1> <<5:4, 5:4>>.
<<"U">>
2> <<Int1:2, Int2:6>> = <<128>>.
<<128>>
(foo@Vaio)3> Int1.
2
(foo@Vaio)4> Int2.
0
5> <<5:4/little-signed-integer-unit:4>>.
<<5,0>>
6> <<5:4/big-signed-integer-unit:4>>.
<<0,5>>
7> <<5:2,5:8>>.
<<65,1:2>>
```

Note how `<<5:4,5:4>>` returns `<<"U">>`. The integer 5 represented in four bits is equivalent to 0101. In our binary, we put two of them together, 01010101, which is the integer 85, denoting the ASCII value of U. Put 85 in a list, and you get back the string notation containing the capital letter U. Writing `<<"Hello">>` is the same as writing `<<$H,$e,$l,$l,$o>>`, or its ASCII equivalent, `<<72,101,108,108,111>>`.

Pattern-Matching Bits

You can use the same syntax in pattern-matching binaries, and in particular, you can use size and type qualifications in pattern matching. When types are omitted, the default type is an integer:

```
1> A = 1.
1
2> Bin = <<A, 17, 42:16>>.
<<1,17,0,42>>
3> <<D:16,E,F/binary>> = Bin.
<<1,17,0,42>>
4> [D,E,F].
[273,0,<<"*">>]
```

As you can see here, the way a binary sequence is unpacked can be quite different from the way it is constructed. This is one of the mechanisms by which complex protocol handling, where you encode and decode frames, is made easy in Erlang:

```
5> Frame = <<1,3,0,0,1,0,0,0>>.
<<1,3,0,0,1,0,0,0>>
6> <<Type, Size, Bin:Size/binary-unit:8, _/binary>> = Frame.
<<1,3,0,0,1,0,0,0>>
7> Type.
1
8> Size.
3
9> Bin.
<<0,0,1>>
```

It is possible to match bitstrings of any length, as in the following:

```
10> <<X:7/bitstring,Y:1/bitstring>>  = <<42:8>>.
<<"*">>
11> X.
<<21:7>>
12> Y.
<<0:1>>
```

There are a number of possible pitfalls in pattern matching binaries:

- The system will not understand the expression B=<<1>>, as it will read it as B =< <1>>, so it is important to always separate the symbol << from equality with a space.

- The system will not understand the expression <<X+1:8>>. Parenthesizing the arithmetical expression <<(X+1):8>> does the trick.

- <<X:7/binary,Y:1/binary>> will never match, as each binary sequence in a pattern match must have a length that is a multiple of 8.

- It is not possible to use an unassigned variable as the size of a segment, as in the function definition foo(N, <<X:N, ...>>) -> .... You need to use a defined value here. However, it *is* possible to use a variable to specify segment size if the variable is already defined.

Bitstring Comprehensions

The list comprehension notation [... || X <- List, test, ...] is a powerful way to describe lists that are generated from other lists by "generate and test" methods. The *bitstring comprehension* notation performs a similar role for bitstrings. Here's an example:

```
13> << <<bnot(X):1>> || <<X:1>> <= <<42:6>> >>.
<<21:6>>
```

A bitstring comprehension is delimited by << ... || ... >>, and <= is used (instead of <-) for generators. Most importantly, all bitstring entities are indicated by being enclosed within <<...>>. In the preceding example, the variable X is a bit variable, and will be assigned to the bits 101010 (42 in binary) in turn; each of these is negated in the output by bnot, giving the string 010101, the binary representation of 21.

The principles of bitstring comprehension are the same as for lists: pattern-matching syntax is used in the generator (on the left side of the <= symbol) and results can be described using bit syntax, as explained earlier.

Bit Syntax Example: Decoding TCP Segments

Bit syntax comes into its own when you look at coding and decoding segments for the Transmission Control Protocol (TCP).[†]

TCP segments consist of a header (with a defined structure) followed by data. The header has 10 mandatory fields—one of which consists of eight 1-bit flags—and an optional field. The size of the header (in 32-bit words) is specified by the (4-bit) Data Offset field, from which the size of the optional part of the header can be calculated. This field must be at least 5 (and at maximum, 15).

The following code shows the decode function and two sample packets:

```
decode(Segment) ->
  case Segment of
    << SourcePort:16, DestinationPort:16,
       SequenceNumber:32,
       AckNumber:32,
       DataOffset:4, _Reserved:4, Flags:8, WindowSize:16,
       Checksum:16, UrgentPointer:16,
       Payload/binary>> when DataOffset>4 ->

         OptSize = (DataOffset - 5)*32,
         << Options:OptSize, Message/binary >> = Payload,
         <<CWR:1, ECE:1, URG:1, ACK:1, PSH:1, RST:1, SYN:1, FIN:1>> = <<Flags:8>>,
           %% Can now process the Message according to the
           %% Options (if any) and the flags CWR, ..., FIN.
```

[†] TCP is described in a series of Requests for Comments (RFCs) from the Internet Engineering Task Force. An overview is provided in RFC 4614, "A Roadmap for Transmission Control Protocol (TCP) Specification Documents," at *http://tools.ietf.org/html/rfc4614*.

```
        binary_to_list(Message);

   _ ->
        {error, bad_segment}
   end.

seg1() ->
   << 0:16, 0:16,
      0:32,
      0:32,
      5:4, 0:4, 0:8, 0:16,
      0:16, 0:16,
      "message">>.

seg2() ->
   << 0:16, 0:16,
      0:32,
      0:32,
      7:4, 0:4, 0:8, 0:16,
      0:16, 0:16,
      0:64,
      "message">>.
```

What is so gratifying about this definition is that the pattern in the case statement is a readable (yet formal) definition of what a segment looks like. It begins with two 16-bit words representing the source and destination ports that are followed by 32-bit fields for sequence and acknowledgment number.

So far, we have matched on byte boundaries, but next we match the four variables:

```
DataOffset:4, _Reserved:4, Flags:8, WindowSize:16
```

This gives the DataOffset, four bits that are reserved (and so matched to a "don't care" variable), eight bits of Flags, and so on. After matching some more fields, the remainder of the binary is matched to Payload/binary. The match is also guarded by a check that the DataOffset is indeed at least 5.

The body of the clause also uses pattern matching. In the following statement:

```
<< Options:OptSize, Message/binary >> = Payload,
```

any Options are taken from the front of the Payload; if there are none, Options will be matched to the empty binary <<>> and the Payload will be the Message. In either case, the following pattern match:

```
<<CWR:1, ECE:1, URG:1, ACK:1, PSH:1, RST:1, SYN:1, FIN:1>> = <<Flags:8>>,
```

simultaneously extracts the eight 1-bit flags from the Flags byte.

The two segments, seg1/0 and seg2/0, show that segment construction mirrors the pattern matching in the decode function. This data can be used to test the functionality of the stub decoder shown here.

Bitwise Operators

The bit-level operators in Erlang can be applied to integers, returning integers as results. Table 9-1 lists the bitwise operators.

Table 9-1. Bitwise operators

Operator	Description
band	Bitwise and
bor	Bitwise or
bxor	Bitwise exclusive or
bnot	Bitwise negation
bsl	Bit shift left; the second argument gives the size of the shift
bsr	Bit shift right; the second argument gives the size of the shift

In the following example, the operators are applied to $17 = 10001_2$ and $9 = 1001_2$, among others:

```
1> 9 band 17.
1
2> 9 bor 17.
25
3> 9 bxor 17.
24
4> bnot 9.
-10
5> bnot (bnot 9).
9
6> 6 bsr 1.
3
7> 6 bsl 4.
96
```

Serialization

In this section, we will show how you can serialize binary trees in Erlang. You can represent binary trees with nested tuples: an internal node with {node, ..., ...} and a leaf with {leaf, ...}, as illustrated by the tree in Figure 9-1. One way to serialize the data structure is to pretty-print the structure as a fully bracketed string; for the example in Figure 9-1 this would begin "{node,{node,{leaf,cat ...". You can deserialize a string such as this by parsing it, but this is neither an efficient form of storage nor an efficient deserialization mechanism.

When the structure is serialized, you know the size of the representation that is produced. As a result, you can build this information into the serialization and avoid the structure having to be parsed. Initially, you can think of turning the structure into a stream where before each subtree you record the size of the representation of the tree.

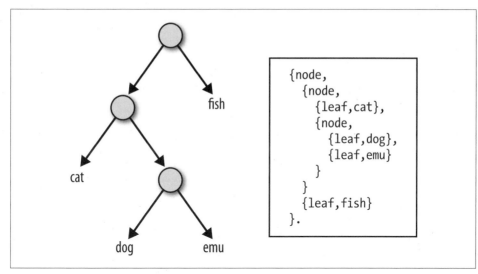

Figure 9-1. Binary tree for serialization

In this example, you would have [11,8,2,cat,5,2,dog,2,emu,2,fish]; here, 11 is the length of the whole representation, 8 is the length of the list [8,2,cat,5,2,dog,2,emu] representing the left subtree of the top node, and so on. Using this length information, it is relatively easy to reconstruct the original tree (we leave this as an exercise for the reader).

However, there is redundant information in that serial form. If you have a segment representing the subtrees at a node, it is enough to know the size of (representation of) the left subtree, as the right subtree is given by whatever remains. So, you can represent the tree in Figure 9-1 by the segment [8,6,2,cat,2,dog,emu,fish]. The following code shows the conversion to this representation:

```
treeToList({leaf,N}) ->
  [2,N];
treeToList({node,T1,T2}) ->
  TTL1 = treeToList(T1),
  [Size1|_] = TTL1,
  TTL2 = treeToList(T2),
  [Size2|List2] = TTL2,
  [Size1+Size2|TTL1++List2].
```

The next code snippet shows the deserialization:

```
deserialize([_|Ls]) ->
  listToTree(Ls).

listToTree([2,N]) ->
  {leaf,N};
listToTree([N]) ->
  {leaf,N};
listToTree([M|Rest] = Code) ->
  {Code1,Code2} = lists:split(M-1,Rest),
```

```
{node,
 listToTree(Code1),
 listToTree(Code2)
}.
```

The essential point here is in the final clause of listToTree, where the length M in the representation is used as the point for splitting the representation before converting the two halves.

Once you have a list representation, you can convert it into a stream of bytes in a straightforward way. You also can remove the intermediate lists from the construction entirely, going directly from tree to bitstream, and vice versa.

References

Remember the frequency server example in Chapter 5? The client sent a request to allocate a frequency and waited for a response of the format {reply, Reply}. How can you be sure this message actually originates from the frequency server, and is not just any other process that decides to send the client a message of this format? The answer is to use *references*.

You create references using the BIF make_ref(), and within the lifetime of a node they are (almost) unique, with values being repeated every 2^{82} calls. They are also unique to a node, and so two references on two different nodes can never be the same.[‡]

You can compare references for equality, and you can use them to match calls and responses within a protocol by labeling a message with a reference, returning that same reference in the response. To do this in the frequency server example, you can use the following code:

```
call(Message) ->
  Ref = make_ref(),
  frequency ! {request, {Ref, self()}, Message},
  receive
    {reply, Ref, Reply} -> Reply
  end.

reply({Ref, Pid}, Message) ->
  Pid ! {reply, Ref, Message}
```

Note how we bind the variable Ref to the reference. When receiving the reply, we include the bound variable Ref, accepting only the messages that pattern-match correctly. This guarantees that the reply originates from the frequency server, as it is the only process to which the client has passed the reference.

---

‡ The only way in practice to get two references that are the same would be to save a reference in a file of Erlang terms, stop the node, and then reload the reference from the file after restarting the node; a reference created in the newly restarted node might then be equal to the saved value, since for each incarnation of the node, references are allocated to be integers starting at 0.

Exercises

Exercise 9-1: Higher-Order Functions

Using funs and higher-order functions, write a function that prints out the integers between 1 and N.

Hint: use `lists:seq(1, N)`.

Using funs and higher-order functions, write a function that, given a list of integers and an integer, will return all integers smaller than or equal to that integer.

Using funs and higher-order functions, write a function that prints out the even integers between 1 and N.

Hint: solve your problem in two steps, or use two clauses in your fun.

Using funs and higher-order functions, write a function that, given a list of lists, will concatenate them.

Using funs and higher-order functions, write a function that, given a list of integers, returns the sum of the integers.

Hint: use `lists:foldl` and try figure out why we prefer to use `foldl` rather than `foldr`.

Exercise 9-2: List Comprehensions

Using list comprehensions, create a set of integers between 1 and 10 that are divisible by three (e.g., `[3,6,9]`).

Using list comprehensions, remove all non-integers from a polymorphic list. Return the list of integers squared: `[1,hello, 100, boo, "boo", 9]` should return `[1, 10000, 81]`.

Using list comprehensions and given two lists, return a new list that is the intersection of the two lists (e.g., `[1,2,3,4,5]` and `[4,5,6,7,8]` should return `[4,5]`).

Hint: assume that the lists contain no duplicates.

Using list comprehensions and given two lists, return a new list that is the symmetric difference of the two lists. Using `[1,2,3,4,5]` and `[4,5,6,7,8]` should return `[1,2,3,6,7,8]`.

Hint: assume that the lists contain no duplicates.

Exercise 9-3: Zip Functions

Define the function `zip`, which turns a pair of lists into a list of pairs:

```
zip([1,2],[3,4,5]) = [{1,3},{2,4}]
```

Using this example, define the function `zipWith` that applies a binary function to two lists of arguments, in lock step:

```
add(X,Y) -> X+Y.
zipWith(Add, [1,2], [3,4,5]) = [4,6]
```

Note that in both cases, the longer of the lists is effectively truncated.

Exercise 9-4: Existing Higher-Order Functions

The Erlang `lists` module contains a number of higher-order functions, that is, functions that take a function as an argument. Write your own definitions of these functions.

Exercise 9-5: Length Specifications in List Comprehensions

Investigate the role of length specifications when making bitstrings from integers. What is the result of `<<42:6>>` and `<<42:5>>`? Contrast this with pattern matching of the following form:

```
<<X:4,Y:2>> = <<42:6>>.
```

Also, investigate pattern matches of the following form:

```
<<C:4,D:4>> = << 1998:6 >>.
<<C:4,D:2>> = << 1998:8 >>.
```

Exercise 9-6: Bitstrings

Using bitstring constructors and pattern matching, give a bit-level implementation of the serialization algorithm in the section "Serialization" on page 208. You will need to think about how much storage is needed for the various items, including the size information that forms part of the sequence.

CHAPTER 10

ETS and Dets Tables

Many practical systems need to store and retrieve large amounts of data within demanding time constraints. For instance, a mobile phone application will need to access and manipulate subscriber details in handling calls as well as in billing and user support. Search times that are proportional to the amount of data being searched are not acceptable in soft real-time systems. Lookup times not only have to be constant, but also have to be very fast!

One of the main composite data types used in programming is a *collection* of items (or elements, or objects). Erlang lists provide one way to implement a collection, but with more than a small number of items in the list, access to elements becomes slow. On average, we need to check through 50% of the elements in a collection to confirm that a given element is present, and we need to look at *all* the elements to verify that a given value is absent.

To handle fast searches, Erlang provides two mechanisms. This chapter introduces Erlang Term Storage (ETS) and Disk Erlang Term Storage (Dets), two mechanisms for memory- and disk-efficient storage and retrieval of large collections of data. Erlang also provides a full database application, Mnesia, which we cover in Chapter 13.

ETS Tables

ETS tables store tuples, with access to the elements given through a *key field* in the tuple. The tables are implemented using hash tables and binary trees, with different representations providing different kinds of collections.

Sets and Bags

In mathematics, *sets* and *bags* are two different kinds of collections. Both contain elements, but the difference is that although a set contains each element only once, a bag can contain duplicate elements. Sets and bags are useful in modeling as well as in programming. Let's look at two examples to see the difference.

You would use a *set* to model the collection of people coming to a birthday party, as all that counts in this example is whether a person is coming. On the other hand, you would use a *bag* to model the presents received at the birthday party,* because in this case, you might receive duplicates of the same item. As a result, you need to store all of the individual presents, not only to ensure that everyone gets a thank you note, but also to keep track of what to sell on eBay.

There are four different kinds of ETS tables, as outlined in the following list. In discussing them and their differences, we'll use the example of an index, where the tuples are pairs containing a word (i.e., a string) and a line number, as in {`"refactorings"`, `4`}. We'll take the first field of the tuple (containing the word) as the key for the rest of this section.

Set
> In a set, each key can occur only once. So, using this kind of table for the index example will mean there can be only one element in the table for each word.

Ordered set
> An ordered set has the same property as the set, but it is stored so that the elements can be traversed following the lexicographical order on the keys. Any entry for `"refactorings"` would precede an entry for `"replay"`, for example. The ordering for different data types is described in Chapter 2.

Bag
> A bag allows multiple entries for the same key, permitting entries such as {`"refactorings"`,`4`} and {`"refactorings"`,`34`}. The elements have to be distinct: in the index example, this means there can be only one entry for a particular word on a particular line.

Duplicate bag
> A duplicate bag allows duplicated elements, as well as duplicated keys, so in the running example, it would be possible for the table to contain the entry {`"refactorings"`,`4`} two or more times.

Referring to the mathematical sidebar "Sets and Bags" on page 213, the Erlang terminology can be seen to reflect how the keys are handled: sets and ordered sets can contain each key only once, whereas bags and duplicate bags give two different variants of multiple occurrences of keys.

Implementations and Trade-offs

The implementations of these collections give a constant lookup time for elements, except in the case of ordered sets, where the access time is logarithmic in the size of the collection. In both cases, this performance is much better than the linear access times for a list representation.

---

* Our birthdays are January 12 and March 26.

Sets, bags, and duplicate bags are stored as *hash tables*, where the position in the table storing the tuple is determined by the value of a function (called a *hash function*) mapping the key of the tuple to the memory location where its contents are stored, as illustrated in Figure 10-1. Assuming our hash table has allocated space for 10 entries, the hash function will return 10 unique memory locations, one for each entry. When we need to insert an additional item, the table gets rehashed (or reorganized), creating space for more entries and returning more unique memory spaces as a result of hashing on the key. This gives us a constant access time from the key to the corresponding tuple(s), but a variable one in case we need to write an entry and rehash the table as a result.[†] (In general, it is possible that two data keys may hash to the same value.)

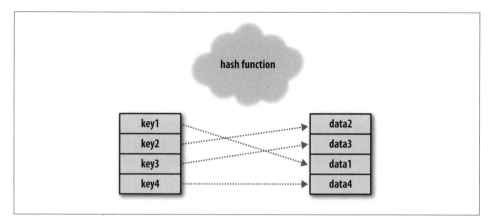

Figure 10-1. Hash table

An ordered set, where the order is given by the built-in order on the key field, is stored as an *AVL balanced binary tree*. This means the length of any of the branches, which determines the time complexity of access, is the logarithm of the size of the collection being stored in the tree. Figure 10-2 shows an example of a balanced binary tree with numerical key values.

If sets and bags give constant time access, why is it necessary to give an ordered set representation that has slower access to particular elements? The answer is that ordered sets allow the collection to be traversed in *key order*, whereas the other representations simply allow the collection to be traversed in *storage order*. In sets, the order depends on the (hidden) details of the hash function, following the order in which they are stored in memory. This will certainly not correspond to any natural order on the data.

The choice of which table type to use depends on the particular application. For instance, in writing out a complete index, it is necessary to traverse the index in

---

[†] You can find more details about the implementation of ETS tables in "A Study of Erlang ETS Table Implementations and Performance," by Scott Lystig Fritchie (Erlang Workshop '03, ACM Press, 2003; *http://doi.acm.org/10.1145/940880.940887*).

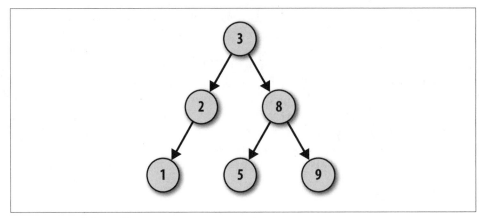

Figure 10-2. A balanced binary tree

alphabetical (key) order, whereas in looking up single words in an online index, it is enough to have access to particular tuples, and an unordered collection is good enough for that.

The implementation of ETS tables in the Erlang distribution is very flexible, allowing key fields to be of any type, including complex data structures. Moreover, it is highly optimized, since it forms the foundation for implementing Erlang's Mnesia database, which we introduce in Chapter 13.

Creating Tables

The `ets` module contains the functions to create, manipulate, and delete ETS tables. A table is created by a call to `ets:new/2`: the first parameter gives the name of the table, and the second consists of a list of options. The function call `ets:new(myTable, Opts)` returns the *table identifier* used to reference the table.

The default setup when an empty list of options is passed to the `ets:new/2` function is to create a set, with the key in position 1, and providing *protected* access to the values of the table. Protected access allows all processes to read the table, but only the owner to write to it. Other options include:

`set, ordered_set, bag, duplicate_bag`
    Including one of these in the options list creates an ETS table of the specified sort.

`{keypos, Pos}`
    This creates a table with the key in position `Pos`.

`public, protected, private`
    A public table is readable and writable by all processes; a private table is readable and writable only by the process that owns the table. We described a protected table earlier.

named_table

> The name of the table is statically registered, and can then be used to reference the table in ETS operations.

Other configuration parameters are described in the `ets` module documentation. You can learn information about a table using `ets:info/1`, passing in the table identifier:

```
1> TabId = ets:new(myTable, []).
15
2> ets:info(TabId).
[{memory,301},
 {owner,<0.31.0>},
 {name,myTable},
 {size,0},
 {node,nonode@nohost},
 {named_table,false},
 {type,set},
 {keypos,1},
 {protection,protected}]
```

If the table is created with the `named_table` option set, you can access it using either the name or the table identifier.

> Even though every table is created with a name in the call to `ets:new/2`, and the name is apparent when viewing the table information, the name *cannot* be used to access the table unless the `named_table` option is enabled.
>
> If this option is not enabled, attempting to access the table in this way will give rise to the `bad argument` runtime error.

The storage used by tables is not recycled automatically when the tables are no longer referenced by the program (i.e., they are not "garbage collected"). Instead, you need to delete the tables manually by calling `ets:delete(TabId)`. However, a table is linked to the process that created it, and if the process terminates, the table is deleted automatically.

> As ETS tables are connected to the process that created them, you need to be extra careful when testing them in the shell. If you cause a runtime error, the shell process will crash and be restarted. As a result, you will lose all of your ETS tables and the data associated with them. Should this happen, use the shell command `f()` to clear all the variables associated with your table references and start again.

Handling Table Elements

You insert elements into a table using `ets:insert/2` and access them by their key using `ets:lookup/2`:

```
3> ets:insert(TabId,{alison,sweden}).
true
4> ets:lookup(TabId,alison).
[{alison,sweden}]
```

In this example, `TabId` is a set, and the insertion of a second element with the `alison` key causes the first element to be overwritten. A common error when dealing with sets is to delete an element before inserting an update. The deletion is superfluous, as the insertion will overwrite the old entry:

```
5> ets:insert(TabId,{alison,italy}).
true
6> ets:lookup(TabId,alison).
[{alison,italy}]
```

If you delete the existing table and re-create the table as a bag, you'll see a different behavior:

```
7> ets:delete(TabId).
true
8> TabId2 = ets:new(myTable,[bag]).
16
9> ets:insert(TabId2,{alison,sweden}).
true
10> ets:insert(TabId2,{alison,italy}).
true
11> ets:lookup(TabId2,alison).
[{alison,sweden},{alison,italy}]
```

The insertion order of elements in bags is preserved. The order of elements in the result here is the order in which they were added, with the oldest element coming first.

Because this ETS table is a bag rather than a duplicate bag, inserting an element where the tuple is identical for a second time has no effect:

```
12> ets:insert(TabId2,{alison,italy}).
true
13> ets:lookup(TabId2,alison).
[{alison,sweden},{alison,italy}]
```

Example: Building an Index, Act I

As an example of ETS tables in use, let's look at the design of an index for a text document; we'll come back to this in later sections when we discuss how the data in the table can be accessed or traversed.

Specification: the document is a text file, and the index is required to show all the lines on which a particular word appears. Words of fewer than four letters are ignored, and all the words in the document are normalized to noncapitalized form in the index. Consecutive lines are shown in the form of a range.

The first part of an index for the preceding paragraph would be as follows:

```
appears      2.
consecutive  3.
document     1,3.
file         1.
form         3-4.
ignored      2.
index        1,3.
   ......
```

The program works in two phases, first collecting the word occurrence data as {Word, LineNumber} pairs from the text file, and then turning that into an index, as shown in Figure 10-3.

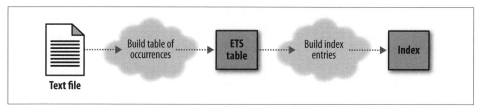

Figure 10-3. Building an index

What type of ETS table should you use to store the word occurrence information?

- If the index is simply used to look up particular words, you could use an *unordered structure*. If the table key is to be the word, duplicate occurrences of a key will be required, as the entries for different lines will have the same key. On the other hand, there is no need to build an index to note repeated entries on the same line: in this case, a *bag* will be the right choice of data structure.

- If the complete index is to be printed in alphabetical order, the table needs to be an *ordered set*. What should the key be in this case?

  — If it is the word, only one entry per word is allowed: in this case, the records stored would need to contain a *list* of line numbers. We leave this option for you to try out in the exercises at the end of the chapter.

  — The option we chose here is to make the pair the key so that entries in the table are tuples of the form {{Word, LineNumber}}; that is, one-field tuples whose only field is itself a pair.

The top-level program for the index shows three separate steps: creating the table, filling the table with data, and building the index from the table. Keep in mind that when reading the example, if you come across new library function calls you do not understand, you should look up the module in the online Erlang documentation:

```
index(File) ->
  ets:new(indexTable, [ordered_set, named_table]),
  processFile(File),
  prettyIndex().
```

Once the file is opened, it can be processed one line at a time by `processLines`, which also keeps track of the current line number:

```erlang
processFile(File) ->
  {ok,IoDevice} = file:open(File,[read]),
  processLines(IoDevice,1).

processLines(IoDevice,N) ->
  case io:get_line(IoDevice,"") of
    eof ->
      ok;
    Line ->
      processLine(Line,N),
      processLines(IoDevice,N+1)
  end.
```

Each line is split into words, using a function from the `regexp` module and a macro definition to hide the details of the punctuation on which words are split:

```erlang
-define(Punctuation,"(\\ |\\,|\\.|\\;|\\:|\\t|\\n|\\(|\\))+").

processLine(Line,N) ->
  case regexp:split(Line,?Punctuation) of
    {ok,Words} ->
      processWords(Words,N) ;
    _ -> []
  end.
```

The `processWords` function inserts the words into the `Table`, after short words have been eliminated, and the remaining words have been normalized using `string:to_lower/1`:

```erlang
processWords(Words,N) ->
  case Words of
    [] -> ok;
    [Word|Rest] ->
      if
        length(Word) > 3 ->
          Normalise = string:to_lower(Word),
          ets:insert(indexTable,{{Normalise , N}});
        true -> ok
      end,
      processWords(Rest,N)
  end.
```

This completes the first stage of building the index; we'll come back to the example after discussing the ways in which you can traverse ETS tables and extract information from them.

Traversing Tables

You already saw how you can find information relating to a single key using `ets:lookup/2`, which will return all the tuples containing that key. In an example such as the index, you need to look at all the data in the table, step by step, one key at a time.

The first key in the table is given by ets:first/1, and given a key, the next key in the table is given by ets:next/2. Assume that we have created and populated the ordered set table in the index example. We would traverse it like this:

```
3> First = ets:first(indexTable).
{"appears",2}
4> Second = ets:next(indexTable,First).
{"consecutive",3}
```

When doing so, we can see the order of the elements is the lexicographical order on the {Word, Number} pairs.

If we were instead to use a bag for the table, with the Word as the key, we would see entirely different behavior after the table has been created:

```
3> First = ets:first(indexTable).
"words"
4> Second = ets:next(indexTable,First).
"form"
```

The order of the keys now is determined by the ordering on the *hash values of the keys*, and not by the order on the keys themselves or by the order in which the values are inserted into the table. Essentially, the ordering is *arbitrary*; the only thing we can depend on is being able to reach all the keys by stepping through them using first and next.

ETS Tables and Concurrent Updates

ETS tables provide very limited support for concurrent updates. What happens if you are traversing an ETS table of type set or a bag using ets:first/1 and ets:next/2 while other processes write and delete elements concurrently?

What happens if another process, when writing a new element, causes a rehash of the table, completely rearranging the order of the entries? In the best of cases, your next/2 call will result in a runtime error. In the worst of cases, the behavior will be undefined in the sense that any element or the '$end_of_table' atom might be returned, or a badarg error might be raised.

If you know other processes will be executing destructive calls to the ETS table while you are traversing it using ets:first/1 and ets:next/2, use the function ets:safe_fixtable/2. It guarantees that during the traversal, you will visit an element only once. If a new element is added after you started your traversal (either by the process traversing the table or by another process), no guarantee is made that this element will be accessed. All other elements, however, will be traversed only once. A process fixes a table by calling ets:safe_fixtable(TableRef, Flag), where Flag is set to true. The flag is released either by setting the flag back to false, or upon process termination. If several processes fix a table, it will remain fixed until all processes have either released it or terminated.

If you fix a table, do not forget to release it, as deleted objects are not removed from the table as long as it remains fixed. This will result not only in the memory used by these deleted objects not being freed, but also in a degradation of the performance of the operations applied on the table.

The last element of an ordered set is given by `ets:last/1`. If `ets:last/1` is applied to any other sort of table, the first element is returned. If `Last` is the final entry in the table, a call to `next` will return `'$end_of_table'`. For the original index example, this would be the result:

```
5> Last = ets:last(indexTable).
{"words",3}
6> ets:next(indexTable,Last).
'$end_of_table'
```

In the next section, we'll discuss how these operations are used in practice.

Example: Building an Index, Act II

In this section, we'll show you how to construct the index for a text file by traversing the ETS table containing the word occurrences. To build the index, you need to write a function that will traverse the table and perform two operations:

- Collecting the entries for a particular word and pairing the word with a list of line numbers, as in `{"form",[4,3]}`
- Pretty-printing each of these tuples as output, with duplicate numbers removed and consecutive numbers shown as a range

The traversal is set up by `prettyIndex/1`, which reads the first field, `{Word,N}`, and builds the tuple `{Word,[N]}`, which contains the information collected so far regarding the `Word`. This partial index entry and the current field, together with the table identifier, are passed to the worker function `prettyIndexNext/3`, which performs the traversal:

```
prettyIndex() ->
  case ets:first(indexTable) of
    '$end_of_table' ->
      ok;
    First ->
      case First of
        {Word, N} ->
          IndexEntry = {Word, [N]}
      end,
      prettyIndexNext(First,IndexEntry)
  end.
```

The function `prettyIndexNext` will read the next record. If no records exist, the current `IndexEntry` is output. If there is a next tuple, `{NextWord, M}`, one of the following things will happen:

- If the NextWord is the same as the word in the IndexEntry, M needs to be added to the list of lines containing the Word, and prettyIndexNext is called recursively.

- If the NextWord is different, the IndexEntry needs to be output before the recursive call with a new partial index entry {NextWord, [M]}.

```
prettyIndexNext(TabId,Entry,{Word, Lines}=IndexEntry) ->
  Next = ets:next(indexTable,Entry),
  case Next of
    '$end_of_table' ->
      prettyEntry(IndexEntry);
    {NextWord, M}  ->
      if
        NextWord == Word ->
          prettyIndexNext(Next,{Word, [M|Lines]});
        true ->
          prettyEntry(IndexEntry),
          prettyIndexNext(Next,{NextWord, [M]})
      end
  end.
```

The definition of prettyEntry is left as an exercise (see Exercise 10-1 at the end of this chapter for hints on how to do this).

Extracting Table Information: match

You've seen how to extract tuples with a given key field from an ETS table, and how to traverse a table; this section shows how you can extract elements from a table by pattern matching. You do this with the ets:match/2 function and the more general ets:select function; in the latter case, the match specification can be given in primitive form, or it can be "compiled" from a function definition using the operation ets:fun2ms/1.

To illustrate match in operation, let's look at an example table containing 3-tuples:

```
1> ets:new(countries, [bag,named_table]).
countries
2> ets:insert(countries,{yves,france,cook}).
true
3> ets:insert(countries,{sean,ireland,bartender}).
true
4> ets:insert(countries,{marco,italy,cook}).
true
5> ets:insert(countries,{chris,ireland,tester}).
true
```

Elements of the table are 3-tuples, and so the patterns matched against the table will also be 3-tuples. The patterns contain three kinds of elements:

- '_', which is a wildcard that will match any value in this position
- '$0' and '$1', which are variables that will match any value in this position
- A *value*, in this case something such as ireland or cook

The result of match is to give a list of results, one for each successful match in the table. The difference between a variable and a wildcard is that the values matched to a variable are reported in the results, whereas the wildcard matches are not. The values matched to the variables are given in a list in ascending order of the variables.

This is best shown through an example:

```
6> ets:match(countries,{'$1',ireland,'_'}).
[[sean],[chris]]
7> ets:match(countries,{'$1','$0',cook}).
[[france,yves],[italy,marco]]
```

In command 6 in the preceding code, the pattern requires that the second field is ireland. There are no restrictions on the other fields, and the value of the first field is reported for each successful match, of which there are two.

In command 7, the third field is required to be cook. Each successful match reports the second and first fields, in that order. Can you predict the results of the following matches? Pay particular attention to the order in which the results are returned:

```
8> ets:match(countries,{'$2',ireland,'_'}).
???
9> ets:match(countries,{'_',ireland,'_'}).
???
10> ets:match(countries,{'$2',cook,'_'}).
???
11> ets:match(countries,{'$0','$1',cook}).
???
12> ets:match(countries,{'$0','$0',cook}).
???
```

You can check your answers by typing the ETS table creation commands (commands 1–5 in the earlier code) into the Erlang shell.[‡] It is possible to return the entire tuple matching a pattern using match_object, and to delete the matching objects by means of match_delete:

```
13> ets:match_object(countries,{'_',ireland,'_'}).
[{sean,ireland,bartender},{chris,ireland,tester}]
14> NewTab = ets:match_delete(countries,{'_',ireland,'_'}).
true
15> ets:match_object(countries,{'_',ireland,'_'}).
[]
```

[‡] Or you can check them here: [[sean],[chris]]; [[],[]]; []; [[yves,france],[marco,italy]]; [].

You need to use match operations with great care, as they can change the real-time behavior of a system. This is because all match operations are implemented as BIFs, and BIFs are executed atomically; a match operation on a large table can therefore stop other processes from executing until the operation has traversed the whole table.

To avoid this problem, it is best to work by table traversal using `first` and `next`, as shown earlier. It might take more time, but it will not disrupt the real-time properties of your system.

Extracting Table Information: select

A *match specification* is an Erlang term that describes a small program. It is a generalization of a pattern, allowing the following:

- Guards to be evaluated over the variables matched
- Return expressions to be more than simply a list of bindings

Here is an example. Working with the `countries` ETS table earlier:

```
16> ets:select(countries,
      [{{'$1','$2','$3'},[{'/=','$3',cook}],[['$2','$1']]}]).
[[ireland,sean],[ireland,chris]]
```

the match specification is a list of 3-tuples, each corresponding roughly to a function clause. Here, there is just one. In that tuple, there are three parts:

`{'$1','$2','$3'}`
> This is a pattern, the same as that used earlier in the `ets:match` function.

`[{'/=','$3',cook}]`
> This is a list of guard expressions, written in prefix form. The single guard here checks the condition that `$3 /= cook`. The match is successful only if each guard evaluates to `true`.

`[['$2','$1']]`
> This is the return expression.

We are introducing match specifications in this chapter not because of their beauty (you must agree with us that they are pretty ugly constructs), but because of their speed and tight connection to the internals of the Erlang runtime system.

As the syntax is cumbersome, support in Erlang for describing match specifications with closures has been implemented. The function `ets:fun2ms/1` takes a `fun` as an argument, describing the comparison (or pattern matching) we want to execute on the ETS table, together with the return values we want the `select` to return. And fortunately for us, `fun2ms` returns a match specification we can use as an argument in our `select` call, relieving us of the need to understand or write match specifications:

```
17> MS = ets:fun2ms(fun({Name,Country,Job}) when Job /= cook ->
                    [Country,Name] end).
[{{'$1','$2','$3'},[{'/=','$3',cook}],[['$2','$1']]}]
18> ets:select(countries, MS).
[[ireland,sean],[ireland,chris]]
```

This allows variables to be named and subexpressions to be written in the usual order when writing a match expression. Note that the fun has to be a literal function, and it cannot be passed to fun2ms as a variable. By *literal function*, we mean a function that is typed in the ets:fun2ms/1 call and not one that is bound to a variable. The fun itself needs to have one argument, which will be a tuple. Finally, if this is used in a module, a header file needs to be included:

```
-include_lib("stdlib/include/ms_transform.hrl").
```

The function ets:select/2 will return all matches for the pattern in the table. You can "batch" the matches by calling ets:select/3, whose third argument is a limit on the number of matches. ets:select/3 returns the matches plus a continuation, which can be called for further matches by passing it to ets:select/1.

We provide the full and gory details of match specifications in Chapter 17. Have a look at it now if you need a detailed explanation or if you are interested in the syntax and semantics of what fun2ms actually returns.

Other Operations on Tables

There are a number of other operations on tables. Here is a list of the most useful of them:

ets:tab2file(TableId | TableName, FileName)
: tab2file/2 dumps a table into a file, returning ok or {error, Reason}.

ets:file2tab(FileName)
: file2tab reads a dumped table back in, returning {ok, Tab} or {error, Reason}.

ets:tab2list(TableId | TableName)
: tab2list returns a list containing all the elements of the table.

ets:i()
: This will list summary information about all the ETS tables visible from the current process.

ets:info(TableId | TableName)
: The info function, discussed earlier, returns the attributes of the given table.

Full details about these and other functions are in the ets module documentation.

Records and ETS Tables

With us having let you in on the internal implementation of Erlang records, by now you should have deduced that it is also possible to insert records into ETS tables. There is a catch, however. Remember that the default key position in ETS tables is the first element of the tuple. In records, that position is reserved for the record type; unless you explicitly state the key position, you will not get the intended behavior. You retrieve the position of the KeyField in a RecordType with the expression #RecordType.Key Field, adding {keypos, #RecordType.KeyField} to the options list of the ets:new/2 call.

Let's try to insert records in ETS tables through the shell. We'll start by defining them, after which we'll create the ETS table and insert and look up a few elements. You might recall from Chapter 7 that when records are used in the shell, they must be either defined, or their definition has to be loaded from the module or include file where it is specified.

```
1> rd(capital, {name, country, pop}).
capital
2> ets:new(countries, [named_table, {keypos, #capital.name}]).
countries
3> ets:insert(countries, #capital{name="Budapest", country="Hungary",
                                  pop=2400000}).
true
4> ets:insert(countries, #capital{name="Pretoria", country="South Africa",
                                  pop=2400000}).
true
5> ets:insert(countries, #capital{name="Rome", country="Italy",
                                  pop=5500000}).
true
6> ets:lookup(countries, "Pretoria").
[#capital{name = "Pretoria",country = "South Africa",
          pop = 2400000}]
7> ets:match(countries, #capital{name='$1',country='$2', _='_'}).
[["Rome","Italy"],
 ["Budapest","Hungary"],
 ["Pretoria","South Africa"]]
8> ets:match_object(countries, #capital{country="Italy", _='_'}).
[#capital{name = "Rome",country = "Italy",
          pop = 5500000}]
9> MS = ets:fun2ms(fun(#capital{pop=P, name=N}) when P < 5000000 -> N end).
[{#capital{name = '$1',country = '_',pop = '$2'},
  [{'<','$2',5000000}],
  ['$1']}]
10> ets:select(countries, MS).
["Budapest","Pretoria"]
```

Look at how we matched the record at commands 7 and 8. To ensure that changes of record fields will not impact operations on ETS tables where those fields are not used, an expression of the format #capital{name = '$1', country = "Italy", _ = '_'} is passed as a pattern to the match functions. The construct _ = Expression replaces all unbound record fields with the Expression. In our example, we picked '_', meaning "all fields that are not explicitly mentioned" are matched to a wildcard, '_'.

Visualizing Tables

The Erlang system comes with a tool for visualizing the current state of ETS and Mnesia tables; tables owned by both the current node and connected nodes are shown when the visualizer is launched by calling `tv:start()`. On launch, a list of the tables is shown, as in Figure 10-4.

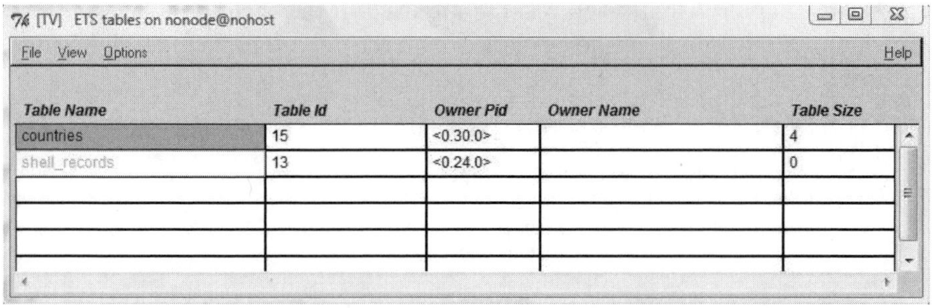

Figure 10-4. The Table Visualizer main window

Clicking one of the tables launches its visualization, as shown in Figure 10-5.

The visualizer allows the contents of the table to be edited, and it will also poll for changes in the table initiated by the program. New tables can also be created from the main visualizer window.

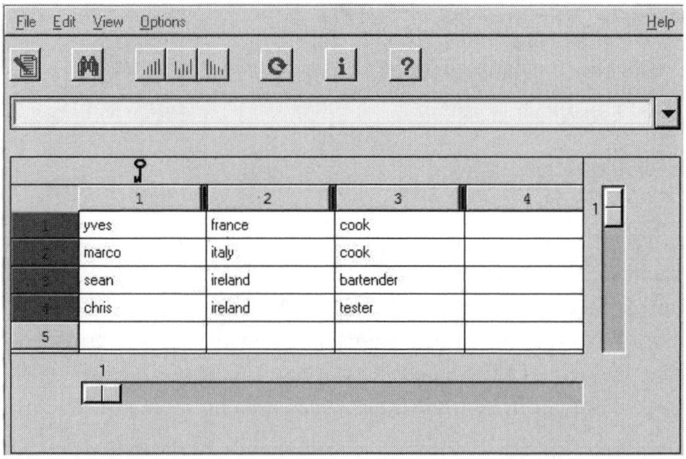

Figure 10-5. Visualizing a table

Dets Tables

Dets tables provide efficient file-based Erlang term storage. They are used together with ETS tables when fast access needs to be complemented with persistency. Dets tables have a similar set of functions to ETS tables, including functions for retrieving, matching, and selecting. But as the calls involve disk seek and read operations, they will be much slower than their counterparts on ETS tables. In the R13 release, the size of a Dets file cannot exceed 2 GB. If you need more than 2 GB of data, you have to fragment it into multiple Dets tables.

Dets table types include `set`, `bag`, and `duplicate_bag`. To use them, you have to open them using `dets:open_file(TableName, Options)`, where the `Options` argument is a list of key value tuples, including the following:

`{auto_save, Interval}`
> Sets the interval at which the table is regularly flushed. Flushing the table means there is no need to repair it if it is not properly closed. `Interval` is an integer in milliseconds (the default is 180,000, or three minutes). If you do not want your file to flush, use the atom `infinity`.

`{file, FileName}`
> Is used to override the default name of the table as a filename and provide a location in which to save the Dets file.

`{repair, Bool}`
> States whether the table should be repaired if it was not properly closed. If repair is needed, setting `Bool` to `true` will trigger the repair automatically, whereas `false` will return the tuple `{error, need_repair}`.

`{type, TableType}`
> Can be `set`, `bag`, or `duplicate_bag`. Ordered sets are currently not supported in Dets tables.

Options used to optimize the table include the following:

`{max_no_slots, Number}`
> Will fragment the table accordingly, optimizing table insertion times. The default value is 2 million entries; the maximum value is 32 million.

`{min_no_slots, Number}`
> Will enhance performance if an estimate is accurate. The default value is 256 entries.

`{ram_file, Bool}`
> Will enhance performance if you need to populate the table with lots of elements. It stores the elements in RAM and spools them to file either when you call `dets:sync(Name)` or when you close the table. The flag is set to `false` by default.

Once you have created the table, you can open it using the call dets:open_file(FileName), where FileName is a string containing the path and name. When using open_file/1, you have to use the reference returned by the function; you will not be able to access the file using its static table name.

Dets tables are closed when the owning process terminates or calls the dets:close(Name) call. If several processes have opened the same table, the table will remain open until all of the processes have either terminated or explicitly closed the table. Not closing a table prior to terminating the Erlang runtime system will result in it being repaired the next time it is opened. This can be a time-consuming task depending on the size of the table.

In the following example, we create a Dets table, experimenting with insertion, selection, and closing the table:

```
1> dets:open_file(food, [{type, bag}, {file, "/Users/Francesco/food"}]).
{ok,food}
2> dets:insert(food, {italy, spaghetti}).
ok
3> dets:insert(food, {sweden, meatballs}).
ok
4> dets:lookup(food, china).
[]
5> dets:insert(food, {italy, pizza}).
ok
6> NotItalian = ets:fun2ms(fun({Loc, Food}) when Loc /= italy -> Food end).
[{{'$1','$2'},[{'/=','$1',italy}],['$2']}]
7> dets:select(food, NotItalian).
[meatballs]
8> dets:close(food).
ok
9> {ok, Ref} = dets:open_file("/Users/Francesco/food").
{ok,#Ref<0.0.0.173>}
10> dets:lookup(Ref, italy).
[{italy,spaghetti},{italy,pizza}]
11> dets:info(Ref).
[{type,bag},
 {keypos,1},
 {size,3},
 {file_size,5920},
 {filename,"/Users/Francesco/food"}]
12> dets:lookup(food, italy).
** exception error: bad argument
     in function  dets:lookup/2
        called as dets:lookup(food,italy)
13> dets:info(Ref).
undefined
```

Pay special attention to dets:open_file/1 in commands 9–12, where table access is by reference instead of by name. Also note command 13; only after the process that opened the table has crashed (or terminated) does the table become unavailable.

ETS and Dets tables are used when a sizable key value store is needed. If your system is in need of transactions, queries, duplication, and persistency, you should be using Mnesia, the database application that comes as part of the OTP middleware. It uses ETS and Dets tables together with the Erlang distribution to provide a relational data model with support for atomic distributed transactions, checkpoints, backups, fragmentation, and runtime schema configuration changes. For more information on Mnesia, see Chapter 13 and the reference guide that comes with the Erlang distribution.

A Mobile Subscriber Database Example

Think of the infrastructure your mobile phone provider needs to manage your account. Every time you send an SMS, your provider must perform a database lookup to ensure that you are a subscriber who is up-to-date with your payments. Every time you authenticate your mobile terminal to start a data session, a similar lookup is required to ensure that you have a data package subscription. Every time you want to use your phone's premium-rated text capability to, for instance, vote for your favorite artist in a televised singing competition, you text the number of the artist you want to vote for and when the mobile operator receives the message, the underlying system needs to perform a subscriber lookup to ensure that your account is enabled to send and receive premium-rated messages. In this example, where the operator will get hundreds of thousands of SMSs in a very short period, the system needs to handle huge bursts without a degradation of service. Similarly, picture a campaign in which the operator sends an SMS to millions of subscribers as quickly as it can; each of these requests requires a lookup before the SMS can be processed to ensure that the recipient is a customer, thus avoiding intranetwork connectivity fees, so not only does this subscriber database need to handle millions of lookups each day, but it also must support massive sustained request bursts. Needless to say, Erlang is the perfect fit for these types of applications. Now that we've looked at ETS and Dets tables in detail, let's use them to build a mobile subscriber database providing an Erlang interface we can use for lookups.

We obviously need to use an ETS table to guarantee a constant and fast lookup time. But because a backup of the data is needed as well, we will mirror the ETS table using a Dets table. Two interfaces will be provided. One is for provisioning the database, where users can be added, updated, and deleted. In this interface, operations have to be fast, as we might be dealing with tens of millions of subscribers, but they do not require soft real-time properties. The second interface will be for the services running in the mobile network, where they have to look up subscriber information. These lookups have to run independently of the provisioning interface so that they will not affect each other, causing bottlenecks during bursts. Assuming the services using the database described earlier are written in Erlang, providing an Erlang API to the user database will further speed up the requests, as no translation of the data is required between systems.

We will store user data in a record of type usr defined in the *usr.hrl* file:

```
%%% File    : usr.hrl
%%% Description : Include file for  user db

-record(usr, {msisdn,            %int()
              id,                %term()
              status = enabled,  %atom(), enabled | disabled
              plan,              %atom(), prepay | postpay
              services = []}).   %[atom()], service flag list
```

The msisdn[§] field refers to the subscriber phone number associated with the GSM SIM card. We will store it as an integer in the ETS table, dropping the leading zero. A number of the format 071234567 is thus represented as 71234567. This representation needs to be used both when provisioning and when doing database lookups.

The id field is the mobile operator's internal subscriber ID. The provisioning interface uses it to manage particular subscribers, and it might differ from the msisdn. As it is not possible to index on secondary keys when using ETS and Dets tables, a separate table containing the subscriber id to msisdn mapping is needed. When starting or restarting the system, the database server can traverse the Dets table, generating the ETS and index mapping table entries which are stored in memory.

The Database Backend Operations

We will implement our example with small development and test iterations. It is a good practice to isolate all the database operations that are directly dependent on Dets and ETS tables in a module of their own, as that will allow you to change and manipulate the storage medium later, without affecting other parts of the system. The day you decide to migrate your subscriber data to Mnesia or any other database, you will be able to do so with very little pain.

Let's call our database module usr_db.erl. The three tables it will manipulate are:

UsrRam
    A named ETS table storing usr records and used for fast access of the subscriber data

UsrIndex
    A named ETS table used to index the subscriber id to the msisdn

UsrDisk
    A Dets table mirroring the usrRam table for redundancy and persistency purposes

---

[§] MSISDN stands for Mobile Subscriber Integrated Services Digital Network Number.

Our first development iteration of the server backend will cover the functionality for opening and closing these tables. As you implement more functions, remember to add them to the export clause of the usr_db module:

```
%%% File     : usr_db.erl
%%% Description : Database API for subscriber DB

-module(usr_db).
-include("usr.hrl").
-export([create_tables/1, close_tables/0]).

create_tables(FileName) ->
    ets:new(usrRam, [named_table, {keypos, #usr.msisdn}]),
    ets:new(usrIndex, [named_table]),
    dets:open_file(usrDisk, [{file, FileName}, {keypos, #usr.msisdn}]).

close_tables() ->
    ets:delete(usrRam),
    ets:delete(usrIndex),
    dets:close(usrDisk).
```

Take particular note of the table properties. The usrRam table needs a key position specified, as it will store usr records. Not doing so will result in the default key position of one being chosen, which in the tuple representation of records maps to the record name. The table will have the default type *set*, making every key unique. It is *protected*, allowing all processes to read the usr records, but only the owner process to manipulate them. The usrIndex table will store a tuple mapping a usr id to an msisdn.

The usrDisk file is a Dets table, with the filename set as an option, allowing us to use different filenames for test purposes. As it will mirror the usrRam table with usr records, it also needs its key position specified:

```
1> c(usr_db).
{ok,usr_db}
2> usr_db:create_tables("UsrTabFile").
{ok,usrDisk}
3> ets:info(usrIndex).
[{memory,308},{owner,<0.29.0>},{name,usrIndex},{size,0},{node,nonode@nohost},
 {named_table,true},{type,set},{keypos,1},{protection,protected}]
4> ets:info(usrRam).
[{memory,308},{owner,<0.29.0>},{name,usrRam},{size,0},{node,nonode@nohost},
 {named_table,true},{type,set},{keypos,2},{protection,protected}]
5> dets:info(usrDisk).
[{type,set},{keypos,2},{size,0},{file_size,5432},{filename,"UsrTabFile"}]
6> usr_db:close_tables().
ok
7> dets:info(usrDisk).
undefined
8> ets:info(usrRam).
undefined
9> ets:info(usrIndex).
undefined
```

So far, so good; it all seems to work. Let's now include two functions to insert a new subscriber in the database, creating an entry in the index and ETS tables together with a backup in the Dets table. We will call the functions add_usr/1 if the subscriber is being provisioned for the first time, and update_usr/1 if the subscriber already exists and we are updating only its status, plan, or services:

```
add_usr(#usr{msisdn=PhoneNo, id=CustId} = Usr) ->
    ets:insert(usrIndex, {CustId, PhoneNo}),
    update_usr(Usr).

update_usr(Usr) ->
    ets:insert(usrRam, Usr),
    dets:insert(usrDisk, Usr),
    ok.
```

Study the preceding functions. Notice that when calling update_usr/1, we are over-writing the existing table entries. In doing so, we are making the assumption that the customer id does not change and that the user already exists. If these preconditions are not met, the database will become corrupt, as the usrIndex table will not map the subscriber id to the msisdn anymore. In Erlang, you should trust your internal interfaces. It is the responsibility of the calling process to guarantee that these preconditions are met. We will tell you how when we write these functions.

Now that we can insert data, we should also be able to look it up, using either the msisdn or the subscriber id. The functions to do this are called lookup_id/1 and lookup_msisdn/1. If we use the id, we need a local get_index/1 call, mapping the id to the msisdn:

```
lookup_id(CustId) ->
    case get_index(CustId) of
        {ok,PhoneNo}     -> lookup_msisdn(PhoneNo);
        {error, instance} -> {error, instance}
    end.

lookup_msisdn(PhoneNo) ->
    case ets:lookup(usrRam, PhoneNo) of
        [Usr] -> {ok, Usr};
        []    -> {error, instance}
    end.

get_index(CustId) ->
    case ets:lookup(usrIndex, CustId) of
        [{CustId,PhoneNo}] -> {ok, PhoneNo};
        []                 -> {error, instance}
    end.
```

Let's test these functions from the shell. In doing so, we need to create the ETS tables and reopen the Dets table. As we already created the Dets table in a previous call, the file with no entries should still be there. As a result, it will be opened instead of being created.

We also need to load the usr record definition in the shell, as we will be reading and writing records. Injecting 100,000 subscribers in all three tables, an operation which might take a few seconds as it will be dealing with heavy I/O when writing the Dets entries to files, should provide us with sufficient data to ensure that the database is production worthy for a smaller mobile operator:

```
1> c(usr_db).
{ok,usr_db}
2> rr("usr.hrl").
[usr]
3> usr_db:create_tables("UsrTabFile").
{ok,usrDisk}
4> usr_db:lookup_id(1).
{error,instance}
4> Seq = lists:seq(1,100000).
[1,2,3,4,5,6,7,8,9,10,11,12,13,14,15,16,17,18,19,20,21,22,
 23,24,25,26,27,28,29|...]
5> Add = fun(Id) -> usr_db:add_usr(#usr{msisdn = 700000000 + Id,
                                        id = Id,
                                        plan = prepay,
                                        services = [data, sms, lbs]})
          end.
#Fun<erl_eval.6.13229925>
6> lists:foreach(Add, Seq).
ok
7> ets:info(usrRam).
[{memory,2214643}, {owner,<0.29.0>}, {name,usrRam}, {size,100000},
 {node,nonode@nohost}, {named_table,true}, {type,set}, {keypos,2},
 {protection,protected}]
8> {ok, UsrRec} = usr_db:lookup_msisdn(700000001).
{ok,#usr{msisdn = 700000001,id = 1,status = enabled,
         plan = prepay,
         services = [data,sms,lbs]}}
9> usr_db:update_usr(UsrRec#usr{services = [data, sms], status = disabled}).
ok
10> usr_db:lookup_msisdn(700000001).
{ok,#usr{msisdn = 700000001,id = 1,status = disabled,
         plan = prepay,
         services = [data,sms]}}
```

Now, let's kill the shell process. All tables owned by this process will be terminated, and as a result, the usr entries and indexes stored in memory will be lost. We re-create the tables using create_tables/1:

```
11> exit(self(), kill).
** exception exit: killed
12> usr_db:lookup_msisdn(700000001).
** exception error: bad argument
     in function  ets:lookup/2
        called as ets:lookup(usrRam,700000001)
     in call from usr_db:lookup_msisdn/1
13> usr_db:create_tables("UsrTabFile").
{ok,usrDisk}
14> usr_db:lookup_msisdn(700000001).
```

```
{error,instance}
15> dets:lookup(usrDisk, 700000001).
[#usr{msisdn = 700000001,id = 1,status = disabled,
      plan = prepay,
      services = [data,sms]}]
```

What's happening here? Have we missed something? The database is not being restored in the RAM-based tables, and we're getting an {error, instance} return value. The subscriber information is accessible only if we read it directly from the Dets table. Simple: we've opened the Dets usrDisk table, giving us access to all of the entries we had prior to terminating the shell. We've also re-created the usrIndex and usrRam tables, but in doing so, we never populated them with a RAM copy of the usr records and the usr id to msisdn mapping. We need a restore backup function that traverses the Dets table and uses its contents when creating the index and RAM entries. Luckily, the dets module exports a function called traverse that we can use:

```
restore_backup() ->
    Insert = fun(#usr{msisdn=PhoneNo, id=Id} = Usr) ->
                ets:insert(usrRam, Usr),
                ets:insert(usrIndex, {Id, PhoneNo}),
                continue
             end,
        dets:traverse(usrDisk, Insert).
```

As arguments, the traverse function takes the Dets table name and a fun that is applied to every element. The fun we pass creates an entry in both the usrRam and usrIndex tables, restoring our subscriber database to its original state:

```
16> c(usr_db).
{ok,usr_db}
17> usr_db:restore_backup().
[]
18> usr_db:lookup_msisdn(700000001).
{ok,#usr{msisdn = 700000001,id = 1,status = disabled,
         plan = prepay,
         services = [data,sms]}}
19> usr_db:lookup_id(1).
{ok,#usr{msisdn = 700000001,id = 1,status = disabled,
         plan = prepay,
         services = [data,sms]}}
```

Finally, we need an efficient way to traverse the whole database, clearing all of the subscribers who have terminated their accounts, not paid their bill, or taken their number to another network provider. We will recognize them through their status field, set to disabled.

We will traverse the mobile subscribers using the first/1 and next/2 calls on the RAM copy of the usrRam table. Do you recall us mentioning that when traversing tables, we need to lock the tables using the safe_fixtable/2 call, as destructive operations during the traversal might cause a runtime error, or even worse, an undefined behavior? Using safe_fixtable/2 guarantees that we will traverse each and every element only once without being affected by any destructive operations executed after starting the traversal. It is a good idea to wrap the traversal operation in a `catch`, as we need to guarantee the release table in the event of a runtime error. Not doing so will cause a memory leak if the exception is caught elsewhere. The reason for using a catch, and not a try catch, is that we are not interested in the return value of the `loop_delete_disabled/1` call, we just want to make sure it does not crash the process:

```erlang
delete_disabled() ->
    ets:safe_fixtable(usrRam, true),
    catch loop_delete_disabled(ets:first(usrRam)),
    ets:safe_fixtable(usrRam, false),
    ok.

loop_delete_disabled('$end_of_table') ->
    ok;
loop_delete_disabled(PhoneNo) ->
    case ets:lookup(usrRam, PhoneNo) of
      usr{status=disabled, id = CustId}] ->
        delete_usr(PhoneNo, CustId);
      _ ->
        ok
    end,
    loop_delete_disabled(ets:next(usrRam, PhoneNo)).
```

With `delete_disabled/0` in place, we have all of the database backend functionality we need. We set one subscriber's status to disabled in command 9 earlier, so traversing the table and deleting entries with the *disabled* status should result in 99,999 subscribers. Let's test it, and once we're satisfied that it works, we'll start defining and implementing the server code itself:

```erlang
20> c(usr_db).
{ok,usr_db}
21> usr_db:delete_disabled().
ok
22> ets:info(usrRam).
[{memory,2214621}, {owner,<0.182.0>}, {name,usrRam}, {size,99999},
 {node,nonode@nohost}, {named_table,true}, {type,set}, {keypos,2},
 {protection,protected}]
```

After having added all of the functions, the `export` clause of the `usr_db` module should look like this:

```erlang
-export([create_tables/1, close_tables/0, add_usr/1, update_usr/1, delete_usr/1,
         lookup_id/1, lookup_msisdn/1, restore_backup/0, delete_disabled/0]).
```

The Database Server

The database server exports three sets of functions. The first is used for operation and maintenance purposes to start and stop the server; the second is used to interface the customer service software to provision the database with users and the services they are entitled to use; and the third is an interface used by the services to request information on the users.

Next, we describe the interface to the database server: in describing the functions we use the Erlang type notation, covered in more detail in Chapter 18. For example, `delete_usr(CustId) -> ok | {error, timeout}` says that `delete_usr` has one argument (which is a customer identifier), and it returns either the atom `ok` or the tuple `{error, timeout}`. The vertical bar (`|`) is used to separate alternatives in the return type.

To begin, we will look at the operation and maintenance functions to start and stop the server:

`start() -> ok`
> Starts the database server using *usrDb* as a default Dets filename. It is there for testing purposes.

`start(FileName) -> ok`
> Is used to override the *usrDb* filename with any valid string. `start/0` and `start/1` are synchronous calls and return `ok` only when the Dets file has been traversed and its contents written to ETS tables.

`stop() -> ok | {error, Reason}`
> Deletes the ETS tables and closes the Dets file, terminating the database server. It returns the result of the `dets:close/1` call.

The customer service software will use the following functions to provision the users and the services they are allowed to use. All of the functions that change the database state are synchronously executed by the database server to guarantee data consistency:

`add_usr(PhoneNum, CustId, Plan) -> ok | {error, timeout}`
> Is used to provision users. `PhoneNum` and `CustId` are integers, and `Plan` is one of the atoms `prepay` or `postpay`.

`delete_usr(CustId) -> ok | {error, timeout}`
> Is used to delete a particular user and all of its associated indexes.

`set_service(CustId, Service, Flag) -> ok | {error, instance | timeout}`
> Is used to add or delete a `Service` for a particular subscriber. `Service` is a list of atoms, including (but not limited to) `data`, `lbs`, and `sms`. The atom `data` confirms the user has subscribed to a data plan, `sms` allows the user to send and receive premium-rated SMSs, and `lbs` allows third parties to execute location lookups on this particular user. `Flag` denotes the atom `true` or `false`, depending on whether the `Service` is being added or deleted.

`set_status(CustId, Status) -> ok | {error, instance | timeout}`
> Will set the status to `enabled` or `disabled` for a particular `CustId`.

`delete_disabled()-> ok | {error, timeout}`
> Will traverse the table and delete all users whose status is set to `disabled`.

`lookup_id(CustId) -> {ok, #usr{}} | {error, instance}`
> Will look up a user based on its `CustId` and return a record of type `usr`, as defined in the *usr.hrl* include file. This function does not change the user data, and as a result, it is executed in the scope of the calling process, and not by the server.

Finally, we have the functions used by the service applications. Both of these functions are executed in the scope of the calling process, as the data needed is available through the protected ETS tables. No messages are exchanged between the client and the process, avoiding a bottleneck during heavy load and decreasing the overall response time. As most of the services will have access only to the msisdn and not to usr id, the msisdn is used as a key:

lookup_msisdn(PhoneNo)) -> {ok, #usr{}} | {error, instance}

> Will look up a user based on its PhoneNo and return a record of type usr. The main use for this function is for service applications to use the data in the record to determine whether the user is eligible for a particular service. For example, a prepay customer might have to be charged to ensure that he has good credit before being allowed to send or receive a premium-rated SMS. Or we might want to send an SMS with latitude and longitude only if a user has the sms and lbs flags set.

service_flag(PhoneNo, Service) -> true | false | {error, instance | disabled}

> Will check whether a user exists and has an enabled status. If so, it will traverse the list of services to determine whether the subscriber is allowed to use this Service in a particular request. This is a variant of the lookup_msisdn/1 call, where logical checks are done in the usr module.

Note how we have abstracted the server loop in the code, handling all of the messages in separate functions where we pattern-match on the first argument of each call. All client/server communication has also been abstracted through the call/1 and reply/2 functions. Starting the server is a synchronous operation, and only when the Dets table has been loaded and mirrored in the usrIndex and usrRam tables do the start/0 and start/1 calls return. The timeout has been set to 30 seconds. Depending on the server load and the size of the database, this value might have to be fine-tuned. Real timeout values are usually determined when stress-testing the system:

```
%%% File     : usr.erl
%%% Description : API and server code for cell user db

-module(usr).
-export([start/0, start/1, stop/0, init/2]).
-export([add_usr/3, delete_usr/1, set_service/3, set_status/2,
        delete_disabled/0, lookup_id/1]).
-export([lookup_msisdn/1, service_flag/2]).

-include("usr.hrl").
-define(TIMEOUT, 30000).

%% Exported Client Functions
%% Operation & Maintenence API

start() ->
  start("usrDb").

start(FileName) ->
  register(?MODULE, spawn(?MODULE, init, [FileName, self()])),
```

```erlang
    receive started-> ok after ?TIMEOUT -> {error, starting} end.

stop() ->
  call(stop).

%% Customer Service API

add_usr(PhoneNum, CustId, Plan) when Plan==prepay; Plan==postpay ->
  call({add_usr, PhoneNum, CustId, Plan}).

delete_usr(CustId) ->
  call({delete_usr, CustId}).

set_service(CustId, Service, Flag) when Flag==true; Flag==false ->
  call({set_service, CustId, Service, Flag}).

set_status(CustId, Status) when Status==enabled; Status==disabled->
  call({set_status, CustId, Status}).

delete_disabled() ->
  call(delete_disabled).

lookup_id(CustId) ->
  usr_db:lookup_id(CustId).

%% Service API

lookup_msisdn(PhoneNo) ->
  usr_db:lookup_msisdn(PhoneNo).

service_flag(PhoneNo, Service) ->
  case usr_db:lookup_msisdn(PhoneNo) of
    {ok,#usr{services=Services, status=enabled}} ->
      lists:member(Service, Services);
    {ok, #usr{status=disabled}} ->
      {error, disabled};
    {error, Reason} ->
      {error, Reason}
    end.

%% Messaging Functions

call(Request) ->
  Ref = make_ref(),
  ?MODULE! {request, {self(), Ref}, Request},
  receive
    {reply, Ref, Reply} -> Reply
  after
    ?TIMEOUT -> {error, timeout}
    end.

reply({From, Ref}, Reply) ->
  From ! {reply, Ref, Reply}.

%% Internal Server Functions
```

```erlang
init(FileName, Pid) ->
  usr_db:create_tables(FileName),
  usr_db:restore_backup(),
  Pid ! started,
  loop().

loop() ->
  receive
    {request, From, stop} ->
      reply(From, usr_db:close_tables());
    {request, From, Request} ->
      Reply = request(Request),
      reply(From, Reply),
      loop()
  end.

%% Handling Client Requests

request({add_usr, PhoneNo, CustId, Plan}) ->
  usr_db:add_usr(#usr{msisdn=PhoneNo,
                      id=CustId,
                      plan=Plan});

request({delete_usr, CustId}) ->
  usr_db:delete_usr(CustId);

request({set_service, CustId, Service, Flag}) ->
  case usr_db:lookup_id(CustId) of
    {ok, Usr} ->
      Services = lists:delete(Service, Usr#usr.services),
      NewServices = case Flag of
        true  -> [Service|Services];
        false -> Services
        end,
          usr_db:update_usr(Usr#usr{services=NewServices});
      {error, instance} ->
        {error, instance}
    end;

request({set_status, CustId, Status}) ->
  case usr_db:lookup_id(CustId) of
    {ok, Usr} ->
      usr_db:update_usr(Usr#usr{status=Status});
    {error, instance} ->
      {error, instance}
    end;

request(delete_disabled) ->
  usr_db:delete_disabled().
```

Testing the usr server in the shell could yield something similar to this:

```erlang
1> c(usr).
{ok,usr}
2> rr("usr.hrl").
```

```
[usr]
3> usr:start().
ok
4> usr:add_usr(700000000, 0, prepay).
ok
5> usr:set_service(0, data, true).
ok
6> usr:lookup_id(0).
{ok,#usr{msisdn = 700000000,id = 0,status = enabled,
        plan = prepay,
        services = [data]}}
7> usr:set_status(0, disabled).
ok
8> usr:service_flag(700000000,lbs).
{error,disabled}
```

When reviewing the database server code base, try to break it with negative test cases and corrupt data. Using the interface provided in the usr module, are you able to corrupt the data by changing the customer ID without updating the usrIndex table? See whether you manage to detect weak spots in the code that might cause a runtime error and result in the server terminating.

There are a number of ways to build more efficient collections in a pure functional style, many of which are described in *Purely Functional Data Structures* by Chris Okasaki (Cambridge University Press).

Exercises

Exercise 10-1: Pretty-Printing

This exercise asks you to complete the definition of prettyEntry from the index example earlier in this chapter.

In defining prettyEntry, you might find it useful to define these functions:

accumulate/1
>   This function should take a list of line numbers, in descending order, and produce a list containing ranges as well as removing duplicates. For instance:

>   accumulate([7,6,6,5,3,3,1,1]) = [{1},{3},{5,7}]

prettyList/1
>   This function will print the output of accumulate so that on the list [{1},{3}, {5,7}] the output is 1,3,5-7..

pad/2
>   This function, called with number N and string Word, will return the string padded to the right with spaces to give it length N (assuming Word is not longer than N).

Exercise 10-2: Indexing Revisited

How would you modify the index program so that the ETS table is an ordered set keyed on the Word field of tuples of the form {Word, LNs}, where LNs is a list of line numbers on which the word occurs?

Exercise 10-3: ETS Tables for System Logging

An ETS table can be used to log traces from a communication system, over which analysis and error reporting can be done. Take the example of a simple messaging system in which each message is expected to receive a single acknowledgment. Messages can be identified by the time at which they are sent using the BIF now(); assume that messages are timestamped with their send time so that they can be matched with their acknowledgment.

Design an ETS table or tables to contain messages and acknowledgments, and using these tables write a program to do the following:

- Check that each message receives a unique acknowledgment.
- Monitor the average time taken for acknowledgments to be received over a sliding one-second send window.

CHAPTER 11

Distributed Programming in Erlang

To write a fault-tolerant system, you need at least two computers[*] and you need to distribute your program across them. Distributed systems lie at the heart of modern computing. In server-side programming, it is the exception rather than the rule to see a single computer performing a task of any difficulty; instead, a number of computers (or processors) will together provide a robust, efficient, and scalable platform upon which applications can be built.

Erlang distribution is built into the language, and from the user's point of view, it can be completely transparent: processes are accessed by a pid, and this may equally well refer to a process on the local computer or a process on a system on the other side of the world. In this chapter, we will look at the theory behind distributed systems and see how it is applied to Erlang-based systems.

Distributed Systems in Erlang

The essence of distributed systems is to provide *in a transparent way* a service of some kind through a number of computers, processors, or cores linked together by a network. A service can be specific, such as the storage provided by a distributed filesystem or database, or more general, as in a distributed operating system that provides all the facilities of a general-purpose OS across a network of computers. Distribution can be seen in tightly coupled parallel processors, but more clearly in the loosely coupled grids of e-science systems. Erlang provides *distributed programming* facilities so that Erlang systems can be run across *networked Erlang nodes*.

Take an installation of Ejabberd, an Erlang open source Jabber-based instant messaging (IM) server. Its implementation is distributed across a cluster of two or more Erlang nodes. These nodes, residing on the same or separate machines, help each other by sharing the message and event loads. Should one of the nodes terminate because of a software or hardware error, or simply because of lack of memory, the other nodes take over the traffic, hiding the fault from the end user. In the worst case, end users might

---

[*] At least two, according to Joe Armstrong, but three if you ask Leslie Lamport.

believe they experienced a network glitch when the socket reconnects to the new node, but all they would notice are other users signing out and in.

The Erlang Web framework, an open source application for Erlang-based web applications, uses distribution for scalability and reliability. A typical cluster consists of frontend and backend nodes. The frontend nodes contain the web servers (running in the Erlang node), a cache layer, and a layer handling XML parsing for inbound requests. It also contains the functionality for handling the dynamic generation of XHTML. Two or more backend nodes contain the databases and all of the glue and logic needed to generate the dynamic content. The real load will be on the frontend, as it handles the socket connections and most of the parsing. To scale the system, all you need to do is add more hardware and frontend nodes, increasing the backend support only when necessary. Should any of the nodes fail the load balancers will automatically redirect the traffic to the nodes that are still alive.

If you want to scale a system in Erlang by distributing functionality across a number of nodes, one thing you need to consider is how the load might be balanced across the nodes. It would be possible to allocate tasks to nodes at random or to use a round-robin approach; either process works well with tasks of a similar size. Otherwise, you need to estimate the size of the tasks to be distributed. Finally, you could use a master-slave model where tasks are delegated as required.

Whatever approach you use, it is crucial to monitor the system behavior and to adapt the distribution strategy—either in real time or via code updates—to respond to the system's changing requirements.

Another example of distributed systems is in one of the first flagship Erlang products, the AXD301 ATM switch. The smallest Erlang cluster consists of two nodes, a call setup node and an operation and maintenance (O&M) node. If the O&M node fails, a *failover* occurs, and the O&M applications are restarted on the call setup node. When the O&M node comes back up, through automated recovery or manual intervention, a *takeover* occurs, and the O&M applications are migrated back to the original node.

Failures in the call setup node are considered critical, as they affect ATM traffic. If a call setup node terminates, a failover would move the call setup applications to the O&M node. Data distribution ensures that any calls whose setups were initiated before the failover are not lost. They are picked up by the new call setup application running on the O&M node. When the original call setup node is restarted, to ensure that traffic is not disrupted and no call setup requests are lost, a takeover of the O&M applications results in the O&M functionality being migrated to the newly restarted node and the call setup functionality remaining on what was formerly the O&M node.

Concurrency is central to all distributed systems, since computation and communication can proceed in parallel across the processors and networks comprising the system. Central to the challenges of distributed systems is robustness in the event of failure. This is memorably summarized by one of the pioneers in the field, Leslie Lamport:

A distributed system is one in which the failure of a computer you didn't even know existed can render your own computer unusable.

But if you get it right, there are a number of advantages in building a distributed system:

- It will provide *performance* that can be scaled with demand. A typical example here is a web server: if you are planning a new release of a piece of software, or you are planning to stream video of a football match in real time, distributing the server across a number of machines will make this possible without failure.

  This performance is given by *replication* of a service—in this case a web server—which is often found in the architecture of a distributed system.

- Replication also provides *fault tolerance*: if one of the replicated web servers fails or becomes unavailable for some reason, HTTP requests can still be served by the other servers, albeit at a slower rate. This fault tolerance allows the system to be more *robust* and reliable.

- Distribution allows transparent access to *remote resources*, and building on this, it is possible to *federate* a collection of different systems to provide an overall user service. Such a collection of facilities is provided by modern e-commerce systems, such as the Amazon.com website.

- Finally, distributed system architecture makes a system *extensible*, with other services becoming available through remote access.

Telecom systems need to reflect all of this. But looking at the bigger picture, they are not the only ones. Trading systems, retail banking systems, air and railway traffic control systems, and web services are just some of the areas where highly transactional, mission-critical systems have benefited from Erlang distribution.

Distributed Computing in Erlang: The Basics

An Erlang *node* is an executing Erlang runtime system that has been given a name. Multiple nodes can run on a single host, but they can also be running on different host computers, too, as shown in Figure 11-1, where three nodes are running on the hosts STC and FCC in their respective subnetworks.

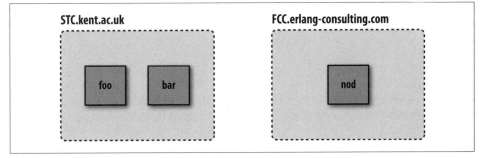

Figure 11-1. Three nodes running on two hosts

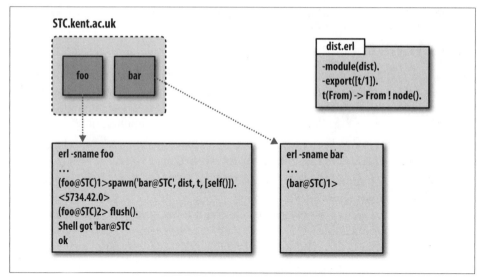

Figure 11-2. A first example: two nodes on one host

As a first example, we'll look at two nodes running on the same machine, STC, as shown in Figure 11-2. To run a node, the `erl` command needs to be given the `sname` flag (the `name` flag can also be used; we'll discuss this shortly). For example:

```
erl -sname foo
Erlang (BEAM) emulator version 5.6.3 [source] [64-bit] [smp:2]
  [async-threads:0] [kernel-poll:false]

Eshell V5.6.3  (abort with ^G)
(foo@STC)1>
```

Note that the prompt in the Erlang shell displays the name of the node, as well as its host computer: foo@STC. This is called the *(unique) identifier* of the node.

A similar command, `erl-sname bar`, will set up a second node on the STC system.

To understand what happens next, you need to look at the module `dist.er1`. This contains the function `t`:

```
t(From) -> From ! node().
```

The function takes as a parameter a pid, `From`, and in its single action it sends a message to the process with that pid. The message is the result of calling `node()`, which returns the identifier of the node where it is called.

Next, we'll look at the user input typed at foo@STC. This uses the `spawn/4` function, whose first argument is the node where the spawn should take place. The remaining arguments are as for `spawn/3`: the module, function, and initial arguments. The effect of this is as follows:

1. The process is spawned at the node bar@STC and starts executing function t with the pid of the shell running at foo@STC as the argument.

2. The effect of this is to send the value of node(), the identifier of the current node, which here will be bar@STC, to the pid; t then terminates.

3. This can be tested by flushing the remaining messages at foo@STC, which shows that it has been sent the identifier bar@STC.

This example shows the *transparency* of communication. The command to send the message between the two nodes has exactly the same form as in the nondistributed case: Pid!Message. Moreover, the messages from one node to another will be delivered in the same order they are sent. The only difference between this and the nondistributed case is that a remote node may become unavailable.

 Sending a message to a named process is different from the nondistributed case, in that *naming is local to each node*. So, to send a message to the process named frequency at the node foo@STC the form {frequency,foo@STC}!Message is used.

Node Names and Visibility

As we said before, a node is an executing Erlang system, and a node is said to be *alive* if it can communicate with other nodes; this is another way of saying that the node is named, and so can take part in communication.

The function erlang:is_alive() will test whether the local runtime system is alive, and as you can see from the following example, it is possible to change the live status of a running runtime system using functions from the net_kernel module, as well as finding out the name of the current node using the node/0 BIF. Try it out:

```
1> erlang:is_alive().
false
2> net_kernel:start([foo]).
{ok,<0.33.0>}
(foo@STC.local)3> erlang:is_alive().
true
(foo@STC.local)4> node().
'foo@STC.local'
(foo@STC.local)5> net_kernel:stop().
ok
6> erlang:is_alive().
false
```

Each live node has to be named: these names must be unique on that host, but can be duplicated across different hosts. The name/host pair, called the *identifier* of the node, is used to uniquely identify the node in the network.

Names take two forms. You already saw the first one; the second one is new:

Short names: `erl –sname foo ...`

> The `sname` will name a host on the local network, and takes the form `name@host` (e.g., `foo@STC`).

Long names: `erl –name foo ...`

> The name gives the *full IP address* of the host: this could be `foo@192.168.1.11`, or (on the local network) `foo@STC.local`. As you can see in command 4 of the preceding example, using the `net_kernel:start` functions to start a distributed node results in the node being given a long name.

Nodes with long names can communicate only with other nodes with long names; similarly for nodes with short names.

> To use hostnames such as `server.kent.ac.uk`, rather than raw IP addresses such as `192.168.1.11`, it is necessary to resolve hostnames to IP addresses. A domain name system (DNS) server does this, but without access to a DNS server, names can be resolved locally using information contained in a hosts file. The details of how to do this vary across different platforms. Consult the documentation for your particular operating system for more information on how it works.

Communication and Security

For two nodes to communicate, not only must both of them be alive, but also they must share some information contained in an atom called the *secret cookie*. Each node has a single cookie value at any time, and nodes sharing the same value can communicate.

Each node can be started with an explicit cookie value, as in the following:

```
erl -sname foo -setcookie blah
```

If no value is set on launch, the Erlang runtime system will pick up the value stored in the file *.erlang.cookie*. If the file does not exist, it will be created in the home directory of the user's account. A randomly generated secret cookie value will be stored in it. As a result, nodes created on the same user account will share the same cookie value by default. If you have been experimenting with distributed Erlang without setting a cookie, look for the *.erlang.cookie* file. You can edit it to whatever value you want.

To show secret cookies and distribution in action, we have repeated the example from Figure 11-2 with nodes running on separate host computers on the same network; this is shown in Figure 11-3.

Note in Figure 11-3 how the two nodes are explicitly started with the same cookie value; this value will override any values contained in an *.erlang.cookie* file on either host.

Distributing the Erlang code: A warning

Your first attempt to distribute the Erlang code may fail: why? The following call:

```
(foo@STC)1> spawn('bar@FCC', dist, t, [self()]).
```

is on the node foo at host STC, but the spawned code is executed on FCC. The module dist.erl will not be executable on FCC unless it is there, as it will not be transferred from STC to FCC automatically.

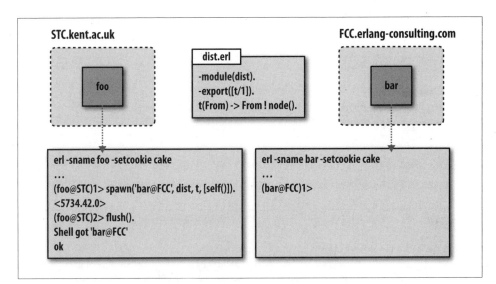

Figure 11-3. Example with nodes on different hosts

Moreover, the call will still fail unless there is a compiled version of the code in the code search path on the remote Erlang node on host FCC.

Erlang Distribution and Security

When a distributed node on your machine gets connected to a remote node through a shared cookie, the owner of the remote node gains the same user access rights as the account on which your local Erlang node is running.

This might be acceptable in a closed telecom system or a banking system running behind a firewall, but if you are running the node on your personal account, anyone connected to it would be able to read and delete files, execute commands, and hijack your machine. Although the caller of the following function:

```
spawn(YourNode, os, cmd, ["rm -rf *"])
```

might find it funny, you might not enjoy the trail of peace and tranquility the call leaves behind.[†]

[†] Unless your account was in need of some serious housekeeping.

As a result, you should *never* publish your Erlang node name and cookie to anyone, unless you've adapted the net kernel to cater for security issues or you explicitly trust the person not to do anything malicious.

Communication and Messages

The most elementary communication is for one node to test whether it can communicate with another, a process informally known as *pinging* the node (see Figure 11-4).

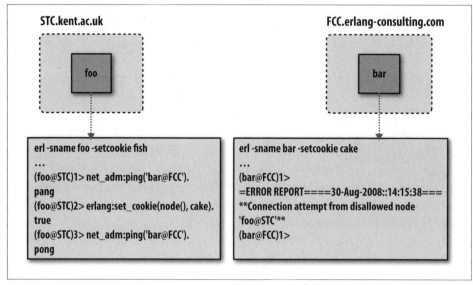

Figure 11-4. Pinging a node

Figure 11-4 shows two nodes initialized with different cookie values, which will prevent their communication. The foo node attempts to communicate using net_adm:ping/1: the pang response shows that this fails. The pinged node also gives an error report to signal the connection attempt, by way of warning that a potential security problem has occurred. After changing the cookie for foo to cake, the ping is successful, indicated by the pong result; such a successful attempt is not signaled at the bar node.

Next, we'll look at an example in which a call to spawn/4 registers a process on a remote node, and then communicates with it; this is illustrated in Figure 11-5.

The first command in the foo node spawns the dist:s/0 process. The effect of this is to register the loop loop/0 under the name server. The effect of this loop is to repeatedly receive messages of the form {M, Pid} and to return the message M to the Pid. In the second command at foo, the message hi is sent together with the Pid of the shell running at foo. This is received by server, and the hi message is returned to foo; you can see this in the inbox of foo as a result of the flush() of the inbox.

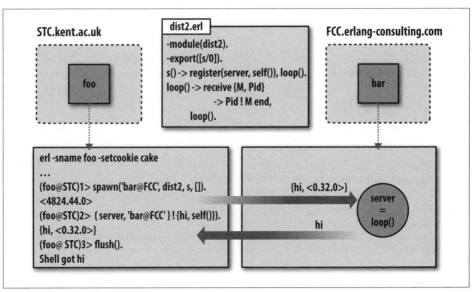

Figure 11-5. Communication with a registered process on another node

Node Connections

Distributed Erlang nodes are able to communicate with each other, provided that they share the same cookie information, but you haven't seen how the connections between the nodes are set up. That is because the Erlang runtime system sets up the connection automatically to a node when it is first referred to. This might be through the net_adm:ping/1 call or by sending a message to a registered process on it. Information about nodes is, by default, shared between connected nodes so that if A knows about B, and B about C, then A will also find out about C.

Each node has a cookie value at any one time. Security in distributed Erlang is based on sharing cookie information: what happens when cookie values are changed? The following interaction shows this in action:

```
(foo@STC)1> net_adm:ping('bar@STC').
pang
(foo@STC)2> erlang:set_cookie(node(),cake).
true
(foo@STC)3> net_adm:ping('bar@STC').
pong
(foo@STC)4> erlang:set_cookie(node(),fish).
true
(foo@STC)5> net_adm:ping('bar@STC').
pong
```

The nodes foo and bar initially have different cookies, hence the negative reply to the first command. In command 2, the cookie of foo is set to that of bar, and so in command 3, the ping is successful, because a connection can be established. In command 4, the cookie is changed back to its original value, but command 5 shows that the two nodes remain connected despite the fact that the two nodes now have different cookies.

This "inclusive" model of connection may not be what is required, and using the facilities of the net_kernel module it is possible to control connections by hand. It is also possible to use the erl command with the flag -connect_all false to avoid nodes from globally connecting to each other.

The net_kernel process at each node coordinates operations at a distributed Erlang node. BIFs such as spawn/4 are converted by the net_kernel to messages that are sent to the net_kernel on the remote node. The net_kernel process also handles authentication by cookies. Because the net_kernel is simply another Erlang process, it is possible for a user to modify it to provide a different behavior, such as changing the authentication scheme or not allowing processes from another node to spawn processes.

Even the most security-unconscious readers will have realized that basing your security on secret cookies alone is not very reliable. As telecom clusters tend to run behind firewalls, enhancing security in its distribution model has never been an issue. In the early days, cookies were in fact sent across the network unencrypted!

Considering the low level of security in distributed Erlang, how can you build a secure distributed system in Erlang? There are two answers to this question:

- If you are building a distributed system for scalability and robustness, it's likely that you are working in a closed and secure network environment. In this case, the Erlang distribution model directly supports what you require in a transparent and effective way.
- If you want to build a geographically distributed system, it is best to communicate between nodes using existing secure mechanisms, such as SSL over TCP/IP. The Erlang distribution has library support for many protocols, including secure protocols such as SSL. We'll cover the fundamentals of how to communicate using TCP/IP in Erlang in Chapter 15.

You can enhance security by writing your own net_kernel process, giving the process whatever behavior and level of security you might require.

Hidden nodes

When nodes get connected, they start monitoring each other, creating a fully meshed network. If you have four Erlang nodes, the fully meshed network would result in six TCP/IP connections among the nodes. Using the formula N * (N - 1) / 2, we quickly compute that 10 nodes require 45 connections. Not only would this result in an overhead of the monitoring messages being sent among the nodes, but also we might not have wanted to connect all of these nodes to each other in the first place. Going up to 100 or more nodes makes the situation even worse, especially if many of these nodes have no relation to each other.

The solution is to use *hidden nodes* and explicitly set up the connections where necessary. You start a hidden node by starting Erlang with the following:

```
erl -sname foo -hidden
```

Once started, connect to other nodes using the `net_kernel:connect(NodeName)` call. Using the `nodes/0` BIF will not return any hidden nodes. To view them, you would have to call `nodes/1` with an atom as the argument: calling `nodes(hidden)` will list the hidden nodes that you are connected to, and `nodes(connected)` will give an aggregated list of all nodes, both hidden and not hidden.

In the following example, we start three nodes, naming them `alpha`, `beta`, and `gamma`, where `gamma` is a hidden node. Having started them, we connect to `beta@STC` and `gamma@STC` from `alpha@STC`. Once the three nodes have been started, the result is as follows:

```
(alpha@STC)1> net_kernel:connect('beta@STC').
true
(alpha@STC)2> net_kernel:connect('gamma@STC').
true
(alpha@STC)3> nodes().
['beta@STC']
(alpha@STC)4> nodes(hidden).
['gamma@STC']
(alpha@STC)5> nodes(connected).
['beta@STC', 'gamma@STC']
```

Once we've executed all of the commands in `alpha`, we can inspect how the node connectivity has spread to `beta` and `gamma`. Let's have a look:

```
UNIXSHELL> erl -sname beta
Erlang (BEAM) emulator version 5.5
Eshell V5.5 (abort with ^G)
(beta@STC)1> nodes().
['alpha@STC']
(beta@STC)2> nodes(connected).
['beta@STC']
```

As you can see in the preceding code, the hidden node `gamma` does not appear. In `gamma` itself, no nodes are visible unless you view them with the `hidden` or `connected` flag. When doing so, the only visible node is `alpha`, as the information on `beta` was not spread when the connection was established:

```
UNIXSHELL> erl -sname gamma -hidden
Erlang (BEAM) emulator version 5.5
Eshell V5.5 (abort with ^G)
(gamma@STC)1> nodes().
[]
(gamma@STC)2> nodes(hidden).
['alpha@STC']
```

Hidden nodes can be used as gateways connecting smaller distributed clusters together. It is a technique which allows hundreds of nodes to be loosely connected in a grid, without the overhead of them having to monitor each other. Hidden nodes are also commonly used for operation and maintenance, as well as for trace nodes, where the node does not carry any traffic, and is not required for the system to operate, but has to be able to retrieve information and interact with remote nodes.

Remote Procedure Calls

The classic construct in distributed computing is the *remote procedure call* (RPC) in which a local call to a procedure is replaced by a call to the same procedure running on a remote node. RPC is simple to implement in Erlang, and moreover, the Erlang implementation avoids many of the pitfalls of RPC in other languages.

In the basic Erlang implementation of RPC, the following (local) function call,

```
Val = fac(N)
```

is replaced by a message to send and receive, as shown here and illustrated in Figure 11-6:

```
remote_call(Message, Node) ->
  {facserver, Node} ! {self(), Message},
  receive
    {ok, Res} ->
      Res
  end.
```

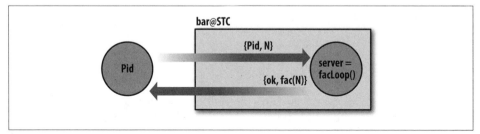

Figure 11-6. Remote procedure call

In the following code, the `facserver` process is on the `bar@STC` node and runs in the `facLoop/0` loop function:

```
server() ->
  register(facserver,self()),
  facLoop().

facLoop() ->
  receive
    {Pid, N} ->
      Pid ! {ok, fac(N)}
```

```
    end,
    facLoop().
```

The main difference between a local and a remote call is the fact that a remote node may go down. You can deal with this in a number of different ways. For example, you can add a timeout in the client code:

```
remote_call(Message, Node) ->
    {facserver, Node} ! {self(), Message},
    receive
      {ok, Res} ->
        Res
      after 1000 ->
        {error, timeout}
    end.
```

If no reply is received within one second, the tuple {error, timeout} is returned. You have to be careful when using timeouts, as the message might still be received after the timeout and stored in the process mailbox. The remote server might be extremely busy or the network may be highly congested. If you do not flush the message, the next time remove_call/2 is invoked and a new request to the factorial server is sent, you'll end up retrieving the first message in the queue containing the reply from the previous call.

Alternatively, it is possible to link to the server process, so if that fails, the client process will also fail. You can do this by launching the server process using spawn_link/4 rather than spawn/4:

```
setup() ->
    process_flag(trap_exit, true),
    spawn_link('bar@STC',myrpc,server,[]).
```

If the remote process terminates, you will receive the usual 'EXIT' signal. If the network connection between the two nodes goes down, the network kernel sends an 'EXIT' signal with reason noconnection.

Finally, it is possible to *monitor* whether a node is alive; the monitor_node(Node,Bool) BIF will switch this on/off for Node according to the Boolean flag Bool, as in the following code. When monitoring is active, the message {nodedown, Node} will be sent to the monitoring process:

```
remote_call(Message, Node) ->
    monitor_node(Node,true),
    {facserver, Node} ! {self(), Message},
      receive
        {ok, Res} ->
          monitor_node(Node,false),
          Res;
        {nodedown, Node} ->
          {error, node_down}
      end.
```

Don't forget to demonitor your node when you are done, because calling monitor_node(Node, true) will generate a nodedown message for each time the BIF was called.

The rpc Module

The rpc library module provides implementations of services that are similar to remote procedure calls, as well as facilities for broadcast and parallel evaluation of RPC calls. The most commonly used function is:

```
rpc:call(Node, Module, Function, Arguments)
```

which executes the function on the remote node. The Module must be in the code search path on the remote node, and the nodes must either be connected or share the cookie. The result is the return value of the call, or, upon failure, {badrpc, Reason}.

If your applications are going to be distributed, spend some time reading through the manual pages of the rpc module and become familiar with it. You will find help functions for synchronous, asynchronous, and blocking calls. There are also calls to broadcast calls, both synchronously and asynchronously to a pool of nodes. You never know when you are going to need these functions, so reviewing them now will avoid your having to reinvent the wheel at a later date.

I/O Group Leaders

Try calling rpc:call(Node, io, format, ["Hello World~n"]) and you will be surprised. Hello World is printed out not on the remote node, but on the local one.

This behavior is explained by a concept called the *group leader*. Every Erlang process has a group leader that is responsible for handling all of the I/O for that process. This group leader process is inherited, resulting in the child processes, including those on remote nodes, sharing the same group leader as their parent. When executing remote procedure calls, the group leader is passed to the remote node with the call.

Group leaders can be reassigned during runtime. The BIF group_leader() returns the process identifier of the group leader of the calling pid while the call group_leader(LeaderPid, Pid) assigns the group leader process LeaderPid to the process denoted by Pid.

Essential Distributed Programming Modules

A number of key modules support distributed programming in Erlang. Some we have already covered, while others are new:

erl

This module contains the `erl` command that starts an Erlang runtime system. You can change the runtime system behavior by setting various flags on launch. These include:

`-connect_all false`

With this flag, the system will not maintain a `global` list of connected nodes, thus preventing global naming.

`-hidden`

This has the effect of launching a hidden node, which is often used for operation and maintenance purposes.

`-name Name/-sname Name`

These flags give the node a long/short name, `Name`.

`-setcookie Cookie`

This flag sets the cookie value for the node to `Cookie`.

erlang

The `erlang` module collects the Erlang BIFs, many of which are auto-imported and can thus be called without the `erlang:` prefix. Here the ones that are not auto-imported are prefixed with the `erlang` module:

`disconnect_node(Node)`

This will disconnect the `Node` passed as an argument.

`erlang:get_cookie()`

This returns the current cookie for the local node if it is alive; otherwise, it returns `nocookie`.

`monitor_node(Node, Flag)`

This turns on/off the monitoring of the node `Node` depending on whether the `Flag` is set to `true` or `false`. There is also a variant, `monitor_node/3`, which is not autoimported.

`node()`

This returns the name of the local node, `Name@Host`, or it returns `non ode@nohost` if it is not alive.

`node(Arg)`

This returns the node where `Arg` is located: `Arg` can be a pid, a reference, or a port.

`nodes()`

This returns a list of visible nodes in the system, excluding the local node. `nodes(Type)` will return a list of particular nodes, where `Type` is the atom `hidden` or `connected`.

erlang:set_cookie(Node, Cookie)
> This sets the cookie at Node to be Cookie.

spawn(Node, Module, Function, ArgumentList)
> This performs spawn(Module, Function, ArgumentList) on the node Node. spawn_link/4 is similarly analogous to spawn_link/3.

net_kernel
> This module contains the infrastructure for manually starting, stopping, connecting, and monitoring nodes. These functions will be called automatically by the runtime system, but can also be used in modified ways by the user.

net_adm
> This module contains various useful functions, including ping (described earlier) and functions to examine the local hosts file, among others.

The epmd Process

When running the distributed Erlang examples in this chapter, you might have noticed an OS thread running a command called epmd. This is a part of the Erlang runtime system that acts as a *port mapper deamon* for Erlang distributed nodes. One epmd daemon process is started per machine, regardless of the number of distributed Erlang nodes running on it. The daemon will listen for incoming connection requests on port 4369, mapping them to the listening port of the node that is being connected to. If not already running, epmd is automatically launched when you start your first distributed Erlang node. Starting it manually, however, allows you to pass a set of commands and configuration parameters.

You will find the epmd command useful when troubleshooting problems relating to distribution, configuring Erlang distribution to work through firewalls, or trying to simulate busy networks. The executable is located in the Erlang root directory, together with the binaries of the virtual machine. Flags which can be passed to it include the following:

-help
> Prints a list of debugging commands. These commands are not always listed in the manual pages.

-port PortNumber
> Changes the listening port. This is useful when dealing with particular ports in firewalls.

-names
> Lists the names of the local nodes. This is useful when running Erlang as a background process without a shell looking for name conflicts.

-daemon
> Starts epmd as a daemon process.

`-kill`

> Kills the epmd process. Connected processes remain connected, but new attempts to connect on that host will fail. Restarting epmd will result in the loss of information regarding all connected nodes. New nodes will be able to connect to each other, but old ones will not.

`-packet_timeout`

> Sets the number of seconds a connection can be inactive before epmd times out and closes the connection. Connections are kept open by a keepalive; if there is no other traffic, tic messages are sent and acknowledged by a tok.

`-delay_accept and -delay_write`

> Are used in testing environments to simulate busy servers and network congestion.

Distributed Erlang Behind Firewalls

When running distributed Erlang nodes behind firewalls, you need to open the port on which epmd listens. By default, this is port 4369, but you can change it to whatever port you please, as long as it is consistent in your node cluster. You also need to open the ports that the individual nodes use to connect to each other. You can specify the node range by running the following commands:

```
application:set_env(kernel, inet_dist_listen_min, 9100)
application:set_env(kernel, inet_dist_listen_max, 9105)
```

These commands force Erlang to use ports from 9100 to 9105 for distribution. You can replace these values with whatever range you want.

Exercises

Exercise 11-1: Distributed Associative Store

Design a distributed version of an associative store in which values are associated with tags. It is possible to store a tag/value pair, and to look up the value(s) associated with a tag. One example for this is an address book for email, in which email addresses (values) are associated with nicknames (tags).

Replicate the store across two nodes on the same host, send lookups to one of the nodes (chosen either at random or alternately), and send updates to both.

Reimplement your system with the store nodes on other hosts (from each other and from the frontend). What do you have to be careful about when you do this?

How could you reimplement the system to include three or four store nodes?

Design a system to test your answer to this exercise. This should generate random store and lookup requests.

Exercise 11-2: System Monitoring

Design a system to monitor the behavior of your distributed store systems under test conditions. This system—which could be another node in the overall system—should log throughput and load-balancing information. How does the system behave when it becomes overloaded?

CHAPTER 12

OTP Behaviors

In previous chapters, we introduced patterns that recur when you program using the Erlang concurrency model. We discussed functionality common to concurrent systems, and you saw that processes will handle very different tasks in a similar way. We also emphasized special cases and potential problems that have to be handled when dealing with concurrency.

For example, picture a project with 50 developers spread across several geographic locations. If the project is not properly coordinated and no templates are provided, how many different client/server implementations might the project end up with? Even more dangerous, how many of these implementations will handle special borderline cases and concurrency-related errors correctly, if at all? Without a code review, can you be sure there is a uniform way across the system to handle server crashes that occur after clients have sent a request to the server? Or guarantee that the response from a request is indeed the response, and not just any message that conforms to the internal message protocol?

OTP behaviors address all of these issues by providing library modules that implement the most common concurrent design patterns. Behind the scenes, without the programmer having to be aware of it, the library modules ensure that errors and special cases are handled in a consistent way. As a result, OTP behaviors provide a set of standardized building blocks used in designing and building industrial-grade systems. The subject of OTP behaviors and their related middleware is vast. In this chapter, we provide the overview you need to get started.

Introduction to OTP Behaviors

OTP behaviors are a formalization of process design patterns. They are implemented in library modules that are provided with the standard Erlang distribution. These library modules do all of the generic process work and error handling. The specific code, written by the programmer, is placed in a separate module and called through a set of predefined callback functions.

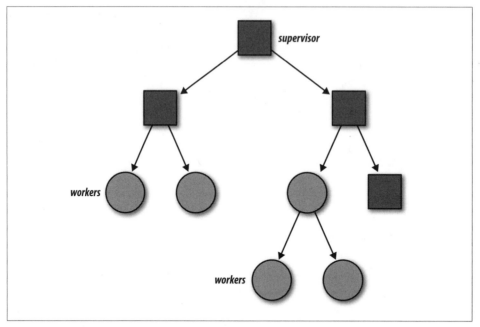

Figure 12-1. Supervision tree in an application

OTP behaviors include *worker* processes, which do the actual processing, and *supervisors*, whose task is to monitor workers and other supervisors. Worker behaviors, often denoted in diagrams as circles, include servers, event handlers, and finite state machines. Supervisors, denoted in illustrations as squares, monitor their children, both workers and other supervisors, creating what is called a *supervision tree* (see Figure 12-1).

Supervision trees are packaged into a behavior called an *application*. OTP applications not only are the building blocks of Erlang systems, but also are a way to package reusable components. Industrial-grade systems consist of a set of loosely coupled, possibly distributed applications. These applications are part of the standard Erlang distribution or are specific applications developed by you, the programmer.

Do not confuse OTP applications with the more general concept of an application, which usually refers to a more complete system that solves a high-level task. Examples of OTP applications include the Mnesia database, which we cover in Chapter 13; an SNMP agent; or the mobile subscriber database introduced in Chapter 10, which we will convert to an application using behaviors later in this chapter. An OTP application is a reusable component that packages library modules together with supervisor and worker processes. From now on, when we refer to an application, we will mean an OTP application.

The behavior module contains all of the generic code. Although it is possible to implement your own behavior module, doing so is rare because the behavior modules that come as part of the Erlang/OTP distribution will cater to most of the design patterns you would use in your code. The generic functionality provided in a behavior module includes operations such as the following:

- Spawning and possibly registering the process
- Sending and receiving client messages as synchronous or asynchronous calls, including defining the internal message protocol
- Storing the loop data and managing the process loop
- Stopping the process

Although the behavior module is provided, the programmer has to develop the callback module (see Figure 12-2). We introduced the concept of callback modules in Chapter 5. A callback module contains all of the specific code required to deliver the desired functionality. The specific code is invoked through a callback interface that is standardized for each behavior.

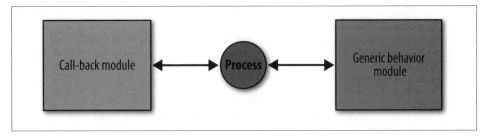

Figure 12-2. Splitting the code into generic and specific modules

The *loop data* is a variable that will contain the data the behavior needs to store in between calls. After the call, an updated variant of the loop data is returned. This updated loop data, often referred to as the *new loop data*, is passed as an argument in the next call. Loop data is also commonly referred to as the *behavior state*.

The functionality to be included in the callback module to deliver the specific behavior required includes the following:

- Initializing the process loop data, and, if the process is registered, the process name.
- Handling the specific client requests, and, if synchronous, the replies sent back to the client.
- Handling and updating the process loop data in between the process requests.
- Cleaning up the process loop data upon termination.

There are many advantages to splitting the code into generic behavior libraries and specific callback modules:

- Because many of the special cases and errors that might occur are already handled in the solid, well-tested behavior library, you can expect fewer bugs in your product.

- For this reason, and also because so much of the code is already written for you, you can expect to have a shorter time to market.

- It forces the programmer to write code in a way that avoids errors typically found in concurrent applications.

- Finally, your whole team will come to share a common programming style. When reading someone else's code while armed with a basic comprehension of the existing behaviors, no effort is required to understand the client/server protocol, looking for where and how processes are started or terminated, or how the loop data is handled. All of it is managed by the generic behavior library. Instead of having to focus on how everything is done, you can focus on what is being done *specifically* in this case, as coded in the callback module.

In the sections that follow, we will look at some of the most important behaviors—including generic servers and supervisors—and how to package them into applications.

Generic Servers

Generic servers that implement client/server behaviors are defined in the `gen_server` behavior that comes as part of the standard library application. In this chapter, you will use the mobile customer database example from Chapter 10 to understand how the callback principle works. If you do not remember the example, take a quick look at it before proceeding.

We will rewrite the `usr.erl` module, migrating it from an Erlang process to a `gen_server` behavior. In doing so, we will not touch the `usr_db` module, keeping the backend database as it is. When working your way through the example, if you are interested in the details, have the manual pages for the `gen_server` module at hand.

Starting Your Server

With the `gen_server` behavior, instead of using the `spawn` and `spawn_link` BIFs, you will use the `gen_server:start/4` and `gen_server:start_link/4` functions.

The main difference between `spawn` and `start` is the *synchronous* nature of the call. Using `start` instead of `spawn` makes starting the worker process more deterministic and prevents unforeseen race conditions, as the call will not return the pid of the worker until it has been initialized. You call the functions as follows (we show two variants for each of the two functions):

```
gen_server:start_link(ServerName, CallBackModule, Arguments, Options)
gen_server:start(ServerName, CallBackModule, Arguments, Options)
```

```
gen_server:start_link(CallBackModule, Arguments, Options)
gen_server:start(CallBackModule, Arguments, Options)
```

In the preceding calls:

ServerName
> Is a tuple of the format {local, Name} or {global, Name}, denoting a local or global Name for the process if it is to be registered. If you do not want to register the process and instead reference it using its pid, you omit the argument and use the start_link/3 or start/3 call instead.

CallbackModule
> Is the name of the module in which the specific callback functions are placed.

Arguments
> Is a valid Erlang term that is passed to the init/1 callback function. You can choose what type of term to pass: if you have many arguments to pass, use a list or a tuple; if you have none, pass an atom or an empty list, ignoring it in the callback function.

Options
> Is a list that allows you to set the memory management flags fullsweep_after and heapsize, as well as tracing and debugging flags. Most behavior implementations just pass the empty list.

The start functions will spawn a new process that calls the init(Arguments) callback function in the CallbackModule, with the Arguments supplied. The init function must initialize the LoopData of the server and has to return a tuple of the format {ok, LoopData}. LoopData contains the first instance of the loop data that will be passed between the callback functions. If you want to store some of the arguments you passed to the init function, you would do so in the LoopData variable.

The obvious difference between the start_link and start functions is that start_link links to its parent and start doesn't. This needs a special mention, however, as it is an OTP behavior's responsibility to link itself to the supervisor. The start functions are often used when testing behaviors from the shell, as a typing error causing the shell process to crash would not affect the behavior. All of the start and start_link variants return {ok, Pid}.

Before going ahead with the example, let's quickly review what we have discussed so far. You start a gen_server behavior using the gen_server:start_link call. This results in a new process that calls the init/1 callback function. This function initializes the LoopData and returns the tuple {ok, LoopData}.

In our example, we call start_link/4, registering the process with the same name as the callback module, calling the MODULE macro. We pass one argument, the filename of the Dets table. The options list is kept empty:

```
start_link(FileName) ->
  gen_server:start_link({local, ?MODULE}, ?MODULE, FileName, []).

init(FileName) ->
```

```
usr_db:create_tables(FileName),
usr_db:restore_backup(),
{ok, null}.
```

Although the supervisor process might call the start_link/4 function, the init/1 callback is called by a different process: the one that was just spawned. We don't really need the LoopData variable in our server, as the ETS and Dets tables are named. Nonetheless, a value still has to be included when returning the {ok, LoopData} structure, so we'll get around it by returning the atom null. Had the ETS and Dets tables not been named_tables, we would have passed their references here.

Do only what is necessary and minimize the operations in your init function, as the call to init is a synchronous call that prevents all of the other serialized processes from starting until it returns.

Passing Messages

If you want to send a message to your server, you use the following calls:

```
gen_server:cast(Name, Message)
gen_server:call(Name, Message)
```

In the preceding calls:

Name
> Is either the *local* registered name of the server or the tuple {global, Name}. It could also be the process identifier of the server.

Message
> Is a valid Erlang term containing a message passed on to the server.

For *asynchronous* message requests, you use cast/2. If you're using a pid, the call will immediately return the atom ok, regardless of whether the gen_server to which you are sending the message is alive. These semantics are no different from the standard Name ! Message construct, where if the registered process Name does not exist, the calling process terminates.

Upon receiving the message, gen_server will call the callback function handle_cast(Message, LoopData) in the callback module. Message is the argument passed to the cast/2 function, and LoopData is the argument originally returned by the init/1 callback function. The handle_cast/1 callback function handles the specifics of the message, and upon finishing, it has to return the tuple {noreply, NewLoopData}. In future calls to the server, the NewLoopData value most recently returned will be passed as an argument when a message is sent to the server.

If you want to send a *synchronous* message to the server, you use the call/2 function. Upon receiving this message, the process uses the handle_call(Message, From, LoopData) function in the callback module. It contains specific code for the particular server, and having completed, it returns the tuple {reply, Reply, NewLoopData}. Only now does the call/3 function synchronously return the value Reply. If the process you are sending a message to does not exist, regardless of whether it is registered, the process invoking the call function terminates.

Let's start by taking two functions from our service API; we will provide the whole program later. They are called by the client process and result in a synchronous message being sent to the server process registered with the same name as the callback module. Note how we are validating the data on the client side. If the client sends incorrect information, it terminates.

```
set_status(CustId, Status) when Status==enabled; Status==disabled->
    gen_server:call(?MODULE, {set_status, CustId, Status}).

delete_disabled() ->
    gen_server:call(?MODULE, delete_disabled).
```

Upon receiving the messages, the gen_server process calls the handle_call/3 callback function dealing with the messages in the same order in which they were sent:

```
handle_call({set_status, CustId, Status}, _From, LoopData) ->
  Reply = case usr_db:lookup_id(CustId) of
    {ok, Usr} ->
      usr_db:update_usr(Usr#usr{status=Status});
    {error, instance} ->
      {error, instance}
  end,
  {reply, Reply, LoopData};

handle_call(delete_disabled, _From, LoopData) ->
  {reply, usr_db:delete_disabled(), LoopData}.
```

Note the return value of the callback function. The tuple contains the *control atom* reply, telling the gen_server generic code that the second element of the tuple is the Reply to be sent back to the client. The third element of the tuple is the new LoopData, which, in a new iteration of the server, is passed as the third argument to the handle_call/3 function; in both cases here it is unchanged. The argument _From is a tuple containing a unique message reference and the client process identifier. The tuple as a whole is used in library functions that we will not be discussing in this chapter. In the majority of cases, you will not need it.

The gen_server library module has a number of mechanisms and safeguards built in that function behind the scenes. If your client sends a synchronous message to your server and you do not get a response within five seconds, the process executing the call/2 function is terminated. You can override this by using the following code:

```
gen_server:call(Name, Message, Timeout)
```

where Timeout is a value in milliseconds or the atom infinity. The timeout mechanism was originally put in place for deadlock prevention purposes, ensuring that servers that accidentally call each other are terminated after the default timeout. The crash report would be logged, and hopefully would result in a patch. Most applications will function appropriately with a timeout of five seconds, but under very heavy loads, you might have to fine-tune the value and possibly even use `infinity`; this choice is very application-dependent. All of the critical code in Erlang/OTP uses `infinity`.

Other safeguards when using the `gen_server:call/2` function include the case of sending a message to a nonexisting server or a server that crashes before sending its reply. In both cases, the calling process will terminate. In raw Erlang, sending a message that is never pattern-matched in a `receive` clause is a bug that can cause a memory leak.

What do you think happens if you do a call or a cast to your server, but do not handle the message in the `handle_call/3` and `handle_cast/2` calls, respectively? In OTP, when a call or a cast is called, the message will always be extracted from the process mailbox and the respective callback functions are invoked. If none of the callback functions pattern-matches the message passed as the first argument, the process will crash with a function clause error. As a result, such issues will be caught in the early stages of the testing phase and dealt with accordingly.

Stopping the Server

How do you stop the server? In your `handle_call/3` and `handle_cast/2` callback functions, instead of returning `{reply, Reply, NewLoopData}` or `{noreply, NewLoopData}`, you can return `{stop, Reason, Reply, NewLoopData}` or `{stop, Reason, NewLoopData}`, respectively. Something has to trigger this return value, often a stop message sent to the server. Upon receiving the `stop` tuple containing the `Reason` and `LoopData`, the generic code executes the `terminate(Reason, LoopData)` callback.

The `terminate` function is the natural place to insert the code needed to clean up the `LoopData` of the server and any other persistent data used by the system. In this example, it would mean closing the ETS and Dets tables. The `stop` call does not have to occur within a synchronous call, so let's use `cast` when implementing it:

```
stop() ->
  gen_server:cast(?MODULE, stop).

handle_cast(stop, LoopData) ->
  {stop, normal, LoopData}.

terminate(_Reason, _LoopData) ->
  usr_db:close_tables().
```

Remember that `stop/0` will be called by the client process, while the `handle_cast/2` and `handle_call/2` are called by the behavior process. In the `handle_cast/2` callback, we return the reason `normal` in the `stop` construct. Any reason other than `normal` will result in an error report being generated.

With thousands of generic servers potentially being spawned and terminated every second, generating error reports for every one of them is not the way to go. You should return a *nonnormal* value only if something that should not have happened occurs and you have no way to recover. A socket being closed or a corrupt message from an external port is not a reason to generate a nonnormal termination. On the other hand, corrupt internal data or a missing configuration file is.

If your server crashes because of a runtime error, `terminate/2` will be called. But if your behavior receives an `EXIT` signal from its parent, `terminate` will be called only if you are trapping exits. Watch out for this special case, as we've been caught by it many times, especially when starting the behavior from the shell using `start_link`.

 Use of the behavior callbacks as library functions and invoking them from other parts of your program is an extremely bad practice. For example, you should never call `usr_db:init(FileName)` from another module to create and populate your database. Calls to behavior callback functions should originate only from the behavior library modules as a result of an event occurring in the system, and *never* directly by the user.

The Example in Full

Here is the `usr.erl` module from Chapter 10, rewritten as a `gen_server` behavior:

```
%%% File    : usr.erl
%%% Description : API and gen_server code for cellphone user db

-export([start_link/0, start_link/1, stop/0]).
-export([init/1, terminate/2, handle_call/3, handle_cast/2]).
-export([add_usr/3, delete_usr/1, set_service/3, set_status/2,
         delete_disabled/0, lookup_id/1]).
-export([lookup_msisdn/1, service_flag/2]).
-behavior(gen_server).

-include("usr.hrl").

%% Exported Client Functions
%% Operation & Maintenance API

start_link() ->
  start_link("usrDb").

start_link(FileName) ->
  gen_server:start_link({local, ?MODULE}, ?MODULE, FileName, []).

stop() ->
  gen_server:cast(?MODULE, stop).

%% Customer Services API

add_usr(PhoneNum, CustId, Plan) when Plan==prepay; Plan==postpay ->
  gen_server:call(?MODULE, {add_usr, PhoneNum, CustId, Plan}).

delete_usr(CustId) ->
  gen_server:call(?MODULE, {delete_usr, CustId}).

set_service(CustId, Service, Flag) when Flag==true; Flag==false ->
  gen_server:call(?MODULE, {set_service, CustId, Service, Flag}).
```

```erlang
set_status(CustId, Status) when Status==enabled; Status==disabled->
  gen_server:call(?MODULE, {set_status, CustId, Status}).

delete_disabled() ->
  gen_server:call(?MODULE, delete_disabled).

lookup_id(CustId) ->
  usr_db:lookup_id(CustId).

%% Service API

lookup_msisdn(PhoneNo) ->
  usr_db:lookup_msisdn(PhoneNo).

service_flag(PhoneNo, Service) ->
  case usr_db:lookup_msisdn(PhoneNo) of
    {ok,#usr{services=Services, status=enabled}} ->
      lists:member(Service, Services);
    {ok, #usr{status=disabled}} ->
      {error, disabled};
    {error, Reason} ->
      {error, Reason}
  end.

%% Callback Functions

init(FileName) ->
  usr_db:create_tables(FileName),
  usr_db:restore_backup(),
  {ok, null}.

terminate(_Reason, _LoopData) ->
  usr_db:close_tables().

handle_cast(stop, LoopData) ->
  {stop, normal, LoopData}.

handle_call({add_usr, PhoneNo, CustId, Plan}, _From, LoopData) ->
  Reply = usr_db:add_usr(#usr{msisdn=PhoneNo,
                              id=CustId,
                              plan=Plan}),
  {reply, Reply, LoopData};

handle_call({delete_usr, CustId}, _From, LoopData) ->
  Reply = usr_db:delete_usr(CustId),
  {reply, Reply, LoopData};

handle_call({set_service, CustId, Service, Flag}, _From, LoopData) ->
  Reply = case usr_db:lookup_id(CustId) of
    {ok, Usr} ->
      Services = lists:delete(Service, Usr#usr.services),
      NewServices = case Flag of
        true  -> [Service|Services];
        false -> Services
```

```
        end,
        usr_db:update_usr(Usr#usr{services=NewServices}));
    {error, instance} ->
        {error, instance}
  end,
  {reply, Reply, LoopData};

handle_call({set_status, CustId, Status}, _From, LoopData) ->
  Reply = case usr_db:lookup_id(CustId) of
    {ok, Usr} ->
        usr_db:update_usr(Usr#usr{status=Status});
    {error, instance} ->
        {error, instance}
  end,
  {reply, Reply, LoopData};

handle_call(delete_disabled, _From, LoopData) ->
  {reply, usr_db:delete_disabled(), LoopData}.
```

Running gen_server

When testing the gen_server instance in the shell, you get exactly the same behavior
as when you used the server process that you coded yourself. However, the code is
more solid, as deadlocks, server crashes, timeouts, and other errors related to concurrent programming are handled behind the scenes:

```
1> c(usr).
/Users/Francesco/otp/usr.erl:11: Warning: undefined callback
 function code_change/3 (behaviour 'gen_server')
/Users/Francesco/otp/usr.erl:11: Warning: undefined callback
 function handle_info/2 (behaviour 'gen_server')
{ok,usr_db}
2> c(usr_db).
{ok,usr_db}
3> rr("usr.hrl").
[usr]
4> usr:start_link().
{ok,<0.86.0>}
5> usr:add_usr(700000000, 0, prepay).
ok
6> usr:set_service(0, data, true).
ok
7> usr:lookup_id(0).
{ok,#usr{msisdn = 700000000,id = 0,status = enabled,
         plan = prepay,
         services = [data]}}
8> usr:set_status(0, disabled).
ok
9> usr:service_flag(700000000,lbs).
{error,disabled}
10> usr:stop().
ok
```

Did you notice the -behavior(gen_server) directive in the module? This tells the compiler that your module is a gen_server callback module, and as a result, it has to expect a number of callback functions. If all callback functions are not implemented, you will get the warnings you noticed as a result of the compile operation in the first command line. Don't write your code to avoid these warnings. If your server has no asynchronous calls, you will obviously not need a handle_cast/2. Ignore the warnings.

 British or Canadian readers: don't despair or shake your heads! You are welcome to use the U.K. English spelling in your directive: -behaviour(gen_server). The compiler is bilingual and can handle both U.S. and U.K. English.

What happens if you send a message to the server using raw Erlang message passing of the form Pid!Msg? It should be possible, as the gen_server is an Erlang process capable of sending and receiving messages like any other process. Don't be shy; try it:

```
11> {ok, Pid} = usr:start_link().
{ok,<0.119.0>}
12> Pid ! hello.
hello

=ERROR REPORT==== 24-Jan-2009::18:08:07 ===
** Generic server usr terminating
** Last message in was hello
** When Server LoopData == null
** Reason for termination ==
** {'function not exported',[{usr,handle_info,[hello,null]},
                             {gen_server,handle_msg,5},
                             {proc_lib,init_p,5}]}
** exception exit: undef
     in function  usr:handle_info/2
        called as usr:handle_info(hello,null)
     in call from gen_server:handle_msg/5
     in call from proc_lib:init_p/5
```

Oops! Something did not go according to plan. Look at the error and try to figure out what happened. Use of Pid!Msg does not comply with the internal OTP message protocol. Upon receiving a message that is not compliant, the gen_server process tries to call the function usr:handle_info(hello, null), where hello is the message and null is the loop data.

The callback function handle_info/2[*] is called whenever the process receives a message it doesn't recognize. These could include "node down" messages from nodes you are monitoring, exit signals from processes you are linked to, or simply messages sent using the ...!... construct. If you are expecting such messages but are not interested in them, add the following definition to your callback module, and don't forget to export it:

---

[*] Did you notice it in the compiler warning in the example?

```
handle_info(_Msg, LoopData) ->
    {noreply, LoopData}.
```

If, on the other hand, you do want to do something with the messages, you should pattern-match them in the first argument of the call. If your server is not expecting nonOTP-compliant messages, don't add the handle_info/2 call, which ignores incoming messages, "just in case." Doing so is considered defensive programming, which will probably make any fault you are hiding hard to detect.

One of the downsides of OTP is the layering that the various behavior modules require. This will affect performance. In the attempt to save a few microseconds from their calls, developers have been known to use the Pid ! Msg construct instead of a gen_server cast, handling their messages in the handle_info/2 callback.

Don't do this! You will make your code impossible to support and maintain, as well as losing many of the advantages of using OTP in the first place. If you are obsessed with saving microseconds, try to hold on and optimize only when you know your program is not fast enough. We discuss optimizations in Chapter 20 and will cover there what really affects the performance of your code.

Before we look at the next behavior, here is a summary of the exported gen_server API, the resulting callback functions, and their expected return values:

Setup
The following calls:

```
start(Name, Mod, Arguments, Opts)
start_link(Name, Mod, Arguments, Opts),
```

where Name is an optional argument, spawn a new process. The process will result in the callback function init(Arguments) being called, which should return one of the values {ok, LoopData} or {stop, Reason}. If init/1 returns {stop, Reason} the terminate/2 "cleanup" function will *not* be called.

Synchronous communication
Use call(Name, Msg) to send a synchronous message to your server. It will result in the callback function handle_call(Msg, From, LoopData) being called by the server process. The expected return values include the following:

```
{reply, Reply, NewLoopData}
{stop, Reason, Reply, NewLoopData}.
```

Asynchronous communication
If you want to send an asynchronous message, use cast(Name, Msg). It will be handled in the handle_cast(Msg, LoopData) callback function, returning either {noreply, NewLoopData} or {stop, Reason, NewLoopData}.

Non-OTP-compliant messages

Upon receiving non-OTP-compliant messages, `gen_server` will execute the `handle_info(Msg, LoopData)` callback function. The function should return either {`noreply, NewLoopData`} or {`stop, Reason, NewLoopData`}.

Termination

Upon receiving a `stop` construct from one of the callback functions (except for `init`), or upon abnormal process termination when trapping exits, the `terminate(Reason, LoopData)` callback is invoked. In `terminate/2`, you would typically undo things you did in `init/1`. Its return value is ignored.

Supervisors

The supervisor behavior's task is to monitor its children and, based on some preconfigured rules, take action when they terminate. The children that make up the supervision tree include both supervisors and worker processes. Worker processes are OTP behaviors including `gen_server`, `gen_fsm` (supporting finite state machine behavior), and `gen_event` (which provides event-handling functionality).

Worker processes have to link themselves to the supervisor behavior and handle specific system messages that are not exposed to the programmer. This is different from the way in which one process links to another in raw Erlang, and because of this, we cannot mix the two mechanisms. For this reason, it is not possible to add Erlang processes to the supervision tree in the form you know them. So, for the remainder of this section, we will stick to describing supervision within the OTP framework.

You start a supervisor using the `start` or `start_link` function:

```
supervisor:start_link(ServerName, CallBackModule, Arguments)
supervisor:start(ServerName, CallBackModule, Arguments)
supervisor:start_link(CallBackModule, Arguments)
supervisor:start(CallBackModule, Arguments)
```

In the preceding calls:

ServerName

Is the *name* to be registered for the supervisor, and is a tuple of the format {`local, Name`} or {`global, Name`}. If you do not want to register the supervisor, you use the functions of arity two.

CallbackModule

Is the name of the module in which the `init/1` callback function is placed.

Arguments

Is a valid Erlang term that is passed to the `init/1` callback function when it is called.

Note that the supervisor, unlike the `gen_server`, does not take any options. The `start` and `start_link` functions will spawn a new process that calls the `init/1` callback function. Upon initializing the supervisor, the `init` function has to return a tuple of the following format:

```
{ok, {SupervisorSpecification, ChildSpecificationList}}
```

The *supervisor specification* is a tuple containing information on how to handle process crashes and restarts. The *child specification* list specifies which children the supervisor has to start and monitor, together with information on how to terminate and restart them.

Supervisor Specifications

The supervisor specification is a tuple consisting of three elements describing how the supervisor should react when a child terminates:

```
{RestartStrategy, AllowedRestarts, MaxSeconds}
```

The restart strategy determines how other children are affected if one of their siblings terminates. It can be one of the following:

one_for_one
> Will restart the child that has terminated, without affecting any of the other children. You should pick this strategy if all of the processes at this level of the supervision tree are not dependent on each other.

one_for_all
> Will terminate all of the children and restart them. You should use this if there is a strong dependency among all of the children regardless of the order in which they were started.

rest_for_one
> Will terminate all of the children that were started after the child that crashed, and will restart them. This strategy assumes that processes are started in order of dependency, where spawned processes are dependent only on their already started siblings.

What will happen if your process gets into a cyclic restart? It crashes and is restarted, only to come across the same corrupted data, and as a result, it crashes again. This can't go on forever! This is where `AllowedRestarts` comes in, by specifying the maximum number of abnormal terminations the supervisor is allowed to handle in `MaxSeconds` seconds. If more abnormal terminations occur than are allowed, it is assumed that the supervisor has not been able to resolve the problem, and it terminates. The supervisor's supervisor receives the exit signal and, based on its configuration, decides how to proceed.

Finding reasonable values for `AllowedRestarts` and `MaxSeconds` is not easy, as they will be application-dependent. In production, we've used anything from ten restarts per second to one per hour. Your choice will have to depend on what your child processes do, how many of them you expect the supervisor to monitor, and how you've set up your supervision strategy.

Child Specifications

The second argument in the structure returned by the `init/1` function is a list of child specifications. Child specifications provide the supervisor with the properties of each of its children, including instructions on how to start it. Each child specification is of the following form:

```
{Id, {Module, Function, Arguments}, Restart, Shutdown, Type, ModuleList}
```

In the preceding code:

Id

> Is a unique identifier for a particular child within a supervisor. As a child process can crash and be restarted, its process identifier might change. The identifier is used instead.

> The supervisor uses the tuple `{Module, Function, Arguments}` to start the child process. The supervisor has to eventually call the `start_link` function for the particular OTP behavior, and return `{ok, Pid}`.

Restart

> Is one of the atoms `transient`, `temporary`, or `permanent`. Transient processes are never restarted. Temporary processes are restarted only if they terminate abnormally, and permanent processes are always restarted, regardless of whether the termination was normal or nonnormal.

Shutdown

> Specifies how many *milliseconds* a behavior that is trapping exits is allowed to execute in its `terminate` callback function after receiving the `shutdown` signal from its supervisor, either because the supervisor has reached its maximum number of allowed child restarts or because of a `rest_for_one` or `one_for_all` restart strategy.

> If the child process has not terminated by this time, the supervisor will kill it unconditionally. `Shutdown` will also take the atom `infinity`, a value which should always be chosen if the process is a supervisor, or the atom `brutal_kill`, if the process is to be killed unconditionally.

Type

> Specifies whether the child process is a `worker` or a `supervisor`.

ModuleList

> Is a list of the modules that implement the process. The release handler uses it to determine which processes it should suspend during a software upgrade. As a rule of thumb, always include the behavior callback module.

In some cases, child specifications are created dynamically from a config file. In most cases, however, they are statically coded in the supervisor callback module. The init/1 function is the only callback function that needs to be exported.

It can be easy to insert syntactical and semantic errors in child specification lists, as they tend to get fairly complex. The help function check_childspecs/1 in the supervisor module takes a list of child specifications and returns ok or the tuple {error, Reason}. An example of a child specification for the mobile subscriber database will follow in the next section. To ensure that you understand what is happening, map all of the entries to their respective fields in the child specification structure.

Supervisor Example

In this example, the usr_sup module is a supervisor behavior, supervising one child that is the usr example of a gen_server from earlier in the chapter.

We'll start the supervisor using the start_link/0 call. Note that we've omitted the option of passing a filename for the Dets tables, as it was originally included for test purposes. Pay particular attention to the child and the supervisor specifications returned by the init/1 function:

```
-module(usr_sup).
-behavior(supervisor).

-export([start_link/0]).
-export([init/1]).

start_link() ->
  supervisor:start_link({local, ?MODULE}, ?MODULE, []).

init(FileName) ->
  UsrChild = {usr,{usr, start_link, []},
              permanent, 2000, worker, [usr, usr_db]},
  {ok,{{one_for_all,1,1}, [UsrChild]}}.
```

Now you can try it out from the shell. Do not test only positive cases; also try to kill the child and ensure that it has been restarted. Finally, kill the server more than MaxRestart times in MaxSeconds (twice in one second in this example), to see whether the supervisor terminates:

```
13> c(usr_sup).
{ok,usr_sup}
14> usr_sup:start_link().
{ok,<0.149.0>}
15> whereis(usr).
<0.150.0>
```

```
16> exit(whereis(usr), kill).
true
17> whereis(usr).
<0.156.0>
18> usr:lookup_id(0).
{ok,#usr{msisdn = 700000000,id = 0,status = disabled,
         plan = prepay,
         services = [data]}}
19> exit(whereis(usr), kill).
true
20> exit(whereis(usr), kill).
** exception exit: shutdown
```

 When a process terminates, all of the ETS tables that it created are destroyed. If you want ETS tables to survive process restarts without incurring the overhead of dealing with Dets tables or the filesystem, a trick is to let your supervisor create the tables in its init/1 function, rather than in the processes spawned.

Dynamic Children

So far, we have looked only at static children. What if you need a supervisor that dynamically creates a child whose task is to handle a specific event, take care of the task, and terminate when completed? It could be for every incoming instant message (IM) or buddy update coming into your IM server. You can't specify these children in your init callback function, as they are created dynamically. Instead, you need to use the calls to functions supervisor:???_child/2:

```
supervisor:start_child(SupervisorName, ChildSpec)
supervisor:terminate_child(SupervisorName, Id)
supervisor:restart_child(SupervisorName, Id)
supervisor:delete_child(SupervisorName, Id).
```

In the preceding calls:

SupervisorName
    Is either the process identifier of the supervisor or its registered name

ChildSpec
    Is a single child specification tuple, as described in the section "Child Specifications" on page 278

Id
    Is the unique child identifier defined in the ChildSpec

Of particular importance in the ChildSpec tuple is the child Id. Even after termination, the ChildSpec will be stored by the supervisor and referenced through its Id, allowing processes to stop and restart the child. Only upon deletion will the child specification be permanently removed.

If you've been skimming through the manual page for the supervisor behavior, you probably realize that it does not export a **stop** function. As supervisors are never meant to be stopped by anyone other than their parent supervisors, this function was not implemented.

You can easily add your own **stop** function by including the following code in your supervisor callback module. However, this will work only if **stop** is called by the parent:

```
stop() -> exit(whereis(?MODULE), shutdown).
```

If your supervisor is not registered, use its pid.

Applications

The *application* behavior is used to package Erlang modules into reusable components. An Erlang system will consist of a set of loosely coupled applications. Some are developed by the programmer or the open source community, and others will be part of the OTP distribution. The Erlang runtime system and its tools will treat all applications equally, regardless of whether they are part of the Erlang distribution.

There are two kinds of applications. The most common form of applications, called *normal applications*, will start the supervision tree and all of the relevant static workers. *Library applications* such as the Standard Library, which come as part of the Erlang distribution, contain library modules but do not start the supervision tree. This is not to say that the code may not contain processes or supervision trees. It just means they are started as part of a supervision tree belonging to another application.

In this section, we will cover all the functionality needed to encapsulate the mobile subscriber system into an OTP application, starting its top-level supervisor. When done, this application will behave like any other normal application. And don't forget, when we talk about applications in this chapter, we mean OTP applications.

Applications are loaded, started, and stopped as one unit. A resource file associated with every application not only describes it, but also specifies its modules, registered processes, and other configuration data. Applications have to follow a particular directory structure which dictates where beam, module, resource, and include files have to be placed. This structure is required for many of the existing tools, built around behaviors, to function correctly. To find out which applications are running in your Erlang runtime system, you use **application:which_applications()**:

```
1> application:which_applications().
[{stdlib,"ERTS  CXC 138 10","1.15.2"},
 {kernel,"ERTS  CXC 138 10","2.12.2"}]
```

The Standard Library and the Kernel are part of the basic Erlang applications and together form the minimal OTP subset when starting the runtime system. The first item in the application tuple is the application name. The second is a description string, and the third is the application version number. If you are wondering what the description string in the preceding example means, you are not alone. It is the internal Ericsson product numbering scheme.

We will show you where to configure the description of your applications later in this chapter.

Directory Structure

In your Erlang shell, type code:get_path(). You did this when we were explaining how to manipulate the code search path in the code server. What you probably did not realize at the time was that each code path was pointing to a specially structured directory of an OTP application.

Let's pick the Inets application and inspect its contents in more detail. In Mac OS X, the path for this particular installation of Erlang would be as follows:

```
/usr/local/lib/erlang/lib/inets-5.0.12/
```

In other operating systems, just `cd` to the *lib* directory from the Erlang root directory, typically something like */usr/local/lib/erlang/lib* or *C:/Program Files/erl5.6.2/lib/*, and look for the latest Inets release. Among all the subdirectories in an application, the following ones comprise an OTP release of the application in question:

src
Contains the source code of all the Erlang modules in the application.

ebin
Contains all of the compiled beam files and the application resource file; in this example, it's *inets.app*.

include
Contains all the Erlang header files (`hrl`) intended for use outside the application. By using the following directive:

```
-include_lib("Application/include/Name.hrl")
```

where `Application` is the application directory name *without* the version number (in the example it would be `inets`) and `Name.hrl` is the name of the include file, the compiler will automatically pick up the version of the application pointed to by the code search path.

priv
Is an optional directory that contains necessary scripts, graphics, configuration files, or other non-Erlang-related resources. You can access it without knowing the application version by using the `code:priv_dir(Application)` call.

You will notice that Inets (and other) applications may have a few more directories, including *docs* and *examples*. These have no effect on the system during runtime, and are there just for convenience. In some applications, you might not find the *priv* directory. If you do not use it, omitting it is not a problem, even if it might not be considered a good practice by some. In live systems, the only mandatory directory is *ebin*. This is because you probably don't want to include your source code when shipping your system to clients!

It is common to use scripts to create these directory structures, and to use make files which, having compiled your code, move the beam files to the *ebin* directory. How you set this up depends on the operating systems, build systems, repositories, and many other non-Erlang-related dependencies in your application. Although it might be feasible to set this up manually for small projects, you will probably want to use templates and automate the task for larger projects.

The Application Resource File

The application resource file, also known as the *app file*, contains information on your application resources and dependencies. Move into the *ebin* subdirectory of the Inets application and look for the *inets.app* file. This is the resource file of the Inets application. On closer inspection, you will notice that all other applications also have an *inets.app* file. The application resource file consists of a tuple where the first element is the application tag, the second is the application name, and the third is a list of features.

Let's go through the features individually. Note that for space considerations, we've omitted some of the modules in the example:

```
{application,inets,
 [{description,"INETS  CXC 138 49"},
  {vsn,"5.0.5"},
  {modules,[inets,inets_sup,inets_app,inets_service,
           %% FTP
           ftp, ftp_progress,ftp_response,ftp_sup,
           %% HTTP client:
           http,httpc_handler,httpc_handler_sup,httpc_manager,
           %% TFTP
           tftp,tftp_binary,tftp_engine,tftp_file,tftp_lib,tftp_sup
       ]},
  {registered,[inets_sup, httpc_manager]},
  {applications,[kernel,stdlib]},
  {mod,{inets_app,[]}}]}.
```

In the preceding code, the `description` is a string that is displayed as a result of calling the `application:which_application/0` function. The `vsn` attribute is a string denoting the version of the application. This should be the same as the suffix of the application directory. In larger build systems, the application version is usually updated automatically through proprietary scripts executed when committing your code.

The `modules` tag lists all the modules that belong to this application. The purpose of listing them is twofold. The first is to ensure that all of them are present when building the system and that there are no name clashes with any other applications. The second is to be able to load them either at startup or when loading the application. For every module, there should be a corresponding beam file. To ensure that there are no registered name clashes with other applications, we list all of the `registered` processes in this field. Clashes in module and registered process names are detected by the release-handling tools used when creating your boot file. We will look at boot files in the next section. Just including them in the application resource files will have no effect unless these tools are used.

Most applications will have to be started after other applications on which they depend. Your application will not start if the applications in the `applications` list included in your *app* file are not already started. *kernel* and *stdlib* are the basic standard applications on which every other application depends. After that, the particular dependencies will be based on the nature of the application.

Finally, the `mod` parameter is a tuple containing the callback module and the arguments passed to the `start/2` callback function.

Not necessary to the Inets application, but certainly important to applications in general, are *environment variables*. The `env` tag indicates a list of key-value tuples that can be accessed from within the application using calls to the following functions:

```
application:get_env(Tag)
application:get_all_env().
```

To access environment variables belonging to other applications, just add the application `Name` to either function call, as in the following:

```
application:get_env(Name,Tag)
application:get_all_env(Name).
```

The application resource file *usr.app* of our mobile subscriber service database would contain four modules, two registered processes, and dependencies on the *stdlib* and *kernel* applications. Let's also add the filename for the Dets table among the environment variables:

```
{application, usr,
 [{description, "Mobile Services Database"},
  {vsn, "1.0"},
  {modules, [usr, usr_db, usr_sup, usr_app]},
  {registered, [usr, usr_sup]},
  {applications, [kernel, stdlib]},
  {env, [{dets_name, "usrDb"}]},
  {mod, {usr_app,[]}}]}.
```

Starting and Stopping Applications

You start and stop applications using the following commands:

```
application:start(ApplicationName).
application:stop(ApplicationName).
```

In the preceding code, `ApplicationName` is an atom denoting the name of your application.

The *application controller* loads the environment variables belonging to the application, as well as starts the top-level supervisor through a set of callback functions. When calling start/1, the start(StartType, Arguments) function in the application callback module is invoked. StartType is usually the atom normal, but if you are dealing with distributed applications,[†] you might come across the start types takeover and failover. Arguments is a value of any valid Erlang data type, which together with the callback module is defined in the application resource file.

Start has to return the tuple {ok, Pid} or {ok, Pid, Data}. Pid denotes the process identifier of the top-level supervisor. Data is a valid Erlang data type used to store data that is needed when terminating the application.

If you stop your application, the top-level supervisor is sent a shutdown message. This results in the termination of all of its children in reverse startup order, propagating the exit path through the supervision tree. Once the supervision tree has terminated, the callback function stop(Data) is called in the application callback module. Data was originally returned in the {ok, Pid, Data} construct of the start/2 callback function. If your start/2 function did not return any data, just ignore the argument. Should you want a callback function to be called before terminating the supervision tree, export the function prep_stop(Data) in your callback module.

So, armed with all of the preceding information, how would you package your usr server database into an application, what would the directory structure look like, and what are the contents of the app file?

Let's start with the application callback file. We export the start/2 and stop/1 functions:

```
-module(usr_app).
-behaviour(application).
-export([start/2, stop/1]).

start(_Type, StartArgs) ->
    usr_sup:start_link().
stop(_State) ->
    ok.
```

As you can see, the application callback module is relatively simple. Although we have not done it in our example, it is not uncommon to join the supervisor and application behavior modules into one. You would have the two -behaviour directives next to each other, and if there is no conflict with the callback functions, the compiler will not issue any warnings.

This leaves one minor change to be made in the usr.erl module, where we read the environment variable in the start_link/0 call:

[†] We do not cover distributed applications in this chapter. For more information on them, you will need to consult the OTP documentation.

```
start_link() ->
    {ok, FileName} = application:get_env(dets_name),
    start_link(FileName).
```

With all of this in place, all that remains is our application directory structure, placing the relevant files in there:

```
usr-1.0/src/usr.erl
            usr_db.erl
            usr_sup.erl
            usr_app.erl
        /ebin/usr.beam
              usr_db.beam
              usr_sup.beam
              usr_app.beam
              usr.app
        /priv/
        /include/usr.hrl
```

Let's compile all the modules and take them for a test run. Move the beam files to the *ebin* directory, and make sure they are accessible by telling the system about the path to them. You can do that either with the `erl -pa Dir` directive when starting Erlang, or directly in the shell using `code:add_path(Dir)`.

In the following interaction, we start the application and run a few operations on the customer settings before stopping it. In doing so, we check that the **supervisor** and **gen_server** processes no longer exist:

```
1> code:add_path("usr-1.0/ebin").
true
2> application:start(usr).
ok
3> application:start(usr).
{error,{already_started,usr}}
4> usr:lookup_id(10).
{error,instance}
5> application:get_env(usr, dets_name).
{ok,"usrDb"}
6> application:stop(usr).

=INFO REPORT==== 27-Jan-2009::22:14:33 ===
    application: usr
    exited: stopped
    type: temporary
ok
6> whereis(usr_sup).
undefined
```

Note how we retrieved the `dets_name` environment variable from the environment. In our usr example, we are calling the function from within the application, and as a result, we do not need to specify the application name. Look through the manual page of the application module and experiment with the various options for retrieving application environment variables to get a better understanding of what is available.

The Application Monitor

The application monitor is a tool that provides an overview of all running applications. Upon launching it with the `appmon:start()` call, you are presented with a list of all the applications running on all distributed nodes. The various menus allow you to manipulate the node and presentation, and the bar on the left shows the load on the node under scrutiny (see Figure 12-3).

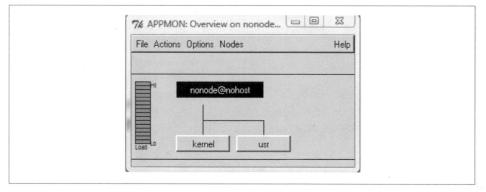

Figure 12-3. The application monitor window

In Figure 12-3, note how the *stdlib* application is not shown. Only applications with a supervision tree appear. Double-clicking the application opens a new window with a view of its supervision tree (see Figure 12-4). The menus and buttons allow you to manipulate the various processes. The top processes linking to `usr_sup` are part of the application controller. They are the ones that start, monitor, and stop the top-level supervisor.

Release Handling

From our behaviors, we've created a supervision tree. The supervision tree is packaged in an application that can be loaded, started, and stopped as one entity. Erlang systems consist of a set of loosely coupled applications specified in a release file. This includes the basic Erlang installation you have been running. From your Erlang root directory, enter the *releases* directory, followed by one of the release subdirectories. In our example, it is *R12B* (see Figure 12-5).

In it, you will find a list of release files, indicated by the *.rel* suffix, as shown in Figure 12-5. Pick *start_clean.rel* and inspect it:

```
{release, {"OTP  APN 181 01","R12B"}, {erts, "5.6.2"},
 [{kernel,"2.12.2"},
  {stdlib,"1.15.2"}]}.
```

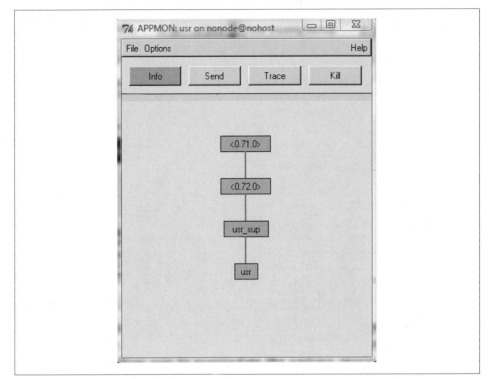

Figure 12-4. The supervision tree viewed in the application monitor

It consists of a tuple where the first element is the `release` tag and the second element is a tuple with a release name and release version number. The third element is a tuple with the version of the Erlang runtime system. The last element in the tuple is a list of applications and their version numbers, defined in the order in which they should be started.

Each application in this list points to an application resource file. When you call the function `systools:make_rel(Name, Options)`, these app files are retrieved and inspected. Module and registered process name conflicts are checked, and if everything matches, *Name.boot* and *Name.script* files are produced.

Name.boot is a binary file containing instructions on loading the application modules and starting the top-level supervisors. The *Name.script* file is a text version of its binary counterpart. The `Options` argument is a list in which the most important data includes `{path, [Dir]}`, which describes any paths to the application *ebin* directories not known by the code server. The paths commonly point to the *ebin* directories of your applications. The `local` directive is another option to `make_rel/2`, stating that the boot file should not assume that all the applications will be found under *lib* in the Erlang root directory. The latter is useful if you want to separate the Erlang installation and your applications.

Figure 12-5. The releases directory

In a deployed system, you will release only the applications that are relevant to your system. These applications include both your applications and the subset of the ones you need from the OTP release. They should all be stored in the *lib* subdirectory of the Erlang root, but you can easily override this recommendation by adding paths in the code search path used by the code server. We are in fact overriding this in our example by using the `local` directive in the options list.

When the boot file has been created, you can start your system using the following command:

```
erl -boot Name
```

This ensures that all of the modules specified in the boot file are loaded and that the applications and their respective supervision trees are started correctly. If anything fails at startup, the Erlang node will not start.

The release file of our mobile subscriber database, *usr.rel*, would include *kernel* and *stdlib*, the two mandatory applications of any OTP release, together with version 1.0 of our usr application:

```
{release, {"Mobile User Database","R1"}, {erts, "5.6.2"},
 [{kernel,"2.12.2"},
  {stdlib,"1.15.2"},
  {usr,"1.0"}]}.
```

Because in the running example we are using the shell where we previously added the path to `usr-1.0/ebin`, we create a boot file which runs with the existing code path. Had we *not* set the path in the shell, we would have had to add the option {`dir`, [`"usr-1.0/ebin"`]} to the release file:

```
7> systools:make_script("usr", [local]).
ok
8> ls().
usr-1.0
usr.boot
usr.rel
usr.script
usrDb
```

We can now start our system using `erl -boot usr`.

> Have a look at the *usr.script* file that was generated when creating the boot file. We will not explain it in this book, as most of its commands should be fairly straightforward. You can edit the files and generate a new boot file using the `systools:script2bootfile/1` call.
>
> Spare a thought for the early pioneers of OTP. In its first release back in 1996, script files had to be generated manually, as the `make_script/2` function had not been implemented!

Other Behaviors and Further Reading

What we described in this chapter should cover a majority of the cases you will come across when using OTP behaviors. However, you can go in more detail when working with generic servers, supervisors, and applications. Behaviors we have not covered but which we briefly introduced in this chapter include finite state machines, event handlers, and special processes. All of these behavior library modules have manual pages that you can reference. In addition, the Erlang documentation has a section on OTP design principles that provides more details and examples of these behaviors.

Finite state machines are a crucial component of telecom systems. In Chapter 5, we introduced the idea of modeling a phone as a finite state machine. If the phone is not being used, it is in state `idle`. If an incoming call arrives, it goes to state `ringing`. This does not have to be a phone; it could instead be an ATM cross-connect or the handling of data in a protocol stack. The `gen_fsm` module provides you with a finite state machine behavior that you can use to solve these problems. States are defined as callback functions that return a tuple containing the next `State` and the updated loop data. You can send events to these states synchronously and asynchronously. The finite state machine callback module should also export the standard callback functions such as `init`, `terminate`, and `handle_info`. As `gen_fsm` is a standard OTP behavior, it can be linked to the supervision tree.

Event handlers and managers are another behavior implemented in the gen_event library module. The idea is to create a centralized point that receives events of a specific kind. Events can be sent synchronously and asynchronously with a predefined set of actions being applied when they are received. Possible responses to events include logging them to file, sending off an alarm in the form of an SMS, or collecting statistics. Each of these actions is defined in a separate callback module with its own LoopData, preserved in between calls. Handlers can be added, removed, or updated for every specific event manager. So, in practice, for every event manager, there could be many callback modules, and different instances of these callback modules could exist in different managers.

Sometimes you might want to add to the supervision tree processes that are not generic OTP behaviors. This might be for efficiency reasons, where you have implemented the process using plain Erlang. You might want to attach to a supervision tree legacy code that was written before OTP was available, or you might have abstracted a design pattern and implemented your own behavior.

Writing your own behaviors is straightforward. The main differences are in how you spawn your processes, and the system calls you need to handle. You should create processes using the proc_lib library, which exports both spawn and start functions. Using the proc_lib function stores the start data of the process, provides the means to start the process synchronously, and generates error reports upon abnormal termination. To be OTP-compliant, processes need to handle system messages and events, yielding the control of the loop to the sys library module. They also need to be linked to their parent, and if they are trapping exits, they need to terminate when the parent terminates. You can find more information on writing your own OTP behaviors in the sys and proc_lib library modules.

Exercises

Exercise 12-1: Database Server Revisited

Rewrite Exercise 5-1 in Chapter 5 using the gen_server behavior module. Use the lists backend database module, saving your list in the loop data. You should register the server and access its services through a functional interface. Exported functions in the my_db_gen.erl module should include the following:

```
my_db_gen:start() ⇒ ok.
my_db_gen:stop() ⇒ ok.
my_db_gen:write(Key,Element) ⇒ ok.
my_db_gen:delete(Key) ⇒ ok.
my_db_gen:read(Key) ⇒ {ok,Element}|{error,instance}.
my_db:match(Element) ⇒ [Key1, ..., KeyN].
```

Hint: if you are using Emacs or Eclipse, use the **gen_server** skeleton template:

```
1> my_db:start().
ok
2> my_db:write(foo, bar).
ok
3> my_db:read(baz).
{error, instance}
4> my_db:read(foo).
{ok, bar}
5> my_db:match(bar).
[foo]
```

Exercise 12-2: Supervising the Database Server

Implement a supervisor that starts and monitors the **gen_server** in Exercise 12-1. Your supervisor should be able to handle five crashes per hour. Your child should be permanent and be given at least 30 seconds to terminate, as it might take some time to close a large Dets file.

Exercise 12-3: The Database Server As an Application

Encapsulate your supervision tree from Exercise 12-2 in an application, setting up the correct directory structure, complete with application resource file.

CHAPTER 13

Introducing Mnesia

Try to picture a cluster of Erlang nodes, distributed over half a dozen computers to which requests are forwarded. Data has to be accessible and up-to-date across the cluster and destructive database operations, even if they are rare, have to be executed in a transaction to avoid inconsistent data as a result of race conditions. You need to be able to add and remove nodes during runtime and provide persistence to ensure a speedy recovery from all possible failure scenarios.

The solution is to merge the efficiency and simplicity of ETS and Dets tables with the Erlang distribution and to add a transaction layer on top. This solution, called *Mnesia*, is a powerful database that comes as part of the standard Erlang distribution. Mnesia is the brainchild of Claes "Klacke" Wikström[*] from the days when he was working at Ericsson's Computer Science Lab. Håkan Mattsson eventually took over and brought Mnesia to the next level, productizing it and adding lots of functionality.

Mnesia can be as easy or as complex as you want it to be. The aim of this chapter is to introduce you to Mnesia and its capabilities without losing you in too many details.

When to Use Mnesia

Mnesia was originally built for integration in distributed, massively concurrent, soft real-time systems with high availability requirements (i.e., telecoms). You *want* to use Mnesia if your system requires the following:

- Fast key-value lookups of potentially complex data
- Distributed and replicated data across a cluster of nodes with support for location transparency
- Data persistency with support for fast data access

---

[*] Klacke is the same person we need to thank for giving us the ASN.1 compiler, the first-generation garbage collector, ETS, Dets, the Erlang Distribution, bit syntax, and YAWS. I am sure he will be thrilled to receive your bug reports.

- Runtime reconfiguration of the table location and table features
- Support for transactions, possibly across a distributed cluster of nodes
- Data indexing
- The same level of fault tolerance as for a typical Erlang system
- Tight coupling of your data model to Erlang data types and to Erlang itself
- Relational queries that don't have soft real-time deadlines

You *do not* want to use Mnesia if your system requires the following:

- Simple key-value lookup
- A storage medium for large binaries such as pictures or audio files
- A persistent log
- A database that has to store gigabytes of data
- A large data archive that will never stop growing

For simple key-value lookups, you can use ETS tables or the `dict` library module. For large binaries, you are probably better off with individual files for each item; for dealing with audit and trace logs, the `disk_log` library module should be your first point of call.

If you are looking to store your user data for the next Web 2.0 social networking killer application that has to scale to hundreds of millions of users overnight, Mnesia might not be the right choice. For *massive* numbers of entries that you want to be able to access readily, you might be better off using CouchDB, MySQL, PostgreSQL, or Berkeley DB, all of which have open source Erlang drivers and APIs available. The upper limit of a Dets table is 2 GB. This means the upper limit of a Mnesia table is 2 GB if the storage type is disc-only copies. For other storage types the upper limit depends on the system architecture. In 32-bit systems the upper limit is 4 GB ($4 * 10^9$ bytes), and in 64-bit systems it is 16 exabytes ($16 * 10^{18}$ bytes). If you need to store larger quantities of data, you will have to fragment your tables and possibly distribute them across many nodes. Mnesia has built-in support for fragmented tables.

While Mnesia might not be the first choice for all of your Web 2.0 user data, it is the perfect choice for caching all of the user session data. Once users have logged on, it can be read from a persistent storage medium and duplicated across a cluster of computers for redundancy reasons. The login might take a little longer while you retrieve the user data, but once it's done, all of the session activities would be extremely fast. When the user logs out or the session expires, you would delete the entry and update the user profile in the persistent database.

That being said, Mnesia has been known to handle live data for tens of millions of users. It is extremely fast and reliable, so using it in the right setting will provide great benefits from a maintenance and operational point of view. Just picture your application's database, the glue and logic alongside the formatting and parsing of data from external

APIs, all running in the same memory space and controlled uniformly by an Erlang system. Your application becomes not only efficient but also easy to maintain.

Configuring Mnesia

Mnesia is packaged as an OTP application. To use it, you usually create an empty schema that is stored on disk. But you can also use Mnesia as a RAM-only database that only keeps its schema in RAM. Having created a schema, you need to start Mnesia and create the tables. Once they are created, you can read and manipulate your data. You need to create your schema and tables only once, usually when installing your system. When you're done, you can just start your system, together with Mnesia, and all persistent data will become available.

Setting Up the Schema

A *schema* is a collection of table definitions that describe your database. It covers which of your tables are stored on RAM, disk, or both, alongside their configuration characteristics and the format of the data they will contain. These characteristics may differ from node to node, as you might want your table to have its disk copies on the operation and maintenance node but have RAM-only copies on the transaction nodes. In Erlang, the schema is stored in a persistent Mnesia table. When configuring your database, you create an empty schema table which, over time, you populate with your table definitions.

To create the schema,[†] start your distributed Erlang nodes and connect them. If you do not want to distribute Mnesia, just start a non-distributed Erlang node. Before doing this it is important to make sure no old schemas exist, as well as ensuring that Mnesia is not started.

In our example, we will be starting the database on two nodes, switch and om:

```
(om@Vaio)1> net_adm:ping(switch@Vaio).
pong
(om@Vaio)2> nodes().
[switch@Vaio]
(om@Vaio)3> mnesia:create_schema([node()|nodes()]).
ok
(om@Vaio)4> ls().
Mnesia.om@Vaio          Mnesia.switch@Vaio      include
lib
ok
```

The mnesia:create_schema(Nodes) command has to be executed on one of the connected nodes only. By creating the list [node()|nodes()], we get all the connected Erlang

---

[†] Those familiar with databases might be surprised that we called this "schema creation," since it more closely resembles creation of the database itself, with the schema—the details regarding which tables the database contains—being created *implicitly* by subsequent table operations to create tables.

nodes, which incidentally are the same nodes on which we want to create the schema tables. You might recall from Chapter 11, where we covered distributed Erlang, that the node() BIF returns the local node, whereas nodes() returns all other connected nodes. The mnesia:create_schema/1 command will propagate to the other *nodes* automatically.

When creating the schema, each node will create a directory. In our case, as both distributed Erlang nodes share the same root directory, their schema directories Mnesia.om@Vaio and Mnesia.switch@Vaio will appear in the same location, where Vaio is the hostname of the computer on which we executed the command. Other contents of the directory will depend on what you've previously done with Erlang, but will not affect your schema.

Had your nodes been on computers that were not connected, only the schema directory for the node running on that computer would have been created. If you do not plan to run Mnesia in a distributed environment, the schema directory name will be Mnesia.nonode@nohost. Just pass [node()] as an argument to the create_schema/1 call.

You can override the location of the root directory by starting Erlang with this directive:

```
erl -mnesia dir Dir
```

replacing Dir with the directory where you want to store your schema.

Starting Mnesia

Once your schema has been created, you start the application by calling the following:

```
application:start(mnesia).
```

If you are using boot scripts as described in Chapter 12, you should include the Mnesia application in your release file. In a test environment where you are not using OTP behaviors, you can also use mnesia:start(). In an industrial project with release handling, separating the applications and starting them individually is considered a best practice.

If you start Mnesia without a schema, a memory-only database will be created. This will not survive restarts, however, so the RAM-only tables will have to be created every time you restart the system. To start Mnesia with a RAM schema, all you need to do is ensure that there is no schema directory for that particular node, and then you can start Mnesia.

You can stop Mnesia by calling either application:stop(mnesia) or mnesia:stop().

Mnesia Tables

Mnesia tables contain Erlang records. By default, the name of the record type becomes the table name. You create a table by using the following function call:

```
mnesia:create_table(Name, Options)
```

In this function call, `Name` is the record type and `Options` is a list of tuples of the format {`Item, Value`}. The following `Items` and `Values` are most commonly used:

{disc_copies, Nodelist}
> Provides the list of nodes where you want disc and RAM replicas of the table.

{disc_only_copies, Nodelist}
> `Nodelist` contains the nodes where you want disc-only copies of this particular table. This is usually a backup node, as local reads will be slow on these nodes.

{ram_copies, Nodelist}
> Specifies which nodes you want to have duplicate RAM copies of this particular table. The default value of this attribute is [`node()`], so omitting it will create a local Mnesia RAM copy.

{type, Type}
> States whether the table is a `set`, `ordered_set`, or `bag`. The default value is `set`.

{attributes, AtomList}
> Is a list of atoms denoting the record field names. They are mainly used when indexing or using query list comprehensions. Please, *do not* hardcode them; generate them using the function call `record_info(fields, RecordName)`.

{index, List}
> Is a list of attributes (record field names) which can be used as secondary keys when accessing elements in the table.

The key position is, by default, the first element of the record. Each instance of a record in a Mnesia table is called an *object*. The key of this object, together with the table name, give you the *object identifier*.

Having created the schema, we want to start Mnesia on all nodes and do the one-off operation of creating the table. This is usually done when installing and configuring the Erlang nodes. For redundancy and performance reasons, we want to run the database on two remote nodes called om and `switch`. On the om node, we want to store the data in RAM and on disk, and on the `switch` node, we want to maintain only a RAM copy. When looking at the example, remember the `rr/1` shell command that reads a file and extracts its record definitions, making them available in the shell:

```
(om@Vaio)5> rr(usr).
[usr]
(om@Vaio)6> Fields = record_info(fields, usr).
[msisdn,id,status,plan,services]
(om@Vaio)7> application:start(mnesia).
 ok
(om@Vaio)8> mnesia:create_table(usr, [{disc_copies, [node()]},
  {ram_copies, nodes()}, {type, set}, {attributes, Fields}, {index, [id]}]).
{atomic,ok}
```

Within this Mnesia example, notice how we create only one table, stating that it has to be indexed and that it must have a RAM copy and file backup. A simplistic explanation of what is happening behind the scenes is evident in our mobile subscriber database backend module using ETS and Dets tables, where we created three tables: one for the disk copy, one for the RAM copy, and one for the index.

There is no need to open Mnesia tables. When starting the Mnesia application, all tables configured in the schema are created or opened. This is a relatively fast, nonblocking operation, done in parallel with the startup of other applications.

For large persistent tables, or tables that were incorrectly closed and whose backup files need repair, other applications might try to access the table even if it has not been properly loaded. Should this happen, the process crashes with the error no_exists. To avoid this, you should call:

```
mnesia:wait_for_tables(TableList, TimeOut)
```

in the initialization phase of your process or OTP behavior, where TableList is a list of Mnesia table names (both *persistent* and *volatile*) used by that process, and TimeOut is either the atom infinity or an integer in milliseconds.

When dealing with large tables containing millions of rows, if you are not using infinity as a timeout, you must ensure that the TimeOut value is at least a few minutes, if not hours, for extremely large, fragmented, disk-based tables. The call wait_for_tables/2 is called independently of starting Mnesia by the process which needs the tables. By pattern matching on the return value, you ensure that the table is loaded. If a timeout occurs, the return value will result in a bad match error, and that is logged. The last thing you want is for the wait_for_tables/2 call to return {timeout, TableList}, ignore this value, and continue assuming that the tables have been properly loaded.

In our mobile subscriber example, we originally needed a process that created and owned the ETS and Dets tables and serialized all destructive (write and delete) operations. That is no longer the case. Mnesia will take care of this for us. We would thus scrap the usr_db.erl module, and instead place all of our functionality in the usr.erl module. We would also remove all the process-related calls, such as start, spawn, init, and stop, or gen_server calls and callbacks if we are using the behavior example. We would instead add the calls create_tables/0 and ensure_loaded/0 and keep the entire client API:

```erlang
-module(usr).
-export([create_tables/0, ensure_loaded/0]).
-export([add_usr/3, delete_usr/1, set_service/3, set_status/2,
         delete_disabled/0, lookup_id/1]).
-export([lookup_msisdn/1, service_flag/2]).

-include("usr.hrl").

%% Mnesia API

create_tables() ->
  mnesia:create_table(usr, [{disc_copies, [node()]}, {ram_copies, nodes()},
                      {type, set}, {attributes,record_info(fields, usr)},
                      {index, [id]}]).
```

```
ensure_loaded() ->
  ok = mnesia:wait_for_tables([usr], 60000).
```

Transactions

As many concurrent processes, possibly located on different nodes, can access and manipulate objects at the same time, you need to protect the data from race conditions. You do this by encapsulating the operations in a fun and executing them in a *transaction*. A transaction guarantees that the database will be taken from one consistent state to another, that changes are persistent and atomic across all nodes, and that transactions running in parallel will not interfere with each other. In Mnesia, you execute transactions using:

```
mnesia:transaction(Fun)
```

where the fun contains operations such as read, write, and delete. If successful, the call returns a tuple {atomic, Result}, where Result is the return value of the last expression executed in the fun. If the transaction fails, {aborted, Reason} is returned. Always pattern-match on {atomic, Result}, as your transactions should never fail unless mnesia:abort(Reason) is called from within the transaction.

Make sure your funs, with the exception of your Mnesia operations, are free of side effects. When executing your fun in a transaction, Mnesia will put locks on the objects it has to manipulate. If another process is holding a conflicting lock on an object, the transaction will first release all of its current locks and then restart. Side effects such as an io:format/2 call, or sending a message, might result in the printout or the message being sent hundreds of times.

Writing

To write an object in a table, you use the function mnesia:write(Record), encapsulating it in a fun and executing it in a transaction. The call returns the atom ok. Mnesia will put a write lock on all of the copies of this object (including those on remote nodes). Attempts to put a lock on an already locked object will fail, prompting the transaction to release all of its locks and start again.

Trying out the functions directly in the shell will give you a better feel for how it all works:

```
(om@Vaio)9> Rec = #usr{msisdn=700000003, id=3, status=enabled,
(om@Vaio)9>             plan=prepay, services=[data,sms,lbs]}.
#usr{msisdn=700000003, id=3, status=enabled,
     plan=prepay, services=[data,sms,lbs]}
(om@Vaio)10> mnesia:transaction(fun() -> mnesia:write(Rec) end).
{atomic,ok}
```

Remember how in the ETS and Dets variants of this example, which first appeared in Chapter 10, you had to create three table entries for every user you inserted in the database? And what if you wanted to distribute this data across multiple nodes? As the usr Mnesia table contains distributed RAM and disk-based copies as well as an index, you only need to do one write. Behind the scenes, Mnesia takes care of the rest for you.

To write or update a record in the mobile subscriber example, you would encapsulate the write operation in the usr.erl module as follows:

```
add_usr(PhoneNo, CustId, Plan) when Plan==prepay; Plan==postpay ->
  Rec = #usr{msisdn = PhoneNo,
             id    = CustId,
             plan  = Plan},
  Fun = fun() -> mnesia:write(Rec) end,
  {atomic, Res} = mnesia:transaction(Fun),
  Res.
```

As add_usr/1 in all of the previous ETS and OTP behavior examples returned ok, you would make the new function backward-compatible.

Reading and Deleting

To read objects, you use the function mnesia:read(OId), where OId is an object identifier of the format {TableName, Key}. This function call will return the empty list if the object does not exist or a list of one or more records if the object exists and the table is a set or bag. You need to execute the function within the scope of a transaction; failing to do so will cause a runtime error.

Note from which node we are now reading the record. It does not make a difference. We just need to make sure Mnesia is started on this node as well, something we did when creating the table.

To delete an object, you can use the mnesia:delete(OId) call from within a transaction. The call returns the atom ok, regardless of whether the object exists.

Our schema was distributed across two nodes. Let's start Mnesia on the second node and look up the entry we wrote on the om node in the previous example:

```
(switch@Vaio)1> application:start(mnesia).
ok
(switch@Vaio)2> usr:ensure_loaded().
ok
(switch@Vaio)3> rr(usr).
[usr]
(switch@Vaio)4> mnesia:transaction(fun() -> mnesia:read({usr, 700000003}) end).
{atomic,[#usr{msisdn = 700000003,id = 3,status = enabled,
              plan = prepay,
              services = [data,sms,lbs]}]}
(switch@Vaio)5> mnesia:read({usr, 700000003}).
** exception exit: {aborted,no_transaction}
     in function  mnesia:abort/1
(switch@Vaio)6> mnesia:transaction(fun() -> mnesia:abort(no_user) end).
{aborted,no_user}
(switch@Vaio)7> mnesia:transaction(fun() -> mnesia:delete({usr, 700000003}) end).
{atomic,ok}
```

```
(switch@Vaio)8> mnesia:transaction(fun() -> mnesia:read({usr, 700000003}) end).
{atomic,[]}
```

As you can see, executing a destructive operation such as write or delete will duplicate
the operation across all nodes. Pay attention to the error in command 5, where we
execute a read outside the scope of a transaction. Also, look at the return value for
command 6, where we abort a transaction.

Indexing

When creating the usr table, one of the options we passed into the call was the tuple
{index, AttributeList}. This will index the table, allowing us to look up and manipulate objects using any of the secondary fields (or keys) listed in the AttributeList. To
use indexes, you have to execute the following call:

```
index_read(TableName, SecondaryKey, Attribute).
```

All of the functions used to provision customer data in our example use the
CustomerId attribute. If you want to delete a subscriber record, you would have to check
its existence, as the function delete_usr/1 returns {error, instance} if the field does
not exist. If the record does exist, you find its primary key and use it to delete the field:

```
delete_usr(CustId) ->
    F = fun() -> case mnesia:index_read(usr, CustId, id) of
                    []    -> {error, instance};
                    [Usr] -> mnesia:delete({usr, Usr#usr.msisdn})
                 end
        end,
    {atomic, Result} = mnesia:transaction(F),
    Result.
```

In a similar fashion, if you wanted to add or remove a service a subscriber is entitled to
use, you would look up the usr record and, if the entry exists, update the status using
the msisdn:

```
set_service(CustId, Service, Flag) when Flag==true; Flag==false ->
    F = fun() ->
            case mnesia:index_read(usr, CustId, id) of
                []    -> {error, instance};
                [Usr] ->
                    Services = lists:delete(Service, Usr#usr.services),
                    NewServices = case Flag of
                                    true  -> [Service|Services];
                                    false -> Services
                                  end,
                    mnesia:write(Usr#usr{services=NewServices})
            end
        end,
    {atomic, Result} = mnesia:transaction(F),
    Result.
```

The same principle applies to enabling and disabling a particular subscriber:

```
set_status(CustId, Status) when Status==enabled; Status==disabled->
   F = fun() ->
        case mnesia:index_read(usr, CustId, id) of
            []    -> {error, instance};
            [Usr] -> mnesia:write(Usr#usr{status=Status})
        end
     end,
   {atomic, Result} = mnesia:transaction(F),
   Result.
```

Note how all of these functions first look up an object, and if it exists, they either delete or manipulate it. When changing the subscriber status or services, there is no risk of any other process deleting the entry between the index_read/3 and the write/1 function calls. This is because both operations are running in a transaction, setting locks on the objects they are manipulating and ensuring that other transactions attempting to access them are kept on hold. As a result, any two transactions on the same object cannot interfere with each other. Keep in mind that the race conditions could occur among processes in different nodes. Let's try the functions we've just defined in the shell and see whether they work:

```
(switch@Vaio)9> usr:add_usr(700000001, 1, prepay).
ok
(switch@Vaio)10> usr:add_usr(700000002, 2, prepay).
ok
(switch@Vaio)11> usr:add_usr(700000003, 3, postpay).
ok
(switch@Vaio)12> usr:delete_usr(3).
ok
(switch@Vaio)13> usr:delete_usr(3).
{error,instance}
(switch@Vaio)14> usr:set_status(1, disabled).
ok
(switch@Vaio)15> usr:set_service(2, premiumsms, true).
ok
(switch@Vaio)16> mnesia:transaction(fun() -> mnesia:index_read(usr, 2, id) end).
{atomic,[#usr{msisdn = 700000002,id = 2,status = enabled,
              plan = prepay,
              services = [premiumsms]}]}
```

If you create a table and want to add or remove indexes during runtime, you can use the schema manipulation functions add_table_index(Tab, Attribute) and del_table_index(Tab, Attribute).

Dirty Operations

Sometimes it is acceptable to execute an operation outside the scope of a transaction without setting any locks. Such operations are known as *dirty* operations. In Mnesia, dirty operations are about 10 times faster than their counterparts that are executed in transactions, making them a very viable option for soft real-time systems. If you can guarantee the consistency, isolation, durability, and distribution properties of your tables, dirty operations will significantly enhance the performance of your program.

Some of the most common dirty Mnesia operations are:

```
dirty_read(Oid)
dirty_write(Object)
dirty_delete(ObjectId)
dirty_index_read(Table, SecondaryKey, Attribute)
```

All of these operations will return the same values as their counterparts executed within a transaction. If you need to implement soft real-time systems with requirements on throughput, transactions quickly become a major bottleneck. In our mobile subscriber example, the time-critical functions are those that are service-related. If you need to send 100,000 SMS messages, where every SMS requires a lookup to ensure that the subscriber not only exists and is enabled, but also is allowed to receive premium-rated SMSs, speed becomes critical. If the subscriber data is changed before or after the dirty read, it would not impact the sending of the SMS, since the functions contain only one nondestructive read operation:

```
lookup_id(CustId) ->
  case mnesia:dirty_index_read(usr, CustId, id) of
    [Usr] -> {ok, Usr};
    []    -> {error, instance}
  end.

%% Service API

lookup_msisdn(PhoneNo) ->
  case mnesia:dirty_read({usr, PhoneNo}) of
    [Usr] -> {ok, Usr};
    []    -> {error, instance}
  end.

service_flag(PhoneNo, Service) ->
  case lookup_msisdn(PhoneNo) of
    {ok,#usr{services=Services, status=enabled}} ->
      lists:member(Service, Services);
    {ok, #usr{status=disabled}} ->
      {error, disabled};
    {error, Reason} ->
      {error, Reason}
  end.
```

A common way to use dirty operations while ensuring data consistency is to serialize all destructive operations in a single process. Although another process might be allowed to execute a dirty read outside the scope of this process, all operations that involve both writing and deleting elements are serialized by sending the request to the process that executes them in the order they are received.

In our Mnesia version of the usr example, we got rid of our central process altogether and used transactions. If we had kept the process, we could have replaced all of the ETS and Dets read, write, and delete operations with Mnesia dirty operations. If we distributed the table across the OM and Switch nodes, however, we would have had to redirect all destructive operations to one of the nodes, as we would otherwise have run the risk of simultaneously updating the same object in two locations.

If you need to use dirty operations in a distributed environment, the trick is to ensure that updates to a certain key subset are serialized through a process on a single node. If your keys are in the range of 1 to 1,000, you could potentially update all even keys on one node and all odd ones on the other, solving the race condition we just described.

Inconsistent Tables

If you want to see how Mnesia tables become inconsistent through the use of dirty operations, start two distributed Erlang nodes on separate computers and make them share a Mnesia table. In one node, type `mnesia:dirty_write/1` with a key and one or more fields, but do not press Enter. In the other node, do the same, keeping the same key but changing the values of the fields. Very quickly, you need to disconnect the network cable between both computers, press Enter in both shells, and reconnect the cable. If you are fast enough, you will reconnect the cable before the TCP/IP connection between the nodes times out.

Read the entry in both nodes, and you will probably discover an inconsistent table, as values will probably differ. What happens upon executing `dirty_write/1` is that you locally update the local object copy, after which it is sent to the remote node. As the TCP/IP connection is temporarily down, this entry gets buffered. The buffering will happen on both nodes, so as soon as you reinsert the cable, you generate a race condition, overwriting the entry in the peer node. Had you been using transactions, this race condition would not have occurred.

Partitioned Networks

One of the biggest problems when using Mnesia in a distributed environment is the presence of *partitioned networks*. Although this problem is not directly related to any distributed transactional database or to Mnesia in particular, sooner or later you are bound to come across it. Assume that you have two Erlang nodes with a shared Mnesia table. If something as minor as a network glitch occurs between the nodes and both copies of the table are updated independently of each other, then when the network comes back up again, you have an inconsistent shared table (Mnesia would have been able to recover if only one node had been updated). Unlike the example with the dirty operations, Mnesia knows the tables are partitioned and will report this event so that you can act on it.

What do you do? Which of the two table copies do you pick? Can you somehow merge the two databases together again? Recovery of databases from partitioned networks is an area of research for which no "silver bullet" solutions have been found. In Mnesia, you can pick the master node by calling the following function:

```
mnesia:set_master_nodes(Table, Nodes).
```

If the network becomes partitioned, Mnesia will automatically take the contents of the master node, duplicating it to the partitioned nodes and bringing them back in sync. All updates during the partitioning done in tables not on the master node are discarded.

The most common Mnesia deployments will have tables replicated on two or three nodes. As soon as you start increasing that number, the risk of partitioned networks increases exponentially. No matter how extensive your testing is, partitioned databases will rarely manifest themselves until you've gone live and your system is under heavy stress. When designing distributed databases, always have a recovery plan from partitioned databases up your sleeve.

Further Reading

We are almost done with our module subscriber database. Only one operation is missing: traversing the list and deleting all disabled subscribers. There are many ways to traverse and search through data in Mnesia. You can use `first` and `next`, query list comprehensions, and even `select` and `match`.

We picked the `mnesia:foldl/3` call for no particular reason other than the fact that it is an interesting function that deserves mention. It behaves just like its counterpart in the `lists` module, but instead of traversing a list, it traverses a table:

```
delete_disabled() ->
  F = fun() ->
    FoldFun = fun(#usr{status=disabled, msisdn = PhoneNo},_) ->
                  mnesia:delete({usr, PhoneNo});
               (_,_) ->
                  ok
            end,
    mnesia:foldl(FoldFun, ok, usr)
  end,
  {atomic, ok} = mnesia:transaction(F), ok.
```

Although we may now be done with our mobile subscriber example, we've barely scratched the surface of what Mnesia has to offer. Some of the most commonly used functionality in industrial systems includes the fragmentation of tables, backups, fall-backs, Mnesia events, and diskless nodes, to mention but a few. All of them are covered in more detail in the Mnesia User's Guide and the Mnesia Reference Manual, both of which are part of the OTP documentation. What we have covered, however, is more than enough to allow you to efficiently get started using Mnesia.

Exercises

Exercise 13-1: Setting Up Mnesia

In this step-by-step exercise, you will create a distributed Mnesia database of Muppets. First, start two nodes:

```
erl -sname foo
erl -sname bar
```

In the first node, declare the Muppet data structure:

```
foo@localhost 1> rd(muppet, {name, callsign, salary}).
```

Next, create a schema so that you can make the tables persistent:

```
foo@localhost2> mnesia:create_schema([foo@localhost, bar@localhost]).
```

Now,you need to start Mnesia on both nodes:

```
foo@localhost 3> application:start(mnesia).
bar@localhost 1> application:start(mnesia).
```

The database is running! Create a distributed table:

```
foo@localhost 4> mnesia:create_table(muppet, [
                          {attributes, record_info(fields, muppet)},
                          {disc_copies [foo@localhost, bar@localhost}]).
```

Note how the `disc_copies` attribute specifies the nodes on which you want to keep a persistent copy. Check that everything looks all right:

```
foo@localhost 5> mnesia:info().
```

Now, look around you and type in your current cast of Muppets:

```
foo@localhost 6> mnesia:dirty_write(#muppet
{name = "Francesco" callsign="HuluHuluHulu", salary = 0}).
```

See how many Muppets you have so far:

```
foo@localhost 7> mnesia:table_info(muppet, size).
```

List their names with a function we have not covered in this chapter, but which you should have picked up when reading the Mnesia manual pages:

```
foo@localhost 8> mnesia:dirty_all_keys(muppet).
```

Excellent; now go to the other node and look up a Muppet:

```
bar@localhost 2> mnesia:dirty_read({muppet, "Francesco"}).
```

Exercise 13-2: Transactions

Write a function that reads a Muppet's salary and increases it by 10%. Use a transaction to guarantee that there are no race conditions.

Exercise 13-3: Dirty Mnesia Operations

Implement the `usr_db.erl` module from Chapter 10 with dirty Mnesia operations. The module should be completely backward compatible with the ETS- and Dets-based solution. Test it from the shell, and when it works, serialize the operations in a process, using the `usr.erl` module. Your `create_tables/1` and `close_tables/0` calls should start and stop the Mnesia application, and the `restore_backup/0` call should implement a `wait_for_tables/2` call. All other functions should return the same values returned in the original example.

CHAPTER 14

GUI Programming with wxErlang

Programming graphical user interfaces (GUIs) is not one of Erlang's touted strengths, but ongoing work has provided Erlang with a cross-platform, state-of-the-art GUI programming system: wxErlang, an Erlang binding of the wxWidgets system.

wxWidgets consists of an extensive C++ library that provides components for building menus, buttons, interactions, text and graphical displays, and much more; wxWidgets also provides a general framework for building cross-platform applications, including support for internationalization, and lower-level facilities such as memory management. Because of the size and complexity of wxErlang, this chapter cannot provide a comprehensive overview of it. Instead, this chapter covers the principles underlying the toolkit and provides a taste of some of its most-used aspects. Our coverage should be enough to get you started and give you the base from which to explore the library in more depth.

This chapter introduces wxWidgets and explains the principles underlying its Erlang binding. After describing the event-handling mechanism in wxErlang, we present a scaled-down blog example in two stages. This chapter concludes with a number of pointers for learning more about wxWidgets and wxErlang, and a series of exercises to improve and extend the running example.

wxWidgets

The wxWidgets open source project was initiated by Julian Smart in the early 1990s, and it is now supported by a team of approximately 20 developers and a wide circle of contributors. wxWidgets is a C++ library, but it has bindings to many other languages, including Haskell, Java, Perl, and Python, giving programmers in these languages direct, high-level access to a state-of-the-art GUI-building toolkit. Recent work led by Dan Gudmundsson and Mats-Ola Persson has given us wxWidgets for Erlang.

Most systems with a GUI will be deployed on multiple platforms (multiple hardware and operating system combinations). To avoid having to write multiple implementations of a system you're developing, you must use a GUI toolkit that will run on multiple platforms. One option is to use a platform-agnostic toolkit, which will give the application the same appearance on multiple platforms. The downside to this approach is that the application is unlikely to share the native look and feel of applications built for any one platform (and we're sure many of you have experienced the frustration of using an application that fails to comply with the UI guidelines of your favorite platform).

Although the wxWidgets toolkit supports multiple platforms, it is designed to support the native look and feel of each one, including various flavors of Windows, Linux, and Mac OS X. This makes the wxWidgets toolkit a perfect fit for use with the platform-independent Erlang language.

wxWidgets has an object-oriented architecture. Each graphical entity is a C++ object and belongs to a C++ class. The classes are related through *multiple inheritance*, with more complex graphical entities, such as a "text entry dialog," inheriting from a number of more fundamental entities. The example in this chapter includes the event handler and dialog classes as well as the `wxWindow` and `wxObject` base classes.

The GUI is *event-driven*: the GUI objects handle events—both user-initiated and internal—by associating the events with (member) functions to be called when the events occur. You can create this association using an *event table*, binding particular events to their handler functions and constructing them *statically* as part of the construction of a class. Alternatively, you can create a *dynamic* association, by connecting handler functions to events on an object-by-object basis and allowing the processing to be modified during an object's lifetime.

Objects in C++ can be created on the stack, but in general, the principal GUI entities will be created on the heap. Any memory allocated to the storage of the GUI entities needs to be deallocated explicitly once the object is no longer accessible. wxWidgets provides some mechanisms to assist with this.

You can find more information about wxWidgets in the extensive online documentation at *http://www.wxwidgets.org/*, and in *Cross-Platform GUI Programming with wxWidgets* (Prentice Hall).

wxErlang: An Erlang Binding for wxWidgets

In binding wxWidgets to Erlang, it is necessary to decide how to render its object-oriented structure and event-handling mechanism within Erlang so that it fits as closely as possible with the design principles underlying the language. This section explains the top-level correspondences, with more specific details coming in subsequent sections.

The wxErlang documentation contains an overview page together with an EDoc page for each module, which gives type information on each function as well as linking into the corresponding pages of the wxWidgets online documentation.

Objects and Types

In the wxErlang binding, each class is represented by a module and each object by an object reference. For example, the wxErlang function call in:

```
File = wxMenu:new(),
```

constructs a new wxMenu object and assigns a reference to this object to the File variable, with the same effect as the call to the constructor in the C++ fragment:

```
wxMenu *File = new wxMenu;
```

To call methods on the object, as in this C++ example:

```
File->Append(NEW,wxT("New\tCtrl-N"));
```

wxErlang provides a three-argument append function, where the first argument to append is the object reference, File:

```
wxMenu:append(File,?NEW,"New\tCtrl-N"),
```

The pattern of wxErlang functions taking an extra "this" or "self" value as the *first* argument is used throughout the binding. In a similar way, the constructors for a class are replaced by new functions of the same arity. Some wxWidgets methods take optional arguments; in wxErlang, all the optional arguments are passed by means of a single unordered property list, as supported by the proplists module.

It is worth reiterating that the File variable in the preceding code is *fixed*: its value is a reference to an object. On the other hand, the object to which it refers is *mutable*; this change is performed by the wxErlang operations—in this case, those in the wxMenu module.

The preceding "append" example illustrates two additional points:

- In the wxWidgets code, strings are wrapped in the wxT constructor; wx strings are assumed to be encoded in UTF-32 in native architecture format. This is the default for ASCII strings in Erlang, but other character sets will need to be handled explicitly.
- The C++ macro mechanism is used heavily in wx, not least in defining object identifiers such as NEW. These are also available in wxErlang through the Erlang preprocessor mechanism with the inclusion of the header file *wx.hrl*.

Finally, although most wxWidgets classes have corresponding wxErlang representations, some of the classes representing data types are mapped directly to Erlang data types. For instance, a wxPoint is represented by a pair, {Xcoord, Ycoord}, and a wxGBPosition by a {Row, Column} tuple. The full details of all data type correspondences are given in the overview section of the wxErlang documentation.

Event Handling, Object Identifiers, and Event Types

The Erlang `wx` binding allows events to be handled in two ways. They can be handled by callback functions as in wxWidgets, or they can be received as Erlang *messages*, thus integrating with the Erlang concurrent programming model. We will cover the latter mechanism here.

To understand the structure of messages, it is first necessary to understand three other aspects of events in `wx`:

Identifiers

Identifiers are integers used to uniquely identify parts of a GUI, such as windows, buttons, menu items, and so forth. wxWidgets contains a collection of standard identifiers for various common elements, such as `wxID_OPEN` and `wxID_ABOUT` (for a File Open menu item and an About box). These are available as macro definitions in wxErlang in *wx.hrl*, and are referenced with `?wxID_ABOUT`.

Use of these standards is encouraged. First, the system can identify default behaviors to associate with particular identifiers, and second, the objects associated with certain identifiers can be treated in a platform-sensitive way. We illustrate this in the blog example later in the chapter.

You can find a full listing of the standard identifiers in the *wx.hrl* header file in the wxErlang distribution, as well as in the wxWidgets documentation and book.

Event types

Events in wxErlang come in different shapes and sizes: menu selections, navigation through trees of commands, and events triggered by the passage of time or by external user gestures with a mouse or keyboard. These different kinds of events are represented by a collection of atoms, including `command_menu_selected`, `enter_window`, `close_window`, and 100+ others, known as the *type* of the event.

Depending on its type, an event will have different kinds of information associated with it. For example, `close_window` will have nothing associated with it (apart from its type), whereas an `enter_window` event will also have associated with it the position of the mouse when the event occurred (among other things). This associated information is presented in a *record*, whose type depends on the type of the event. In every case, the `type` field of the record contains the event type.

The full definition of event types and the associated records is available in the `wxEvtHandler.erl` module and its associated documentation.

Connection

For a graphical component to receive messages it must *connect* to the particular *type* of event; it is also possible to restrict this to a particular set of objects by specifying a range of identifiers.

For example, for a frame (whose reference is in the `Frame` variable) to connect to the events generated by a command being selected in a menu, you can use the following expression:

```
wxFrame:connect(Frame, command_menu_selected)
```

The **connect** operation is available to any class inheriting from `wxEvtHandler`. The implementation of this uses the wxWidgets dynamic event-handler connection mechanism.

With these definitions in place, we can now explain that messages in wxErlang take the following form:

```
#wx{id=Id, obj=Obj, userData=T, event=Rec}
```

where `Id` is an *identifier* for the graphical object receiving the event, `Obj` is a *reference* to the object that established the connection (in the earlier connection example, this would be the `Frame` object), and `Rec` is a record containing information that is dependent on the particular type of the event. In every case, the `type` field of `Rec` contains the type of the message sent.

Messages are processed in the standard way in Erlang, so the message that an "append" menu item has been selected would be treated as follows:

```
receive
  #wx{id=?APPEND, event=#wxCommand{type=command_menu_selected}} ->
     ... handler code for APPEND ...
  ... other messages ...
end
```

Further details about connections, including the way callbacks can be connected and the way events are handled by multiple handlers, is available in the overview of the wxErlang documentation and the wxWidgets online documentation and book.

Putting It All Together

To use wxWidgets, a wxErlang application will need to start and stop a `wx-server`, like so:

```
wx:new().
  ...
wx:destroy().
```

Two processes that have each created a separate server will not be able to share objects. Instead, the environment of a running process can be retrieved using `wx:get_env/0` and set in a new process using `wx:set_env/1`, allowing the processes shared access to their objects.

When **destroy** is invoked, all the memory used by the `wx` application will be reclaimed. It is possible to explicitly reclaim memory allocated to a `Class` widget using `wxClass:destroy/1` on the object. This is particularly recommended for some transitory objects, such as those that represent a dialog, as these would be stack-allocated in the wxWidgets system, but not in wxErlang.

A First Example: MicroBlog

The first example we will develop is minimal: a "micro" blogging application. It is stripped down so that all it can do is display an About box, but it will show you the principles of building a wxErlang application. In the next section, you'll see how to add the blogging functionality.

Our program is given in one file, *microblog.erl*, which begins like this:[*]

```
%% A micro-blog, which sets up a frame with menus, and allows an
%% "about" box to be displayed.

-module(microblog).
-compile(export_all).

-include_lib("wx/include/wx.hrl").

-define(ABOUT,?wxID_ABOUT).
-define(EXIT,?wxID_EXIT)
```

This shows the inclusion of the *wx.hrl* header file (located in an installation-specific position) that contains definitions of the standard identifiers and types. The local macro definitions of ABOUT and EXIT link these to the standard wx identifiers for these menu items. This will allow the application to handle them in a platform-specific way.

The top-level function for our example is start/0:

```
%% Top-level function: create the wx-server, the graphical objects,
%% show the application, process and clean up on termination.

start() ->
    wx:new(),
    Frame = wxFrame:new(wx:null(), ?wxID_ANY, "MicroBlog"),
    setup(Frame),
    wxFrame:show(Frame),
    loop(Frame),
    wx:destroy().
```

This function first creates an instance of the wx-server, and before termination ensures that it is destroyed and memory is reclaimed. The principal graphical object is a wxFrame, created with no parent object (wx:null()), with an arbitrary identifier (?wxID_ANY) and with the title "MicroBlog".

The setup function sets up the graphical objects within the frame, which is then displayed using wxFrame:show before the main processing loop is entered. Here is how the application is set up:

```
%% Top-level frame: create a menu bar, two menus, two menu items
%% and a status bar. Connect the frame to handle events.

setup(Frame) ->
```

---

[*] In releases before R13, this needs to be an explicit include of the *wx.hrl* file.

```
MenuBar = wxMenuBar:new(),
File = wxMenu:new(),
Help = wxMenu:new(),

wxMenu:append(Help,?ABOUT,"About MicroBlog"),
wxMenu:append(File,?EXIT,"Quit"),

wxMenuBar:append(MenuBar,File,"&File"),
wxMenuBar:append(MenuBar,Help,"&Help"),

wxFrame:setMenuBar(Frame,MenuBar),

wxFrame:createStatusBar(Frame),
wxFrame:setStatusText(Frame,"Welcome to wxErlang"),

wxFrame:connect(Frame, command_menu_selected),
wxFrame:connect(Frame, close_window).
```

The setup function creates a menu bar and two menus, File and Help. The About and Exit items are added to the menus, and the menus are appended to the menu bar, which is then set as the menu bar for the frame. A status bar is also added to the frame. Finally, two types of events are connected to the frame for processing: those signaling the choice of a menu item (command_menu_selected) and the close_window event.

Figure 14-1 shows the application in Mac OS X. This has the look and feel of a Mac application, with the menu appearing in the menu bar at the top of the screen, rather than at the top of the main window. The standard Mac OS X menus—Erlang (the application menu), File, Window, and Help—appear, and they contain the standard items seen in all OS X applications; this is without explicitly creating either the Erlang or the Window menu in MicroBlog.

Moreover, the About MicroBlog item appears in the application menu, consistent with the OS X GUI guidelines, despite being appended to the Help menu in setup. wx can do this because the standard identifier ?wxID_ABOUT is used to identify the About menu item. Contrast this with the Windows XP version, where the About menu item appears in the Help menu, as shown in Figure 14-2.

Figure 14-3 shows the effect of selecting About MicroBlog in Windows XP.

The final part of the code gives the main loop function:

```
loop(Frame) ->
  receive
    #wx{id=?ABOUT, event=#wxCommand{}} ->
        Str = "MicroBlog is a minimal WxErlang example.",
        MD = wxMessageDialog:new(Frame,Str,
                        [{style, ?wxOK bor ?wxICON_INFORMATION},
                         {caption, "About MicroBlog"}]),
    wxDialog:showModal(MD),
    wxDialog:destroy(MD),
    loop(Frame);

    #wx{id=?EXIT, event=#wxCommand{type=command_menu_selected}} ->
```

```
    wxWindow:close(Frame,[])
end.
```

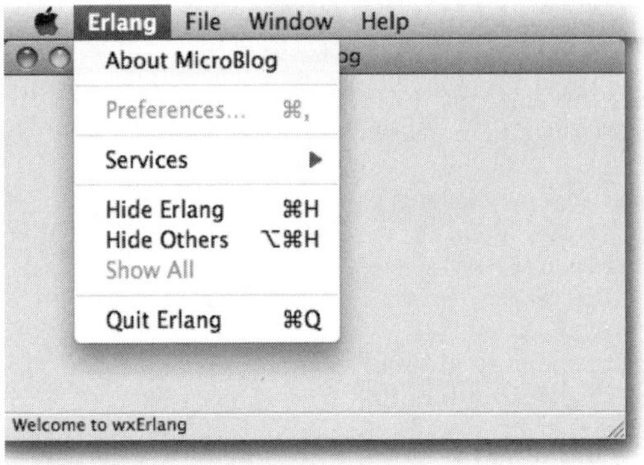

Figure 14-1. MicroBlog in Mac OS X

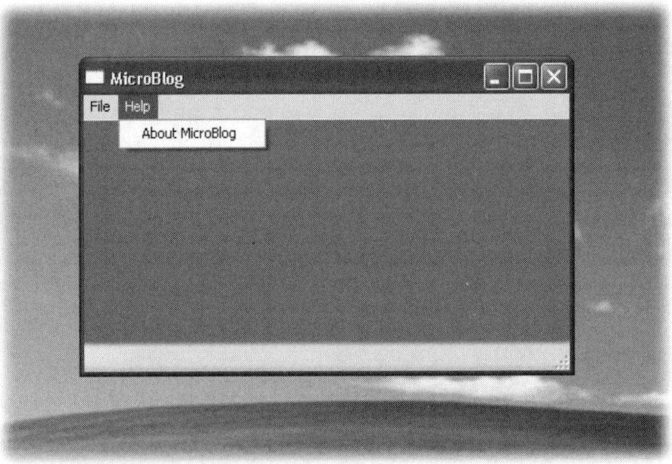

Figure 14-2. MicroBlog in Windows XP

This shows the two kinds of messages to be handled together with the processing code. Selecting the About menu item produces a message dialog, which shows information about the application in a dialog box, closed with an OK button. The dialog is shown modally so that other interactions are halted while the box is displayed; dialogs can also be shown nonmodally. Note also that the dialog MD is explicitly destroyed after it is displayed, allowing memory to be recycled at that point.

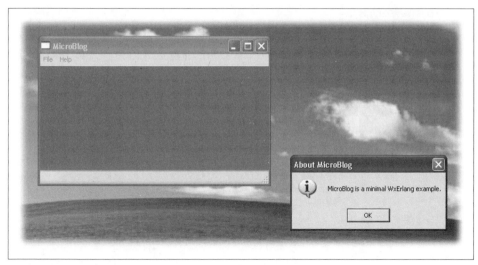

Figure 14-3. About MicroBlog in Windows XP

The MiniBlog Example

This example extends the preceding example to give a basic implementation of a "mini blog"; this forms the basis for a series of suggested extensions that will give you a chance to program in wxErlang for yourself.

The mini blog is a list of dated entries, each occupying a single line, much like a Facebook status message or a Twitter Tweet. As well as the About and Exit options, the GUI provides these operations on the blog:

New
   Creates a new, empty mini blog.

Open
   Opens the blog saved in the *BLOG* file.

Save
   Saves the current blog in the *BLOG* file, overwriting its contents if it already exists.

Add entry
   Adds an entry at the end of the blog. The entry is automatically dated.

Undo latest
   Undoes the latest "add entry"; this can be done recursively.

Figure 14-4 shows a screenshot of the system.

To describe the Erlang code for `MiniBlog` we'll explain how the `MicroBlog` code is modified. The head of the `miniblog.erl` module extends `microblog.erl` with a number of identifier macro definitions, to give a unique identifier to each menu command:

```
-define(APPEND,131).
-define(UNDO,132).
-define(OPEN,133).
-define(SAVE,134).
-define(NEW,135).
```

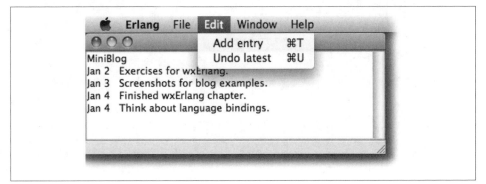

Figure 14-4. The mini blog application

The main function, `miniblog:start/0`, extends the previous function by adding a text
control (`wxTextCtrl`) that contains the entries; this `Text` object is then passed to the
`setup` and `loop` functions:

```
start() ->
  wx:new(),
  Frame = wxFrame:new(wx:null(), ?wxID_ANY, "MiniBlog"),
  Text = wxTextCtrl:new(Frame, ?wxID_ANY,
                        [{value,"MiniBlog"},
                         {style,?wxTE_MULTILINE}]),
  setup(Frame,Text),
  wxFrame:show(Frame),
  loop(Frame,Text),
  wx:destroy().
```

Note that in the construction of the text control the `Frame` is passed as the *parent object*, and also that two optional parameters are passed in the final list argument: the
initial value of the control and a parameter that sets the control to a *multiline* style.

In setting up the GUI, a number of extra menus and menu items are specified, and the
text control is set so that it cannot be edited directly (we've omitted the parts that are
unchanged):

```
setup(Frame,Text) ->
  ...,
  Edit = wxMenu:new(),
  ...,
  wxMenu:append(File,?NEW,"New\tCtrl-N"),
  wxMenu:append(File,?OPEN,"Open saved\tCtrl-O"),
  wxMenu:appendSeparator(File),
  wxMenu:append(File,?SAVE,"Save\tCtrl-S"),
```

```
wxMenu:append(Edit,?APPEND,"Add en&try\tCtrl-T"),
wxMenu:append(Edit,?UNDO,"Undo latest\tCtrl-U"),

wxMenuBar:append(MenuBar,Edit,"&Edit"),
...,

wxTextCtrl:setEditable(Text,false),
....
```

Note that the strings in the menu items contain a *mnemonic* (the letter preceded by an ampersand) and a *shortcut*, preceded by \t in the string. The shortcuts are interpreted in a platform-sensitive way so that the "undo" shortcut becomes ⌘U in Mac OS X.

The main processing loop adds a number of clauses to the **receive** statement:

```
loop(Frame,Text) ->
  receive
    #wx{id=?APPEND, event=#wxCommand{type=command_menu_selected}} ->
      Prompt = "Please enter text here.",
      MD = wxTextEntryDialog:new(Frame,Prompt,
                                 [{caption, "New blog entry"}]),
      case wxTextEntryDialog:showModal(MD) of
        ?wxID_OK ->
          Str = wxTextEntryDialog:getValue(MD),
          wxTextCtrl:appendText(Text,[10]++dateNow()++Str);
        _ -> ok
      end,
      wxDialog:destroy(MD),
      loop(Frame,Text);

    #wx{id=?UNDO, event=#wxCommand{type=command_menu_selected}} ->
      {StartPos,EndPos} = lastLineRange(Text),
      wxTextCtrl:remove(Text,StartPos-2,EndPos+1),
      loop(Frame,Text);

    #wx{id=?OPEN, event=#wxCommand{type=command_menu_selected}} ->
      wxTextCtrl:loadFile(Text,"BLOG"),
      loop(Frame,Text);

    #wx{id=?SAVE, event=#wxCommand{type=command_menu_selected}} ->
      wxTextCtrl:saveFile(Text,[{file,"BLOG"}]),
      loop(Frame,Text);

    #wx{id=?NEW, event=#wxCommand{type=command_menu_selected}} ->
      {_,EndPos} = lastLineRange(Text),
      StartPos = wxTextCtrl:xYToPosition(Text,0,0),
      wxTextCtrl:replace(Text,StartPos,EndPos,"MiniBlog"),
      loop(Frame,Text)
  end.
```

These events are handled by invoking the appropriate operations in the wxTextCtrl module. The system stores the entries literally in the text control, and saves the state in a text file, *BLOG*. Figure 14-5 shows the text entry dialog that is used to input a new blog entry in Mac OS X; note that this comes with standard buttons to accept and to cancel the insertion. Figure 14-6 shows the same text entry dialog in Windows XP.

Figure 14-5. Making a blog entry

Figure 14-6. Making a blog entry in Windows XP

We chose this example to illustrate the basic operation of wxErlang, but you can improve and extend it in a number of ways, some of which we suggest in the exercises at the end of this chapter.

Obtaining and Running wxErlang

wxErlang is part of the standard distribution of Erlang/OTP, and contains information about getting started, either with the prebuilt Mac and Windows binaries or by building the system from source. The documentation of the wxErlang API in EDoc format is also in the distribution.[†]

 When running wxErlang, you need to run Erlang with symmetric multiprocessing enabled via the -smp flag.

The wxErlang distribution contains a number of more substantial examples, including a version of Sudoku, an XRC demo, and wxErlang implementations of etop and erled.

Further documentation for wxWidgets is available online—linked from the wxErlang documentation—and in the book *Cross-Platform GUI Programming with wxWidgets*.

Exercises

wxWidgets is a large and complex GUI toolkit, and this chapter has only scratched the surface of that complexity. These exercises extend the running example from this chapter, and will require you to consult the wxErlang and wxWidgets documentation to learn details of the controls and other widgets to accomplish the tasks set forth.

Exercise 14-1: Selecting the Blog File

The existing system allows the current blog to be saved in a fixed file only. Add controls to allow a user to select a file in which to save her blog file, and to select the file to be opened when a blog is loaded.

When a file is loaded, the system should handle the case where a file does not exist.

Exercise 14-2: Saving Blog Items Separately

The state of the existing system is stored simply as the single block of text contained in the text control. Add to the system a separate backend in which the blog items are stored—together with the date they were written—so that the model of the data and the view of it presented to the user are separated.

[†] If you want to run wxErlang on Erlang releases prior to R13, it is available from the SourceForge website, *http://wxerlang.sourceforge.net*.

Exercise 14-3: Multiple Blogs in Separate Tabs

The current system gives access to a single blog at any particular time. Extend the system so that a number of blogs can be accessed at the same time, through different tabs.

Hint: one mechanism for this is to use a `wxNotebook`.

Exercise 14-4: Extending the Entries—Rich Text

The blog entries here are simply a single line of text. Explore how entries can be multiline and can include styling (e.g., using Rich Text Format).

Exercise 14-5: Tagging Entries

Provide a mechanism by which blog entries can be tagged so that those entries matching particular keywords can be shown.

Exercise 14-6: Multiple Users and Comments

The current system is designed for a single user. Investigate how user identities can be managed (using passwords), and how the system can be extended to accommodate comments on blog entries.

Exercise 14-7: Layout and wxErlang Sizers

To give a complex layout to your answers to Exercises 14-4, 14-5, and 14-6, investigate sizers in wxErlang, which you can use to lay out graphical items without explicitly setting the size of the various widgets involved.

CHAPTER 15

Socket Programming

Although distributed Erlang might be a first step in allowing programs on remote machines to communicate with each other, we sometimes have to rely on lower-level mechanisms and standardized protocols. *Sockets* allow programs written in any language to exchange data on different computers by exchanging byte streams transmitted using the protocols of the Internet Protocol (IP) Suite.

Whereas sockets are used to create a byte-oriented communication stream between programs possibly running on different machines, ports, which we cover in the next chapter, will do the same for programs running on the same machine. Byte streams, which in Erlang can be viewed as either binaries or integer lists, often follow standards and application-level protocols that allow programs written independently of each other to interact with each other.

Examples of socket-based communication include communication between web browsers and servers, instant messaging (IM) clients, email servers and clients, and peer-to-peer applications. The Erlang distribution itself is based on nodes communicating with each other through sockets.

Erlang can hide the raw packets from the user, providing user-friendly APIs to User Datagram Protocol (UDP) and Transmission Control Protocol (TCP). These are contained in the two library modules gen_udp, a connectionless, less reliable, packet-based communication protocol, and gen_tcp, which provides a connection-oriented communication channel. Both of these protocols communicate over IP.

User Datagram Protocol

User Datagram Protocol (UDP) is a connectionless protocol. If a UDP packet is sent, and a socket happens to be listening on the other end, it will pick up the packet. UDP provides little error recovery, leaving it up to the application to ensure packet reception and consistency. UDP packets could take different routes, and as a result could be received in a different order from which they were sent. They can also be lost en route, and as the receiving end does not acknowledge their arrival, their loss happens "silently." Although the protocol might not be reliable, the overhead of using it is small, making it ideal for transmissions in which you would rather drop a packet than wait for it to be re-sent. For example, errors and alarms are often broadcast in the hope that a socket on the other end picks them up.

In Erlang, UDP is implemented in the gen_udp module. Let's get acquainted with it through an example. Start two Erlang nodes on the same host and make sure you execute the commands in the following order:

1. In the first Erlang node, open a UDP socket on port 1234.
2. In the second Erlang node, open a UDP socket on port 1235.
3. Use the socket in the second node to send the binary <<"Hello World">> to the listening socket 1234 on the local host IP address 127.0.0.1.
4. Use the socket in the second node to send the string "Hello World" to the same IP address and listening socket.
5. In the first node, the process that opened (and owns) the socket should have received both of the "Hello World" messages. Retrieve them using the flush() shell command.
6. Close both sockets and thus free the port numbers.

In the Erlang shell on the first node, the commands and output would look like this:

```
1> {ok, Socket} = gen_udp:open(1234).
{ok,#Port<0.576>}
2> flush().
Shell got {udp,#Port<0.576>,{127,0,0,1},1235,"Hello World"}
Shell got {udp,#Port<0.576>,{127,0,0,1},1235,"Hello World"}
ok
3> gen_udp:close(Socket).
ok
```

You should keep in mind that once you've opened the socket, you need to send messages from the second node to the first. In the Erlang shell on the second node, the commands would look like this:

```
1> {ok, Socket} = gen_udp:open(1235).
{ok,#Port<0.203>}
2> gen_udp:send(Socket, {127,0,0,1}, 1234, <<"Hello World">>).
ok
3> gen_udp:send(Socket, {127,0,0,1}, 1234, "Hello World").
ok
4> gen_udp:close(Socket).
ok
```

Play special attention to the format of the UDP messages sent to the process that owns the socket, and the fact that it receives both messages as *lists*, even if the first message was sent as a binary. We will explain all of this when we look at the functions involved in more detail.

If you are trying the example on separate computers, you should replace the local host IP address with the address of the computer to which you want to send messages, and ensure that neither firewall is blocking the relevant ports.

As you can see in Figure 15-1, clients on other hosts send their UDP packets to a listener socket which forwards them to an Erlang process. At any one time, only one process is allowed to receive packets from a particular socket. This process is called the controlling process.

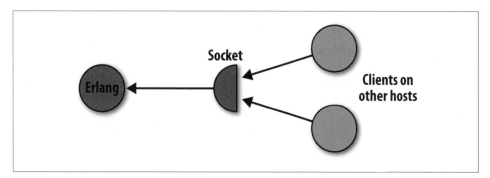

Figure 15-1. UDP listener sockets

To open a socket, on both the client and the server side, you use the following function calls:

```
gen_udp:open(Port)
gen_udp:open(Port, OptionList)
```

The `Port` is an integer denoting the listening port number of the socket. It is used by clients who need to send messages to the socket. The `OptionList` contains configuration options which allow you to override the default values. The most useful parameters include:

`list`
> Forwards all messages in the packet as a list of integers, regardless of how they are sent. It is the default value if no option is chosen.

`binary`
> Forwards all messages in the packet as a binary.

`{header, Size}`
> Can be used if packets are being received as binaries. It splits the message into a list of size `Size`, the header, and the message (a binary). This option was particularly useful before the introduction of bit syntax and pattern matching on binaries, as described in Chapter 9. Repeating the preceding two-node UDP example, but with the first socket opened using the following,
>
> ```
> {ok, Socket} = gen_udp:open(1234,[binary,{header,2}]).
> ```

and sending [0,10|"Hello World"] will result in the first message being received as follows:

```
2> flush().
Shell got {udp,#Port<0.439>,{127,0,0,1},1235,[0,10|<<"Hello World">>]}
ok
```

In the preceding code, the message is split into the (two-integer) header and the message. {active, true} ensures that all the messages received from the socket are forwarded to the process that owns the socket as Erlang messages of the form {udp, Socket, IP, PortNo, Packet}. Socket is the receiving socket, IP and PortNo are the IP address and sending socket number, and Packet is the message itself. This *active mode* is the default value when opening a socket.

{active, false}

Sets the socket to *passive mode*. Instead of being sent, messages from the socket have to be retrieved using the gen_udp:recv/2 and gen_udp:recv/3 calls.

{active, once}

Will send the first message it receives to the socket, but subsequent messages have to be retrieved using the **recv** functions.

{ip, ip_address()}

Is used when opening a socket on a computer that has several network interfaces defined. This option specifies which of the interfaces the socket should use.

inet6

Will set up the socket for IPv6. inet will set it up for IPv4, which is also the default value.

The call to open returns either {ok, Socket} or {error, Reason}, where Socket is the identifier for the socket opened and Reason is one of several POSIX error codes returned as an atom. They are listed in the inet manual page of the Erlang runtime system documentation. The most common errors you will come across are eaddrinuse if the address is already in use, eaddrnotavail if you are using a port in a range your OS has reserved, and eacces if you don't have permission to open the socket.

The gen_udp:close(Socket) call closes the socket and frees the port number allocated to it. It returns the atom ok.

If you want to send messages, you use the following function:

```
gen_udp:send(Socket, Address, Port, Packet)
```

The Socket is the UDP socket on the local machine from which the message is to be sent. The Address can be entered as a string containing the hostname or IP address, an atom containing the local hostname, or a tuple containing the integers making up the IP address. The Port is the port number on the receiving host, and the Packet is the content of the message, as a sequence of bytes, which can be either a list of integers or a binary.

When the socket is opened in passive mode, the connected process has to explicitly retrieve the packet from the socket using these function calls:

```
gen_udp:recv(Socket, Length)
gen_udp:recv(Socket, Length, Timeout)
```

Length is relevant only to the raw transmission mode in TCP, and so it is ignored in this case. If a packet has been received within the timeout, {ok, {Ip, PortNo, Packet}} is returned. If the bytes are not received within Timeout milliseconds {error, timeout} will be returned. If the receiving process calls gen_udp:recv when not in passive mode, expect to see the {error, einval} error, which is the POSIX error code denoting an invalid argument.

The most common use of UDP is in the implementation of Simple Network Management Protocol (SNMP). SNMP is a standard often used to monitor devices and systems across IP-based networks. You can read more about the Erlang SNMP application in the documentation provided with the runtime system.

Transmission Control Protocol

Transmission Control Protocol, or TCP for short, is a connection-oriented protocol allowing peers to exchange streams of data. Unlike UDP, with TCP package reception is guaranteed and packages are received in the same order they are sent. Common uses of TCP include HTTP requests, peer-to-peer applications, and IM client/server connections. Erlang distribution is built on top of TCP. Just as with UDP, neither the client nor the server has to be implemented in Erlang.

On an architectural level, the main difference between TCP and UDP is that once you've opened a socket *connection* using TCP, it is kept open until either side closes it or it terminates because of an error. When setting up a connection, you would often spawn a new process for every request, keeping it alive for as long as the request is being handled.

How does this work in practice? Say you have a listener process whose task is to wait for incoming TCP requests. As soon as a request comes in, the process that acknowledges the connection request becomes the accept process. There are two mechanisms for defining the accept process:

- The first option is to *spawn a new process* which becomes the accept process, while the listener goes back and listens for a new connection request.
- The second option, as shown in Figure 15-2, is to *make the listener process the accept process*, and spawn a new process which becomes the new listener.

If the socket is opened in active mode, the process that owns the socket will receive messages of the form {tcp, Socket, Packet} where Socket is the receiving socket and Packet is the message itself.

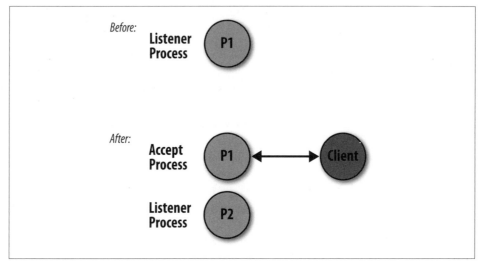

Figure 15-2. The listener and accept processes

If you are working in passive mode, just like with UDP, you need to use the following:

```
gen_tcp:recv(Socket, Length)
gen_tcp:recv(Socket, Length, Timeout)
```

The call will return a tuple of the format {ok, Packet}. In these calls, a nonzero value of Length denotes the number of bytes the socket will wait for before returning the message. If the value is 0, everything available is returned. If the sender socket is closed and fewer than Length bytes have been buffered, they are discarded. The Length option is relevant only if the packet type is raw.

> Using passive mode is a good way to ensure that your system does not get flooded with requests. It is a common design pattern to spawn a new process that handles the request for each message received. In extreme cases under heavy sustained traffic, the virtual machine risks running out of memory as the system gets flooded by requests (and hence, processes). By using sockets in passive mode, the underlying TCP buffer can be used to throttle the requests and reject messages on the client side. The best way to know whether you need to throttle on the TCP level and whether memory is an issue during traffic bursts is through extensive stress testing of your system.

A TCP Example

Let's start with a simple example of how you can use TCP sockets. The client, given a host and a binary, opens a socket connection on port 1234. Using the bit syntax, it breaks the binary into chunks of 100 bytes and sends them over in separate packets.

```
client(Host, Data) ->
    {ok, Socket} = gen_tcp:connect(Host, 1234, [binary, {packet, 0}]),
    send(Socket, Data),
    ok = gen_tcp:close(Socket).
```

You might recall from the description of binaries in Chapter 9 that the expression
<<Chunk:100/binary, Rest/binary>> will bind the first 100 bytes of the binary to Chunk
and what remains to Rest. When the binary contains fewer than 100 bytes, pattern
matching on the first clause of the send/2 call will fail. Whatever remains of the possibly
empty binary will match the second clause, and so its contents are sent to the server,
after which the Socket connection is closed.

```
send(Socket, <<Chunk:100/binary, Rest/binary>>) ->
    gen_tcp:send(Socket, Chunk),
    send(Socket, Rest);
send(Socket, Rest) ->
    gen_tcp:send(Socket, Rest).
```

The server side has a listener process waiting for a client connection. When the request
arrives, the listener process becomes the accept process and is ready to receive binaries
in passive mode. A new listener process is spawned and waits for the next connection
request. The accept process continues receiving data from the client, appending it to a
list until the socket is closed, after which it saves the data to a file.

```
server() ->
    {ok, ListenSocket} = gen_tcp:listen(1234, [binary, {active, false}]),
    wait_connect(ListenSocket,0).

wait_connect(ListenSocket, Count) ->
    {ok, Socket} = gen_tcp:accept(ListenSocket),
    spawn(?MODULE, wait_connect, [ListenSocket, Count+1]),
    get_request(Socket, [], Count).

get_request(Socket, BinaryList, Count) ->
    case gen_tcp:recv(Socket, 0, 5000) of
        {ok, Binary} ->
            get_request(Socket, [Binary|BinaryList], Count);
        {error, closed} ->
            handle(lists:reverse(BinaryList), Count)
    end.

handle(Binary, Count) ->
    {ok, Fd} = file:open("log_file_"++integer_to_list(Count), write),
    file:write(Fd, Binary),
    file:close(Fd).
```

Note how the get_request/3 function receives the binary chunks in batches of 100
bytes. Once all chunks have been received and the socket is closed, you need to reverse
the list in which you stored them, as the first chunk you should be writing is now the
last element of the list. You write the chunks to a file, and when done, you close the
socket, releasing the file descriptors.

To run the example, all you need to do is start the server using `tcp:start()` and the client using the following:

```
tcp:client({127,0,0,1}, <<"Hello Concurrent World">>).
```

You can see that many of the commands are similar to the ones we used in the earlier UDP example. The major difference is the following call:

```
gen_tcp:listen(PortNumber, Options)
```

This starts a listener socket, which then waits for incoming connections. The call takes the same options as the call to `gen_udp:open/2` described earlier, as well as the following TCP-specific ones:

`{active, true}`
Ensures that all messages received from the socket are forwarded as Erlang messages to the process that owns the socket. This active mode is the default value when opening a socket.

`{active, false}`
Sets the socket to passive mode. Messages received from the socket are buffered, and the process must retrieve them through the `gen_tcp:recv/2` and `gen_tcp:recv/3` calls.

`{active, once}`
Will set the socket to active mode, but as soon as the first message is received, it sets it to passive mode so that subsequent messages have to be retrieved using the `recv` functions.

`{keepalive, true}`
Ensures that the connected socket sends keepalive messages when no data is being transferred. As "close socket" messages can be lost, this option ensures that the socket is closed if no response to the keepalive is received. By default, the flag is turned off.

`{nodelay, true}`
Will result in the socket immediately sending the package, no matter how small. By default, this option is turned off and data is instead aggregated and sent in larger chunks.

`{packet_size, Integer}`
Sets the maximum allowed length of the body. If packets are larger than `Size`, the packet is considered invalid.

There are other flags, all of which you can read about in the manual pages of the `gen_tcp` and `inet` modules.

The `gen_tcp:listen/2` call returns immediately. It returns a socket identifier, `Socket`, which is passed to the following functions:

```
gen_tcp:accept(Socket)
gen_tcp:accept(Socket, TimeOut)
```

These calls suspend the process until a request to connect is made to that socket on that IP address. TimeOut is a value in milliseconds resulting in {error, timeout} being returned if no attempt is made to connect to that port. Connections are requested through the following call:

```
gen_tcp:connect(Address, Port, OptionList)
```

The Address is the IP address of the machine to which you are connecting, and Port is the port number of the corresponding socket. The OptionList is similar to the one defined in the gen_tcp:listen/2 call, containing the gen_udp:open/2 options together with the TCP-specific keepalive, nodelay, and packet_size discussed earlier.

As the socket in the example is running in passive mode, you retrieve the socket messages using calls to the functions gen_tcp:recv/1 and gen_tcp:recv/2. Had the sockets been running in active mode, messages would have been sent to the process in the format {tcp, Socket, Packet} and {tcp_error, Socket, Reason}.

You close the socket using the gen_tcp:close(Socket) call. This can be made on either the client or the server side. In either case, the {tcp_closed, Socket} message will be sent to the socket on the other side, effectively closing the socket.

The controlling process is generally the process that established a connection through calling one of gen_tcp:accept or gen_tcp:connect. To redirect messages elsewhere and pass the control to another process, the controlling process has to call gen_tcp:controlling_process(Socket, Pid).

In our previous example, the process calling gen_tcp:accept becomes the controlling process, and we spawned a new listener process. If instead we were to spawn a new process that would become the controlling process, with the listener process remaining the same, the code would look like this:

```
server() ->
    {ok, ListenSocket} = gen_tcp:listen(1234, [binary, {active, false}]),
    wait_connect(ListenSocket,0).

wait_connect(ListenSocket, Count) ->
    {ok, Socket} = gen_tcp:accept(ListenSocket),
    Pid = spawn(?MODULE, get_request, [Socket, [], Count]),
    gen_tcp:controlling_process(Socket, Pid),
    wait_connect(ListenSocket, Count+1).
```

In recent Erlang/OTP releases, it is possible to have multiple acceptors against the same listener socket. This could be expected to give better throughput than spawning a new acceptor each time. We leave this modification of the example as an exercise for you!

The inet Module

The inet module contains generic functions that will work with sockets regardless of whether you are using TCP or UDP. They provide generic access to the sockets as well as useful library functions. Without going into too much detail about what is available, in this section we will demonstrate the most commonly used functions by showing their use in the shell. If you need more information, you can look it up in the `inet` module's manual page. The manual page also contains all of the POSIX error definitions the socket operations will return.

If you need to change your socket options once you've started your socket, you would use the call `inet:setopts(Socket, OptionList)`, where `OptionList` is a list of tagged tuples containing the options described in this chapter together with other, less frequently used ones listed in the `inet` module's manual page.

To retrieve the configuration parameters of an existing socket, you would use `inet:getopts(Socket, Options)` where `Options` is a list of atoms denoting the option values you are interested in retrieving. The function returns a tagged list where, if the underlying operating system or the socket type you are using does not support that particular option, it will be omitted from the result.

```
1> {ok, Socket} = gen_udp:open(1234).
{ok,#Port<0.468>}
2> inet:getopts(Socket, [active, exit_on_close, header, nodelay]).
{ok,[{active,true},{exit_on_close,true},{header,0}]}
```

Sockets will gather statistics about the data they send and receive. Received counters are prefixed with **recv_**, and sent counters with **send_**. They can be retrieved for the following packets:

avg
> The average size of the packets

cnt
> The number of packets that have been sent or received

dvi
> The packet size deviation of bytes sent or received by the socket

max
> The size of the largest package

oct
> The number of bytes sent or received by the socket

In this example, our UDP socket receives four packets and sends none. The output is:

```
3> flush().
Shell got {udp,#Port<0.468>,{127,0,0,1},1235,"Hello World"}
Shell got {udp,#Port<0.468>,{127,0,0,1},1235,"Hello World"}
Shell got {udp,#Port<0.468>,{127,0,0,1},1235,"Hello World"}
Shell got {udp,#Port<0.468>,{127,0,0,1},1235,"Hello World"}
ok
4> inet:getstat(Socket).
{ok,[{recv_oct,44},
     {recv_cnt,4},
     {recv_max,11},
     {recv_avg,11},
     {recv_dvi,0},
```

```
  {send_oct,0},
  {send_cnt,0},
  {send_max,0},
  {send_avg,0},
  {send_pend,0}]}
```

Some of the functions you might find useful and should try in the shell follow. Some of them will return the `hostent` record, defined in the *inet.hrl* include file. Remember that you can load record definitions using the shell command `rr("../lib/kernel-2.13/include/inet.hrl")`.

```
inet:peername(Socket).
inet:gethostname().
inet:getaddr(Host, Family).
inet:gethostbyaddr(Address).
inet:gethostbyname(Name).
```

Finally, a useful command to know, especially if you are having problems trying to open, send, or receive data from a socket, is `inet:i()`. It lists all TCP and UDP sockets, including those that the Erlang runtime system uses as well as those you have created.

In our example, we start a distributed Erlang node. Running the command shows us two sockets—the TCP listener socket waiting for inbound connections, and a socket connected to the `epmd` port mapper daemon:

```
(bar@Vaio)1> inet:i().
Port Module    Recv Sent Owner    Local Address     Foreign Address State
108  inet_tcp 0    0    <0.62.0> *:54843           *:*             ACCEPTING
110  inet_tcp 4    18   <0.60.0> localhost:54844   localhost:4369  CONNECTED
Port Module    Recv Sent Owner    Local Address     Foreign Address State
ok
```

Further Reading

This chapter covered the low-level mechanisms on which to build more complex protocols and layers. The Inets application, which comes as part of the OTP distribution, is a container for IP-based protocol implementations. It includes a web server called Inets as well as HTTP and FTP clients. It also has a Trivial File Transfer Protocol (TFTP) client and server. For more information on the Inets application, refer to its user guide and reference manual.

A part of distributed Erlang is the Secure Sockets Layer (SSL) application, providing encrypted communication over sockets. Erlang's SSL application is based on the open source OpenSSL toolkit. You can read more about this application in the user guides and manuals that come with the Erlang distribution.

If you are interested in reading more about other Internet Protocol implementations, two good books are *Internet Core Protocols* by Eric Hall (O'Reilly) and *TCP Illustrated* by W. Richard Stevens (Addison-Wesley Professional Computing Series).

Exercises

Exercise 15-1: Snooping an HTTP Request

Open a listener socket on your local machine. Start your web browser and send it a request for a web page. Print the contents of the request and study them. How long before the socket connection is closed? What happens if you shut down your browser?

Exercise 15-2: A Simple HTTP Proxy

Change your browser proxy settings to point to your local machine on port 1500.[*] Start a listener socket on that port, and accept any connection coming to it. From your web browser, try to download any web page. The request should be forwarded to your socket connection. Sniff the contents of the request and extract the URL of the web page your browser is trying to load.

Using the HTTP client from the Inets application, retrieve the contents of the page you are trying to load and send it unchanged to the open socket connection. Hint: if you are not behind a proxy or firewall, `http:start()` and `http:request("http://www.erlang.org")` should do the job. Before using them, however, ensure that you read through the HTTP manual pages that come with the Erlang distribution.

Exercise 15-3: Peer to Peer

Write a module that contains code for a peer-to-peer transport layer. You will need a process which, when started, either waits for a socket connection to come in on port 1234 or waits for the function `peer:connect(IpAddress)` to be called. If the latter is called, it will try to connect to port 1234 on that address. Once the connection has been established, you should be able to use the function `peer:send(String)` to send data to your peer. Log what is sent to file and print it to the shell. The functions you should export are:

```
peer:start() -> ok | {error, already_started}
peer:connect(IpAddress) -> ok | {error, Reason}
peer:send(String) -> ok | {error, not_connected}
peer:stop() -> ok | {error, not_started}
```

The tricky part of this exercise, which will require some careful thought, is the fact that your process will be speaking to a copy of itself on another machine. By that, we mean both processes will be running the same code base.

If you are worried about Big Brother watching you, you can encrypt the packets you send using the `crypto` module.

---

[*] Depending on the rights of the user under which you are running your Erlang node, you might not be able to open ports that are either reserved or already taken. If that is the case, pick a higher number.

CHAPTER 16

Interfacing Erlang with Other Programming Languages

It is common for modern computer systems of any size to be built using more than one programming language. Device drivers are typically written in C, and many integrated development environments (IDEs)—such as Eclipse—and other GUI-heavy systems are written in Java or C#. Lightweight web apps can be developed in Ruby and PHP, and Erlang can provide lightweight, fault-tolerant concurrency. If you need to efficiently manipulate or parse strings, Perl or Python is the norm. The library that solves a particular problem for you may not be written in your favorite language, and you must choose whether to use the foreign library or bite the bullet and recode the whole thing in Erlang yourself.[*]

Interlanguage communication is never simple in natural languages or in programming. In natural languages, we must understand the different ways in which the languages work. Do they contain articles? Do they denote gender? Where do the verbs occur in a sentence? We also must understand how words translate. Does the verb *ser* in Portuguese mean the same as "to be" in English, for instance? (It doesn't.) It's the same for programming languages. Which paradigm do they come from? Are the languages functional, object-oriented, concurrent, or structured? Is an integer in Java the same thing as an integer in Erlang? (It isn't!)

Interoperation is not only about interlanguage communication, and Erlang/OTP also supports communication by means of XML, ODBC, CORBA, ASN, and SNMP. These assist in Erlang's growing role as the "distributed glue" joining together single-threaded legacy programs.

---

[*] This is done for the duplicate code detection algorithm in Wrangler, the Erlang refactoring tool. An existing efficient C library is used to identify candidate "clones" in Erlang software.

An Overview of Interworking

Erlang provides a number of mechanisms for interlanguage working: a higher-level model built on *distributed Erlang nodes*, a lower-level model allowing communication with an external program through a *port*, and a mechanism for linking programs into the virtual machine (VM) itself, known as *linked-in drivers*.

The Erlang distributed programming model provides a simple and flexible solution to the high-level question of how other languages can work with Erlang: run the language in other nodes on the same or different machines, making it appear like a distributed Erlang node to which you pass messages back and forth. These nodes provide an environment where foreign programs can be run, yet also communicate with Erlang nodes. To make this communication work, it is possible to use ports or alternatively provide a higher-level model of the Erlang communication primitives in this other language. In either case, there's a question of how to deal with the elementary types, and different degrees of support for translating between complex data in the two languages can be provided.

In this chapter, we'll discuss how to build nodes in Java and in C that can interoperate with Erlang, and we'll introduce `erl_call`, which allows the Unix shell to communicate with a distributed Erlang node, built on the `erl_interface` library. This, together with the `JInterface` Java package, comes with the standard Erlang distribution. These libraries offer stability of code and architecture, at some sacrifice in absolute speed. After this, we will describe how to communicate via ports, and we'll give an example of how to interact with Ruby, using the `erlectricity` library.

Working with Other Languages

We cover Java, C, and Ruby in this chapter. However, it is possible to link Erlang with a number of other programming languages, including the following:

- OTP.NET, which provides a link to the .NET platform, through a port of the `JInterface` code.

- Py-Interface, which is a Python implementation of an Erlang node, allowing communication between Python and Erlang.

- The Perl Erlang-Port, which allows Perl code to communicate with Erlang through a port.

- PHP/Erlang, which aims to be a PHP extension with a simple set of functions for turning a PHP thread into an Erlang C node.

- The Haskell/Erlang-FFI, which enables full bidirectional communication between programs written in Haskell and Erlang. Messages sent from Haskell to Erlang look like function calls, and messages from Erlang to Haskell are delivered to `MVars`.

- An Erlang/Gambit interface, which allows Scheme and Erlang programs to communicate.
- Distel supports the interoperation of Emacs Lisp and Erlang, providing enhancements to the Erlang mode in Emacs.

To gain the greatest efficiency in interoperation, you can define a linked-in driver. The problem is that an erroneous linked-in driver will cause the entire Erlang runtime system to leak memory, hang, or crash, so you should use linked-in drivers with extreme care.

Interworking with Java

The `JInterface` Java package provides a high-level model of Erlang-style processes and communication in Java. You can use this package alone to give Erlang-style concurrency in Java, or you can use it as part of a mixed Java/Erlang distributed system, allowing a Java system to contain Erlang components or vice versa.

A Java package such as `JInterface` consists of a collection of Java classes, mostly beginning with the prefix `Otp`. This section will describe the most common of them, providing examples of passing messages and handling data types when communicating between Erlang and Java. You can find additional information on `JInterface` and its Erlang classes in the "Interface and Communication Applications" section of the Erlang/OTP documentation. The running example for this section is a rework of the remote procedure call (RPC) example in Chapter 11.

Nodes and Mailboxes

We described Erlang nodes in Chapter 11, where we also introduced distributed programming in Erlang. An Erlang node is identified by its name, which consists of an identifier with a hostname (in short or long form); each host can run a number of nodes, if their names are different.

The `OtpNode` class gives the `JInterface` representation of an Erlang node:

```
OtpNode bar = new OtpNode("bar");
```

This creates the Java object `bar`—which we'll call a *node*—that represents the Erlang node bar, running on the host where the statement is executed.

You can create a process on this node by creating a *mailbox*, which is represented by a pid or can be registered to a name. To create the process, use:

```
OtpMbox mbox = bar.createMbox();
```

which gives the process a name on creation. Then, pass in the name as a string on construction:

```
OtpMbox mbox = bar.createMbox("facserver");
```

You can also do this separately from the creation of the mailbox using the following statement:

```
mbox.registerName("facserver");
```

This creates a named process called `facserver`. The process will act as a "factorial server," sending the factorial of the integers that it receives to the processes that have sent the integers.

Once you've named the mailbox, you can access it using its name. If its pid is also required—perhaps by a remote Erlang node—use the `self` method on the mailbox:

```
OtpErlangPid pid = mbox.self();
```

Representing Erlang Types

The `JInterface` package contains a variety of classes that represent, in Java, various Erlang types. Their methods allow the conversion of native Java types to and from these representation types, supporting the conversion of values between the two languages, which is essential for them to work together effectively.

You saw an example of this in the preceding Java statement: the class `OtpErlangPid` gives the Java representation of an Erlang pid. The mapping between Erlang types of atoms, binaries, lists, pids, ports, refs, tuples, and terms is to the corresponding Java class `OtpErlangAtom`, ..., `OtpErlangTuple`, `OtpErlangObject`.

Floating-point types in Erlang are converted to either `OtpErlangFloat` or `OtpErlangDouble`; integral types are converted to `OtpErlangByte`, `OtpErlangChar`, `OtpErlangShort`, `OtpErlangInt`, `OtpErlangUInt`, or `OtpErlangLong`, depending on the particular integral value and sign.

To represent the two special atoms `true` and `false`, there is the `OtpErlangBoolean` class, and Erlang strings—which are lists of integers in Erlang—are described by `OtpErlangString`.

The details of these classes are in the `JInterface` documentation; we will use the classes in the next section as well as in the RPC example.

Communication

Erlang processes send and receive messages, and in `JInterface`, these operations are provided by the `send` and `receive` methods on a mailbox. The interchanged messages are Erlang terms, and are therefore represented by `OtpErlangObject`s in Java. The following Erlang `send` message:

```
Pid ! {ok, M}
```

in the `mbox` process is given by:

```
mbox.send(pid,tuple);
```

where the `pid` variable in Java corresponds to the `Pid` variable in Erlang, and `tuple`[†] represents the Erlang term {ok, M}.

A message is received by:

```
OtpErlangObject o    = mbox.receive();
```

This statement differs from an Erlang `receive` in that it performs no pattern matching on the message. Deconstruction and analysis of the message follow separately; we'll show this in the next example.

Putting It Together: RPC Revisited

The following Erlang code sets up a factorial server on the node called `bar` on the host STC:

```
setup() ->
  spawn('bar@STC',myrpc,server,[]).

server() ->
  register(facserver,self()),
  facLoop().

facLoop() ->
  receive
    {Pid, N} ->
      Pid ! {ok, fac(N)}
  end,
  facLoop().
```

The server receives messages of the form {Pid, N} and sends the result {ok, fac(N)} back to the `Pid`. The next code sample accomplishes the same thing in Java:

```
1  import com.ericsson.otp.erlang.*;  // For the JInterface package
2  import java.math.BigInteger;       // For factorial calculations
3
4  public class ServerNode {
5
6      public static void main (String[] _args) throws Exception{
7
8          OtpNode bar = new OtpNode("bar");
9          OtpMbox mbox = bar.createMbox("facserver");
10
11          OtpErlangObject o;
12          OtpErlangTuple  msg;
13          OtpErlangPid    from;
```

[†] If you've been programming Erlang for a few years and you react at variables such as `pid` and `tuple` not being capitalized, you are not alone. What is important is that in the process, you do not openly question how methods taking atoms as parameters actually work, ensuring that no one picks up on your blunder.

```
14      BigInteger    n;
15      OtpErlangAtom  ok = new OtpErlangAtom("ok");
16
17      while(true) try {
18          o    = mbox.receive();
19          msg  = (OtpErlangTuple)o;
20          from = (OtpErlangPid)(msg.elementAt(0));
21          n    = ((OtpErlangLong)(msg.elementAt(1))).bigIntegerValue();
22          OtpErlangObject[] reply = new OtpErlangObject[2];
23          reply[0] = ok;
24          reply[1] = new OtpErlangLong(Factorial.factorial(n));
25          OtpErlangTuple tuple = new OtpErlangTuple(reply);
26          mbox.send(from,tuple);
27
28      }catch(OtpErlangExit e) { break; }
29  }
30 }
```

In the preceding example, the concurrent aspects are shown in bold in lines 8, 9, 18, and 26; the remaining code is used to analyze, deconstruct, and reconstruct data values, as well as providing the control loop.

The main program starts by running a node **bar**, and a process **facserver** on that node. In the main loop, lines 18–26, a message is received and replied to. The message received is an Erlang term, that is, an **OtpErlangObject**. This is *cast* to an **OtpErlangTuple** in line 19, and from this the pid of the sender (line 20) and the integer being sent (line 21) can be extracted. In line 21, the Erlang value is extracted as a long integer, but is converted to a Java **BigInteger** to allow an accurate calculation of the factorial.

The remainder of the code (lines 22–26) constructs the reply tuple and sends it. Line 22 constructs an array of objects, containing the (representation of the) atom **ok** (line 23) and the return value **factorial(n)** (line 24). This is then converted into a tuple (line 25), before finally being sent back to the client in line 26.

Interaction

To interact with the running Java node, you can use the following code, calling **myrpc:f/1** at the prompt:

```
-module(myrpc).
...
f(N) ->
  {facserver, 'bar@STC'} ! {self(), N},
    receive
      {ok, Res} ->
        io:format("Factorial of ~p is ~p.~n", [N,Res])
    end.
```

This client code is exactly the same as the code that is used to interact with an Erlang node, and a "Turing test"[‡] that sends messages to and from a node should be unable to tell the difference between a Java node and an Erlang node.

The Small Print

In this section, we will explain how to get programs using JInterface to run correctly on your computer.

First, to establish and administer connections between the Java and Erlang nodes it is necessary that epmd (the Erlang Port Mapper Daemon) is running when a node is created. You will recall epmd from Chapter 11. You can run it simply by typing **epmd** (**epmd.exe** on Windows), but you can test whether it is already running by typing the following:

```
epmd -names
```

This will list all the names of the running Erlang nodes on the host. This command is useful for checking whether a node you think should be running actually is running.

The system will create a node with the default cookie if none is supplied when the node is started. This may be OK, but if you need to create a node with a given cookie, use the following:

```
OtpNode bar = new OtpNode("bar", "cookie-value");
```

If a particular port needs to be used, this is the third argument of a three-argument constructor.

Referring back to the program in the section "Putting It Together: RPC Revisited" on page 339, line 1 of the program ensures that the JInterface Java code is imported, but since it is included in the OTP distribution and not in the standard Java, it is necessary to point the Java compiler and runtime to where it is held, which is in the following:

```
<otp-root>/jinterface-XXX/priv/OtpErlang.jar
```

In the preceding code, <otp-root> is the root directory of the distribution, given by typing code:root_dir() within a running node, and XXX is the version number. On Mac OS X the full path is:

```
/usr/local/lib/erlang/lib/jinterface-1.4.2/priv/OtpErlang.jar
```

This value is supplied thus to the compiler:

```
javac -classpath ".:/usr/local/lib/erlang/lib/
   jinterface-1.4.2/priv/OtpErlang.jar" ServerNode.java
```

[‡] The Turing test was proposed by mathematician and computing pioneer Alan Turing (1912–1954) as a test of machine intelligence. The idea, translated to modern technology, is that a tester chats with two "people" online, one human and one a machine: if the tester cannot reliably decide which is the human and which is the machine, the machine can be said to display intelligence.

and to the Java system:

```
java -classpath ".:/usr/local/lib/erlang/lib/
    jinterface-1.4.2/priv/OtpErlang.jar" ServerNode
```

Taking It Further

The `JInterface` library has more extensive capabilities than you have seen so far:

- It is possible to link to and unlink from Java processes using the `link` and `unlink` methods on `OtpMbox`.

- The example relies on connections between nodes being made automatically. You can use the `ping` method on a node to test whether a remote node exists; if it does, a connection is made automatically.

- Arbitrary data can be sent between nodes using binary data, manipulated by the `OtpErlangBinary` class.

- The `OtpConnection` class provides a higher-level mechanism for RPC, just as the `rpc` module does for Erlang.

- Methods on the `OtpConnection` class also support control of tracing.

These and other features are described in more detail in the online documentation.

C Nodes

The `erl_interface` library provides C-side functionality for constructing, manipulating, and accessing C encodings of Erlang binary terms. Also included are functions for dealing with memory allocation in term (de)construction and manipulation, accessing global names, and reporting errors. In a little more detail:

`erl_marshal`, `erl_eterm`, `erl_format`, and `erl_malloc`

> For handling the Erlang term format, including memory management. In particular, these provide conversion to and from C structs similar to Erlang terms, allowing higher-level manipulation of data.

`erl_connect` and `ei_connect`

> For providing a connection with Erlang through a distributed Erlang node.

`erl_error`

> For printing error messages.

`erl_global`

> For providing access to globally registered names.

`registry`

> For providing the facility to store and back up key-value pairs. This provides some of the functionality of ETS tables and can be backed up or restored from a Mnesia table on a linked Erlang node.

In addition, the *Erlang external term format* is a representation of an Erlang term as a sequence of bytes: a binary. You can convert between the two representations in Erlang using the BIFs **term_to_binary/1** and **binary_to_term/1**. We discuss this in more detail in "Port Programs" on page 346.

In this section, we'll revisit the example of the factorial server just given for Java, this time in C, based on the example in the online Interoperability Tutorial for Erlang:

```
1 /* fac.c */
2
3 #include <stdio.h>
4 #include <sys/types.h>
5 #include <sys/socket.h>
6 #include <netinet/in.h>
7
8 #include "erl_interface.h"
9 #include "ei.h"
10
11 #define BUFSIZE 100
12
13 int main(int argc, char **argv) {
14    int fd;                         /* file descriptor of Erlang node */
15
16    int loop = 1;                   /* Loop flag                      */
17    int got;                        /* Result of receive             */
18    unsigned char buf[BUFSIZE];     /* Buffer for incoming message    */
19    ErlMessage emsg;                /* Incoming message               */
20
21    ETERM *fromp, *argp, *resp;     /* Representations of Erlang terms */
22    int res;                        /* Result of the fac call          */
23
24       /* initialize erl_interface (once only) */
25    erl_init(NULL, 0);
26
27       /* initialize the connection mechanism  */
28    if (erl_connect_init(1, "mycookie", 0) == -1)
29      erl_err_quit("erl_connect_init");
30
31       /* connect to a running Erlang node     */
32    if ((fd = erl_connect("blah@STC")) < 0)
33      erl_err_quit("erl_connect");
34
35    while (loop) {
36        /* message received */
37        got = erl_receive_msg(fd, buf, BUFSIZE, &emsg);
38
39 if (got == ERL_TICK) {
40        /* ignore */
41      } else if (got == ERL_ERROR) {
42        loop = 0;
43      } else {
44        if (emsg.type == ERL_REG_SEND) {
45        /* unpack message fields          */
46          fromp = erl_element(1, emsg.msg);
47          argp = erl_element(2, emsg.msg);
```

```
48
49        /* call fac and send result back */
50          resp = erl_format("{ok, ~i}", fac(ERL_INT_VALUE(argp)));
51          erl_send(fd, fromp, resp);
52
53        /* free the term storage used      */
54          erl_free_term(emsg.from); erl_free_term(emsg.msg);
55          erl_free_term(fromp); erl_free_term(argp);
56          erl_free_term(resp);
57 } } } }
58
59 int fac(int y) {
60   if (y <= 0)
61      {return 1;}
62   else
63      {return (y*fac(y-1));};};
64 }
```

The general shape of the C code is similar to the Java node earlier, except that the C code has more lower-level operations, such as the following:

- Library inclusions for C (lines 3–6) and the Erlang interface (lines 8 and 9)
- Allocation of memory for the input buffer (lines 11 and 18)
- Freeing the storage allocated to Erlang terms in the C code (lines 53–56)

The node is set up and connected to an Erlang node in lines 24–33:

- `erl_init(NULL, 0)` initializes the `erl_interface` library, and must be called only once in any program.
- `erl_connect_init(1, "mycookie", 0)` initializes the connection mechanism, including the *identification number* of the node (1 here) and the cookie that it is to use.
- `fd = erl_connect("blah@STC")` connects to the Erlang node `blah@STC` and returns a file descriptor `fd` for the connection.

The `loop` in lines 35–57 will loop forever, reading a message (line 37) using the following line of code:

```
got = erl_receive_msg(fd, buf, BUFSIZE, &emsg);
```

The preceding line of code will receive a message in the buffer `buf` and decode it into an Erlang term, `emsg`. `ERL_TICK` messages that check whether the node is alive are ignored (line 40), and the loop terminates on receiving an `ERL_ERROR` message (line 42). Otherwise, the functional part of the `loop` body will do the following:

- Extract the pid of the message sender, `fromp`, and the payload, `argp` (lines 46 and 47).
- Convert the `argp` into a C integer, pass it to the factorial function, and return the Erlang term {ok, fac(...(argp))} to the `fromp` process (line 51). The `erl_format` call uses a format string to construct Erlang terms in a readable way. To construct the same term manually, you could write:

```
arr[0] = erl_mk_atom("ok");
arr[1] = erl_mk_integer(fac(ERL_INT_VALUE(argp)));
resp   = erl_mk_tuple(arr, 2);
```

- Finally, the storage used for the Erlang terms and subterms is cleaned up (lines 54–56).

To compile this C program, you have to make sure the *erl_interface.h* file and the `liberl_interface.a` and `libei.a` libraries are used. You can do this (on Mac OS X) using the following command:

```
gcc -o fac -I/usr/local/lib/erlang/lib/erl_interface-3.5.9/include \
-L/usr/local/lib/erlang/lib/erl_interface-3.5.9/lib fac.c -lerl_interface -lei
```

In the preceding code, the italicized path gives the path to the latest version of `erl_interface` on your system. This compiles the C code as the executable **fac** in the current directory.

The Erlang code to connect to the C node is given next. In general, the name of the C node is `cN`, where N is the identification number for the node, so we use `c1@STC` here:

```
-module(fac).
-export([call/1]).

call(X) ->
  {any, 'c1@STC2'} ! {self(), X},
  receive
    {ok, Result} ->
      Result
  end.
```

Calling this in the Erlang shell gives the following:

```
% erl -sname "blah" -setcookie "mycookie"
...... at this point the C executable should be called ......
(blah@STC2)1> c(fac).
{ok,fac}
(blah@STC2)2> fac:call(7).
5040
(blah@STC2)3> fac:call(0).
1
```

In the example, where the C node is acting as a client, the Erlang node needs to be launched *first* so that it is already running when the C node attempts to connect to it.

Going Further

The C node you just saw is running as a client: it can make connections to Erlang nodes. It can also run as a server mode; this requires first that the program creates a socket—listening on a particular port number—and then that it publishes the socket by means of **epmd**. This program can then accept connections from Erlang nodes.

The `erl_interface` library provides various other facilities, such as pattern matching on incoming messages, the registry system for storing key-value pairs, and a global naming scheme. All of these, plus the server-style node, are covered in the Interoperability Tutorial and in the user's guide for the `erl_interface` library.

Erlang from the Unix Shell: erl_call

One of the "hidden gems of OTP" is the `erl_call` Unix command, built using `erl_interface` to provide communication with a distributed Erlang node. As well as to start and communicate with a node, you can use this to compile and evaluate Erlang code from the command line. The ability to read from `stdin` allows other scripts to use this, such as those in the CGI bin.

You present arguments and options to the command via a series of flags. The full set is described in the manpage, or summarized by calling `erl_call` with no flags. One of the flags `-n`, `-name`, or `-sname` is required, as these flags are used to specify the name (or short name) of the node to be called. Often, this is accompanied by `-s`, which will start the node if it is not already running.

The `-a` flag is the analog of `apply/3`, with arguments in a similar format, whereas `-e` evaluates what comes from standard input (up to `Ctrl-D`). Here is the `erl_call` command in action:

```
% erl_call -s -a 'erlang date' -n blah
{2009, 3, 21}
% erl_call -s -e -n blah
X=3,
Y=4,
X+Y.
Ctrl-D
{ok, 7}
% erl_call -a 'erlang halt' -n blah
%
```

Port Programs

An Erlang *port* allows communication between an Erlang node and an external program through *binary messages* sent to and from an Erlang process running in the node—known as the *connected process* of the port—and the external program, running in a separate operating system thread (see Figure 16-1). The wxErlang binding to wxWidgets, described in Chapter 14, uses ports.

One of the simplest applications of a port is the `os:cmd/1` function, which can call an operating system command from inside the Erlang shell:

```
1> os:cmd("date").
"Sat 21 Mar 2009 18:11:24 GMT\n"
```

The port is made to behave like an Erlang process that is not trapping exits. Connected processes can link to it, as well as send and receive Erlang messages and exit signals. The mechanism underlying the binary communication depends on the operating

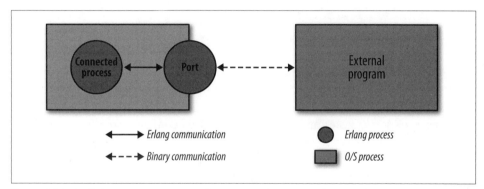

Figure 16-1. An Erlang port, its connected process, and an external program

system: for instance, on Unix-based systems, communication will be through pipes. On the Erlang side, the template for a port-based interaction is given by the following:

```
Port = open_port({spawn, Cmd}, ...),
 ...
port_command(Port, Payload),
 ...
receive
  {Port, {data, Data}} ->
 ...
```

In this fragment, the port is opened by the call to **open_port/2**, returning the *port identifier*, **Port**. Data is *sent* to the **Port** (and on to the external program) through the call to **port_command(Port,...)** in the connected process, and data is *received* from the **Port** in a **receive** clause matching data of the form **{Port, {data, Data}}**.

We've just given you a top-level summary of how ports work. The remainder of this section looks in more detail at the Erlang commands that control ports, as well as the way data is coded and decoded for communication to the external program. Finally, we'll show how you can write external programs in Ruby and C to communicate with Erlang through ports.

Erlang Port Commands

To open an Erlang port, you use the **open_port/2** command:

```
open_port({spawn, Cmd}, Options)
```

This will run the command **Cmd** as an external program; this external program is spawned with the given list of **Options**. Here is a list of the main options available (you can find a complete list in the documentation for the **erlang** module):

{packet, N}
　　This gives the size of the binary packets to be used for this port. **N** can take the value 1, 2, or 4. Under this option, packets are preceded by their size. If variable-sized packets are to be sent, you should use the **stream** option instead.

binary

All I/O from the port comprises binary data objects rather than bytes.

use_stdio

This uses the (Unix) standard input and output for communication with the spawned process; to avoid this, the `nouse_stdio` option is available.

exit_status

This ensures that a message is sent to the port when the external program exits; details are given in the online documentation.

For example, to run a Ruby program, `echoFac.rb`, the following commands need to be executed:

```
Cmd = "ruby echoFac.rb",
Port = open_port({spawn, Cmd}, [{packet, 4}, use_stdio, exit_status, binary]),
```

After these commands are executed, the variable `Port` contains the port identifier for the spawned Ruby program.

The connected process can communicate with the `Port` using `port_command/2`. Executing the following command in the connected process will send the `Data` to the `Port`:

```
port_command(Port, Data)
```

The message sent has the form `{Port, {data, Data}}`.

 A port identifier such as `Port` gives any Erlang process access to the port, and thus to the external program attached to the port. Any process can use this, but it is strongly recommended that instead of direct communication using `Port!...`, which fails if it is called from any process other than the port owner, *all communication should use* `port_command/2`.

To connect a process with pid `Pid` to a port `Port`, the following call must be executed:

```
port_connect(Port, Pid)
```

This can be called by any process, but the old port owner will stay linked to the `Port`; the owner will need to call `unlink(Port)` to remove the link.

To close a port—and therefore to terminate communication with the external program —the connected process needs to execute the command `port_close(Port)`.

To show these examples in action, here is a small Erlang program that calculates the factorial of 23 by sending the argument to Ruby and having it calculated in the Ruby program `echoFac.rb`, based on an echo example in the **erlectricity** library, discussed in the section "Working in Ruby: erlectricity" on page 351.

In the following example, the communication primitives are highlighted:

```
-module(echoFac).
-export([test/0]).

test() ->
```

```
        Cmd = "ruby echoFac.rb",
        Port = open_port({spawn, Cmd}, [{packet, 4}, use_stdio, exit_status, binary]),
        Payload = term_to_binary({fac, list_to_binary(integer_to_list(23))}),
        port_command(Port, Payload),
        receive
          {Port, {data, Data}} ->
            {result, Text} = binary_to_term(Data),
            Blah = binary_to_list(Text),
            io:format("~p~n", [Blah])
        end.
```

Running the test/0 function results in the following behavior:

```
1> echoFac:test().
"23!=25852016738884976640000"
ok
2>
```

We explain the remaining parts of the program in the next section.

Communicating Data to and from a Port

Communication through a port uses binary data; therefore, program data needs to be converted to binary in some way or another before communication, and decoded on receipt. The bit syntax, described in Chapter 9, can be used here, as can a number of coding and decoding functions provided in the erlang module, including the following:

term_to_binary/1

> This converts the argument to a binary data object that encodes its argument according to the Erlang binary term format. This can be communicated through the Port if it is created with the binary option.

binary_to_term/1

> This is the inverse of term_to_binary/1.

list_to_binary/1

> This will return a binary that is composed of the integers and binaries in the argument. For example:

```
> list_to_binary([<<1,2,3>>,1,[2,3, <<4,5>>],4| <<6>>]).
<<1,2,3,1,2,3,4,5,4,6>>
```

binary_to_list/1

> This is the inverse of list_to_binary/1.

Returning to the earlier example:

```
test() ->
  Cmd = "ruby echoFac.rb",
  Port = open_port({spawn, Cmd}, [{packet, 4}, use_stdio, exit_status, binary]),
  Payload = term_to_binary({fac, list_to_binary(integer_to_list(23))}),
  port_command(Port, Payload),
  receive
    {Port, {data, Data}} ->
```

```
        {result, Text} = binary_to_term(Data),
        Blah = binary_to_list(Text),
        io:format("~p~n", [Blah])
    end.
```

The highlighted lines show the conversion in both directions. In creating the Payload, the integer 23 is converted to the string "23" by integer_to_list, and then to a binary <<"23">>, which in turn is paired with the atom fac and the pair is coded as a binary.

In the receive clause, the message from the Port is received in the standard form, {Port, {data, Data}}. The Data is decoded to a term of the form {result, Text}, and then the Text can itself be decoded using binary_to_list. This allows the data to be output to the terminal.

File Descriptors, Ports, and I/O

Imagine an Erlang node that receives HTTP posts with embedded XML that are parsed through a linked-in driver. If you start receiving thousands of requests per second, you will quickly reach the limit of simultaneously allowed open file descriptors.

In your start script (or in the environment running the Erlang process), run the command ulimit -n Max, where Max is the maximum number of simultaneously allowed open file descriptors. Default values and maximum values are OS-dependent. Remember that every port consists of two file descriptors, one for reading and another for writing.

Another optimization is the +A Size flag passed to your erl command when starting the emulator. It will speed your application by increasing the number of asynchronous threads in the virtual machine that handle file I/O. Size is an integer between the default value of 0 and 1,024.

Library Support for Communication

As the example in the preceding section shows, it is pretty tedious to encode and decode the data yourself. Of course, you have the flexibility to choose efficient communication protocols between your external program and an Erlang node, but reinventing an encoding for each program is not the best way to do things.

If you do not want to go down that route, Erlang comes with libraries to support communication with the outside world, and particularly with C and Java. We covered the Java library at the beginning of the chapter; we'll discuss interfacing with Ruby in the remainder of the chapter.

Working in Ruby: erlectricity

`erlectricity` is a Ruby library—a "gem"—that you can download and use with Ruby. It is available as source code from *http://github.com/mojombo/erlectricity*, and you can install it in Ruby using:

```
$ gem install mojombo-erlectricity -s http://gems.github.com
```

You can include the `erlectricity` library in a Ruby program by using a `require` statement at the head of the file.

At the heart of the library is the `receiver.rb` program, which provides the functionality to allow messages to be received by and sent from a Ruby program. This in turn depends on the implementation of ports in `port.rb` and matching in `matcher.rb`; coding and decoding data formats are provided in `encoder.rb` and `decoder.rb` for a variety of different Erlang types.

The library comes with a test suite, as well as examples, including a daemon that links Erlang to Campfire, via the Ruby implementation of the Campfire API, Tinder.

An example using erlectricity

The `erlectricity` library for Ruby provides support for communication with Erlang processes through ports. Here is the Ruby side of the system, the `echoFac.rb` program that communicates with `echoFac.erl`:

```ruby
require 'rubygems'
require 'erlectricity'
require 'stringio'

def fac n
  if (n<=0) then 1 else n*(fac (n-1)) end
end

receive do |f|
  f.when(:fac, String) do |text|
    n = text.to_i
    f.send!(:result, "#{n}!=#{(fac n)}")
    f.receive_loop
  end
end
```

The working part of this program is the `receive` method: on receipt of a message `f` this is matched with {`fac, text`}, where `text` is a `String`. If this match is successful, the text is converted to an integer (by calling the `to_i` method on it) and the following message is sent to the port from which the message came:

```ruby
{:result ,"#{n}!=#{(fac n)}"}
```

In a Ruby string, the construct #{...} surrounds an expression to be evaluated. For example, if the variable n has value 6, then "#{n}!=#{(fac n)}" will be the string "6!=720".

As we said earlier, the result of running echoFac:test() is:

```
1> echoFac:test().
"23!=25852016738884976640000"
ok
```

Linked-in Drivers and the FFI

By default, an external program connecting to Erlang will run in a separate operating system process. This isolates the two parts of the system so that a crash in the external program will not affect the Erlang program, but it has the disadvantage of making it harder to meet certain lower-level, real-time requirements. To meet such requirements, it is possible to run an external program in the same thread as the Erlang system; this is called a *linked-in driver*.

Details of how to build linked-in drivers, including the callback functions that need to be implemented by such drivers, are provided in Chapter 6 of the online Interoperability Tutorial. Communication with linked-in drivers uses ports and the same BIFs as used in port programs.

When using linked-in drivers, your program will execute very, very quickly. But this speed comes at a price, because when things go wrong, they go very, very wrong. A crash or a memory leak in your linked-in driver will result in the Erlang VM crashing. This is in contrast to ports, where the port is closed and an EXIT signal is sent to the connected process.

Use linked-in drivers with extreme care, keep them simple, and integrate them only when performance is critical—for example, when integrating with an external system such as Berkeley DB, or integrating a system call such as sendfile in the Yaws web server.

The Erlang Enhancement Proposal (EEP) process[§] is a mechanism for the Erlang community to propose enhancements to the language, and for the enhancements to be incorporated into the standard distribution. EEP7 is a proposal for a *foreign function interface (FFI)*, which offers the promise of more reliable development of linked-in drivers. In the meantime, there are toolkits, such as the Erlang Driver Toolkit (EDTK) and Dryverl, which aim to support the development of linked-in drivers for Erlang.

[§] *http://www.erlang.org/eeps/*

Exercises

Exercise 16-1: C Factorial via a Port

Write an implementation of the earlier factorial server example in C using a port, rather than a distributed Erlang node.

Exercise 16-2: Factorial Server in Another Language

Reimplement the factorial server example using the interface to one of the following languages: Scheme, Haskell, C#, Python, or PHP.

CHAPTER 17

Trace BIFs, the dbg Tracer, and Match Specifications

Any respectable programming language that has deployments consisting of millions of lines of code running in thousands of installations worldwide must provide built-in low-level tracing mechanisms on which to build tools that can be used for live troubleshooting. Languages that don't provide these tools put a huge burden on developers and support engineers alike, as they have to either develop this infrastructure from scratch themselves or troubleshoot their systems in a black-box environment.

In Erlang, Ericsson's experiences of tracing live telephony switches are reflected in the *trace BIFs*, which, from being part of the first version of the language, have evolved through the years to become the foundation for a set of tools that give full visibility to the changing state of the system and, as a result, drastically reduce bug resolution times and troubleshooting efforts.

Introduction

Imagine you receive a bug report from a live system, where you get a `badmatch` error when pattern matching the result of the call `ets:lookup(msgQ, MsgId)`. You expect your program to pattern-match on an atom denoting a message type, but instead it terminates when coming across the tuple `{error, unknown_msg}`.

You can quickly establish that the database got corrupted and set about trying to find the `ets:insert/2` call that wrote the entry into the ETS table. As messages entering the system are tested at the system boundary, this message should not have made it as far as the `insert` call. You could add a `case` statement ensuring that the error tuple is never inserted in the table, but this would be considered defensive programming, as this message was either tagged incorrectly or should not have made it this far into the system in the first place. Not only that, but you also do not know which particular `ets:insert/2` call caused the problem. In large systems, you would expect to find quite a few of them.

In a huge and complex system, without any knowledge of the module in which the entry was corrupted, you would have to find the tuple {error, unknown_msg}, potentially having to search through millions of lines of code. Once you found the tuple, inserting an io:format/2 statement that prints the error and process information would not solve the problem, as in live systems, processes come and go and millions of entries are inserted and deleted from ETS tables each hour. In addition, because this is a live system with strict revision control, code changes must be tested and approved before being deployed. This option, even if it's tempting, would result in a slow turnaround time. Don't get wound up on release procedures and slow turnaround times from the quality assurance team, however, as you can do better!

In Erlang, the first thing a developer or support engineer would consider doing is to turn on the *trace facility* for all calls to the ets:insert/2 function. You can trace both local and global calls—that is, calls to functions in the same module and in other modules—without having to trace-compile (i.e., recompile) the code. The *trace events* can include the function call itself, its arguments, its result, its calling function, and a timestamp. When tracing is enabled, you can generate a trace event every time the traced function is called. This trace event is either printed in the shell or piped to a socket where a program at the other end receives it, formats it, and stores it in a readable logfile.

But you still have a problem, because in a live system, millions of calls are made to ets:insert/2 every hour. Millions of trace messages would give you lots of unnecessary information and would probably affect performance. You are interested in trace events only if the second argument, the message type, is invalid. So, what do you do? You implement a *match specification* that, for every call, will inspect the parameters passed to it. You write the specification in such a way that you generate a trace event only if your parameter inspection comes across the tuple {error, unknown_msg} instead of a valid message type. In this trace event, you ensure that the *calling function* is displayed, giving you the function where the ets:insert/2 call originated. You can then start tracing this calling function and inspecting its arguments, finding the final clue to the origin of the invalid message, and as a result, locating the bug. What is more, you can do all of this in a live system where thousands of simultaneous transactions are being concurrently executed, without affecting the performance and without having to recompile the code.

If this sounds too good to be true, keep reading. Inspecting arguments in local and global calls, turning process traces on and off based on very specific triggers, and having full visibility over process state changes is possible with no extra effort on behalf of the developer. In Erlang, you achieve these results by using the *dbg tracer* tool, a userfriendly encapsulation of the trace BIFs and match specifications. This is another example of simple but powerful and well-thought-out constructs that reduce development and support efforts and bug turnaround times.

The first section of this chapter covers the low-level trace BIFs, as well as the message-based foundation for tracing in Erlang. This is followed by an introduction to the dbg tracer tool built on the BIFs; although this is probably the tool you will use, the first section will introduce you to the way in which tracing works in Erlang, as well as the terminology used in tracing.

The Trace BIFs

The built-in function `erlang:trace/3` *enables* and *disables* the low-level trace mechanisms in the Erlang runtime system. It provides you with a means to monitor concurrency, code execution, and memory usage. Tools such as the debugger and the process manager use this BIF to collect and display trace events. You prefix the BIF with the name of the `erlang` module that contains it, as it is not autoimported.

The beauty of the trace BIF is that you can use it without having to trace-compile the code. You can use it in live systems, resulting in an incredibly powerful troubleshooting tool that gives full visibility of what is going on.

Trace events are sent as messages of the following format:

```
{trace, Pid, Tag, Data1 [,Data2]}
```

where [,Data2] denotes an optional field dependent on the trace message type. Trace events that are generated include the following:

- Global and local function calls
- Garbage collection and memory usage
- Process-related activities and message passing

At any one time, only one process may receive trace events from another process. This is known as the *tracer process*. The tracer process that receives the trace events is the one that made the call to:

```
erlang:trace(PidSpec, Bool, TraceFlags).
```

In this call, `PidSpec` defines which processes you want to trace. It is either a process identifier or one of the atoms `existing`, `new`, or `all`:

- The `existing` atom enables tracing for all existing processes, but will exclude the tracer process and any process spawned after the call.
- If you pass `new`, the call will trace all processes spawned after the trace BIF call.
- The `all` atom will instead trace all processes created before and after the trace call, excluding the tracer process itself.

Calling `erlang:trace(self(), Bool, TraceFlags)` generates a bad argument error, as the tracer process cannot be traced. The reason the process receiving the trace messages cannot be traced is to avoid infinite cyclic loops. Imagine you are tracing received messages. Every time your process receives a message, a trace message will be generated, resulting in yet another trace message. Moreover, this will happen at a rate that no program would be able to handle.

If you want a process other than the one calling the trace/3 BIF to receive the trace events, you pass the {tracer, Pid} tuple as an element in the TraceFlags list (which we'll described shortly). In this tuple, Pid has to be a process or port identifier.

The second argument to the trace BIF is the atom true or false, specifying whether you want to enable or disable particular aspects of your tracing. These are described by a list of *trace flags*, denoted by atoms[*] and used to specify which trace events you want to generate or suppress, according to the Boolean value of the second argument: true to generate and false to suppress. We describe the trace flags in detail in subsequent sections. The return value of the erlang:trace/3 call is an integer denoting the number of traced processes.

Process Trace Flags

The send flag traces all messages sent by a process, and 'receive' generates events when messages are added to the mailbox of the traced process. Because receive is a reserved word in Erlang, you must enclose it in single quotes, thus generating an atom. Setting the send or 'receive' flag to true generates the following trace events:

```
{trace, Pid, send, Message, To}
{trace, Pid, send_to_non_existing_process, Message, To}
{trace, Pid, 'receive', Message}
```

By now, you must be itching to try the trace BIF in the shell. In the following sections, we will use this program as an example:

```
-module(ping).
-export([start/0, send/1, loop/0]).

start() -> spawn_link(ping, loop, []).

send(Pid) ->
  Pid ! {self(), ping},
  receive  pong -> pong end.

loop() ->
 receive
   {Pid, ping} ->
      spawn(crash, do_not_exist, []),
      Pid ! pong,
      loop()
 end.
```

The start function spawns a child process that waits in a receive-evaluate loop. Upon receiving a ping message, sent as a result of the ping:send/1 call, the child process spawns a new process that immediately terminates abnormally, as the module in which it should be executing does not exist.

---

[*] All trace flags are atoms except the {tracer, Pid} tuple described in the preceding paragraph.

This abnormal termination will result in a crash report being printed. (Remember, spawning a process never fails.[†] Instead, what fails is the newly spawned process, as the function it is supposed to execute is undefined.) Finally, the child process responds to the message by sending a pong message back. It is received in the `ping:send/1` call and is returned as a result of that function.

Running the preceding example in the shell and turning on tracing of all `send` and `receive` events, we get:

```
1> Pid = ping:start().
<0.55.0>
2> erlang:trace(Pid, true, [send, 'receive']).
1
3> ping:send(Pid).
pong
=ERROR REPORT==== 6-Sep-2008::19:16:00 ===
Error in process <0.40.0> with exit value: {undef,[{crash,do_not_exist,[]}]}

4> flush().
Shell got {trace,<0.55.0>,'receive',{<0.39.0>,ping}}
Shell got {trace,<0.55.0>,send,pong,<0.39.0>}
ok
5> erlang:trace(Pid, false, [send, 'receive']).
1
```

Note how commands 2 and 5 return the integer 1, namely the number of processes that are being traced. You can set trace flags on processes at any time, either before they are created, using the `all` or `new` flag, or after they have been created, using the `all` or `existing` flag or the flag's process identifier. And, as the trace messages are being sent to the shell process, which is the process that executed the `trace/3` call, you can retrieve them using the `flush/0` shell command.

At any particular time, a process can be in one of three states (see Figure 17-1). It could be *running* code, it could be *suspended* waiting for its turn to execute, or it could be in a `receive` clause, *waiting* for a message to arrive. A process is said to be *preempted* when its state changes from running to suspended, and *scheduled* when it is moved from the suspended queue ready to execute.

By enabling the `running` flag, you can trace state transfers of the module, function, and arity as well as the pid of the process from when it started running (`in`) to when it stops running (`out`). The trace events generated always come in pairs, and are of the following format:

```
{trace, Pid, in, {M, F, Arity}}
{trace, Pid, out, {M, F, Arity}}
```

Pman, the process manager, uses the `procs` flag to trace process-related events, such as process spawning and termination, process linking, and registration. The trace events generated by the `procs` flag include the following:

[†] Unless you reach the system limit of the maximum number of allowed processes.

```
{trace, Pid, spawn, Pid2, {M, F, Args}}
{trace, Pid, exit, Reason}
{trace, Pid, link | unlink, Pid2}
{trace, Pid, getting_linked | getting_unlinked, Pid2}
{trace, Pid, register | unregister, Pid2}
```

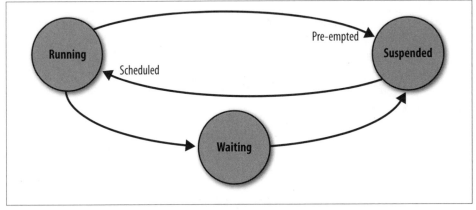

Figure 17-1. Process states

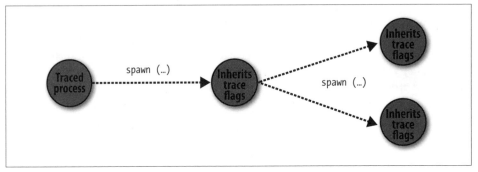

Figure 17-2. The set_on_spawn trace flag

Function calls can be traced using the call flag; we discuss this in more detail in the section "Tracing Calls with the trace_pattern BIF" on page 362.

Inheritance Flags

Use of the set_on_spawn flag specifies that any child process will inherit the flags of its parent, *including* the set_on_spawn flag itself (see Figure 17-2). As a result, any process spawned by the child process will also inherit all of the flags of the child. If instead you use the set_on_first_spawn flag, you specify that any process that is spawned will inherit the flags of its parents *except* the set_on_first_spawn flag. Any process the child will spawn will therefore *not* inherit the flags. In other words, set_on_spawn is *transitive*, whereas set_on_first_spawn is not.

Let's use the ping example to see the set_on_spawn and procs flags in action. Sending the ping message to the process spawns a child that terminates immediately. The tracing process should receive the spawn trace event from the first process and the exit message with the undef runtime error from the second process:

```
1> Pid = ping:start().
<0.31.0>
2> erlang:trace(Pid, true, [set_on_spawn, procs]).
1
3> ping:send(Pid).
pong
=ERROR REPORT==== 6-Sep-2008::19:50:33 ===
Error in process <0.40.0> with exit value: {undef,[{crash,do_not_exist,[]}]}

4> flush().
Shell got {trace,<0.31.0>,spawn,<0.34.0>,{crash,do_not_exist,[]}}
Shell got {trace,<0.34.0>,exit,{undef,[{crash,do_not_exist,[]}]}}
ok
```

The set_on_link and set_on_first_link trace flags are similar to the set_on_spawn flag, but instead they control the flags that are inherited when linking takes place.

Garbage Collection and Timestamps

Memory usage and the amount of time spent garbage collecting are notoriously tricky to predict. The Erlang system uses a generational garbage collection mechanism, described in Chapter 2. The collector has two generations of data: "current" and "old." Current data that survives a garbage collection will become old. The premise behind this approach is that much of the data stored in the heap is short-lived and will not survive to the first garbage collection, whereas data that survives its first collection tends to be much longer-lived. It therefore makes sense to garbage-collect in the current data more frequently than in the old data.

Using the garbage_collection flag, you can receive trace events relating to garbage collection initiation and termination. The events produced are of the following format:

```
{trace, Pid, gc_start, Info}
{trace, Pid, gc_end, Info}
```

where Info is a list of tagged tuples containing the following:

heap_size
    The used part of the current heap

heap_block_size
    The size of the memory block used to store the heap and the stack

old_heap_size
    The used part of the old heap

old_heap_block_size
    The size of the memory block used to store the old heap

stack_size
> The actual stack size

recent_size
> The size of the data that survived the previous garbage collection

mbuf_size
> The message buffer (or mailbox) size

All heap and message buffer sizes provided in the tagged tuples are in words.[‡]

Now, what should you do if you want to calculate the time spent garbage collecting, or if you want to have an accurate record of exactly when a specific trace event is generated? You should use the timestamp flag, which will add a timestamp to all messages. The format is the same as the one returned by the BIF now(), namely {MegaSeconds, Seconds, Microseconds} passed since January 1, 1970. All trace messages will now have the trace_ts tag and will contain an extra field at the end containing the timestamp:

```
1> Pid = ping:start().
<0.31.0>
2> erlang:trace(Pid, true, [garbage_collection, timestamp]).
1
3> ping:send(Pid).
Pong
=ERROR REPORT==== 6-Sep-2008::20:30:41 ===
Error in process <0.40.0> with exit value: {undef,[{crash,do_not_exist,[]}]}

4> flush().
Shell got {trace_ts,<0.31.0>, gc_start,
    [{mbuf_size,0},{recent_size,0},{stack_size,2},{old_heap_size,0},
    {heap_size,231}],{998,578633,990000}}
Shell got {trace_ts,<0.31.0>, gc_end,
    [{mbuf_size,0}, {recent_size,228},{stack_size,2},{old_heap_size,0},
    {heap_size,228}],{998,578633,990001}}
```

The flag cpu_timestamp makes all trace timestamps relative to the CPU time, not the wall clock. To use this flag, the target machine needs to support high-resolution CPU measurements and the Pid argument to the trace BIF must be all.

Tracing Calls with the trace_pattern BIF

You use the erlang:trace_pattern/3 BIF to enable tracing of local and global function calls. You must use this BIF in conjunction with the erlang:trace/3 BIF, called with the call and return_to flags. Tracing will be enabled on the *intersection* of the sets created through the calls to the two trace BIFs (see Figure 17-3). Using trace/3, you define which processes you want to monitor. Using trace_pattern/3, you define the

---

[‡] If you do not recall what we mean by the heap and the old heap, review the sidebar "Memory Management in Erlang" on page 33 in Chapter 2. We covered the process stack in Chapter 3.

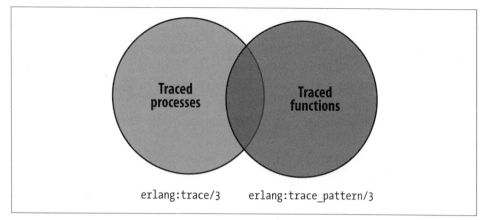

Figure 17-3. The intersection of the trace and trace_pattern BIFs

subset of functions you are tracing. An event will be generated only if a *traced* process executes a *traced* function.

The trace messages generated by the intersection of these two BIFs will be of the following form:

```
{trace, Pid, call, {M, F, Args}}
{trace, Pid, return_to, {M, F, Args}}
```

where `call` is generated when the function `F` in module `M` with argument `Args` is called. The `return_to` event is generated when the call returns and the execution of the function is completed. When tracing recursive calls, the `return_to` event is sent only when the recursive base case has been reached, and not for every iteration of the function. You can use the `return_to` flag only in conjunction with the `call` flag.

Using the `arity` flag in conjunction with the `call` flag, all event tags containing an {M, F, Args} tuple will instead return function arity information, {M, F, Arity}. This is useful if you are not interested in the arguments passed to a function and you want to minimize the size of the messages passed as a result of the call trace.

When calling:

```
erlang:trace_pattern(MFA, Condition, FlagList)
```

you define the functions you want to trace in the first argument, where you specify the module, function, and arity of the calls you want to result in a trace event. The `MFA` format is one of the following:

```
{Module, Function, Arity}
{Module, Function, '_'}
{Module, '_', '_'}
{'_', '_', '_'}
```

Wildcards are allowed using the `'_'` atom, so using {Module, `'_'`, `'_'`} will enable the tracing of *all* calls of *all* arities defined in `Module`. Wildcards need to follow the preceding pattern, however; combinations of the type {`'_'`, Function, `'_'`} are not allowed. The modules we are passing to `erlang:trace_pattern/3` have to be loaded before the BIF is called. Recompiling your code or reloading the module after having set the flags will clear the trace patterns. You need to rerun the trace command BIFs to reenable the tracing.

The `Condition` further controls the tracing. It can take a Boolean value: passing `true` will *enable* tracing of all functions defined in `MFA`, whereas `false` will *disable* it. You can also pass a *match specification*, the same ugly but extremely powerful terms describing a simple program that we discussed in Chapter 10. We will look at them in more detail in the section "Match Specifications: The fun Syntax" on page 374.

Finally, the `FlagList` defines the type of traced function, further filtering what has been defined in `MFA`. Passing `global` traces only exported functions, whereas `local` traces both exported and nonexported calls.

The visibility of your systems gained through these extremely powerful BIFs greatly reduces the time spent troubleshooting, making the turnaround time for patches very short when comparing Erlang to other programming languages. Are you curious and want to experiment with trace function calls from within the shell? Let's look at an example and experiment with the possible range of values you can pass to the BIFs. When doing so, do not forget to prefix the BIF calls with the `erlang` module name. Let's go back to the ping example, explained in the section "Inheritance Flags" on page 360:

```
1> l(ping).
{module,ping}
2> erlang:trace(all, true, [call]).
25
3> erlang:trace_pattern({ping, '_', '_'}, true, [local]).
5
4> Pid = ping:start().
<0.120.0>
5> ping:send(Pid).
pong
=ERROR REPORT==== 4-Apr-2009::19:33:25 ===
Error in process <0.122.0> with exit value: {undef,[{crash,do_not_exist,[]}]}

6> flush().
Shell got {trace,<0.120.0>,call,{ping,loop,[]}}
Shell got {trace,<0.120.0>,call,{ping,loop,[]}}
ok
```

In this example, we are tracing all function calls made within all processes. We intersect them with all *local* function calls in the module `ping`. Remember, this option contains local and nonlocal calls. To understand the calls that we see, let's number the lines in the `ping.erl` module:

```
1 -module(ping).
2 -export([start/0, send/1, loop/0]).
3
4 start() -> spawn_link(ping, loop, []).
5
```

```
 6 send(Pid) ->
 7   Pid ! {self(), ping},
 8   receive  pong -> pong end.
 9
10 loop() ->
11   receive
12     {Pid, ping} ->
13       spawn(crash, do_not_exist, []),
14       Pid ! pong,
15       loop()
16   end.
```

The functions we are tracing will be start/0, send/1, and loop/0; there are other function calls in the module (e.g., spawn_link/3 on line 4, self/0 on line 7), but these are calls to functions that are not defined in the ping module.

What is the effect of calling ping:start()? This starts a process with pid Pid that will execute loop/0, giving us the first call to a function in the traced module, resulting in the first trace message seen as a result of the flush() in command 6. This call is global, as it is called from outside the module as a result of the spawn/3 BIF.

When we then call send(Pid), the code on line 7 sends a message to Pid (which, incidentally, contains a call to self/0), and this message is handled in the body of the loop/0. After receiving the message, it calls spawn/3 and finally loop/0 on line 15. Calling loop/0 is a local call which gives rise to the second trace message found when the flush() was executed.

What would you expect the *global* calls to be, if you do the same trace but replace local with global in command 3, as shown here?

```
erlang:trace_pattern({ping, '_', '_'}, true, [global]).
```

Running the same tests gives you a trace message, namely the one generated as a result of the spawn/3 BIF. But if you are tracing all global calls in all processes, why did ping:send/1 not generate a trace event? It is, after all, a global call. The reason is that *the process receiving the trace messages cannot be traced.* If you want to generate trace events for the process calling the trace BIF, you have to redirect these messages to another process. You do so by passing the {tracer, Pid} option to the trace/3 BIF, specifying the pid of the process you want to send the messages to.

The dbg Tracer

You probably realize that although the trace BIFs are extremely powerful and handy, they are also very low-level and are not very user-friendly. After all, they are there to provide a base on which to build other tools such as the dbg tracer.

The dbg tracer is a text-based debugger providing a user-friendly interface to the trace and trace_pattern BIFs, but using the tracing principles and mechanisms introduced in the preceding section. You can use the dbg tool as a complement to the process manager we discussed in the section "The Process Manager" on page 114 in

Chapter 4, especially when you are tracing on text-based terminals and do not have access to the display or you have to divert the output. The dbg tool has a small impact on system performance, making it a suitable candidate for tracing large live systems.

Getting Started with dbg

The dbg:h() call provides you with a list of helpful functions; to find out more details about any of these functions, you can pass their names to dbg:h/1. The last thing you want to do when under pressure debugging a live system after a support call in the middle of the night is to have to look at the Erlang manual pages. The help function always comes in handy:

```
1> dbg:h().
The following help items are available:
    p, c
        - Set trace flags for processes
    tp, tpl, ctp, ctpl, ctpg, ltp, dtp, wtp, rtp
        - Manipulate trace patterns for functions
    n, cn, ln
        - Add/remove traced nodes.
    tracer, trace_port, trace_client, get_tracer, stop, stop_clear
        - Manipulate tracer process/port
    i
        - Info

call dbg:h(Item) for brief help a brief description
of one of the items above.

ok
2> dbg:h(p).
p(Item) -> {ok, MatchDesc} | {error, term()}
 - Traces messages to and from Item.
p(Item, Flags) -> {ok, MatchDesc} | {error, term()}
 - Traces Item according to Flags.
   Flags can be one of s,r,m,c,p,sos,sol,sofs,
   sofl,all,clear or any flag accepted by erlang:trace/3

ok
```

The call dbg:p(PidSpec, TraceFlags) allows you to specify which processes you want to trace and which trace events you want them to generate. PidSpec can be one of the following:

Pid
  A particular process ID.

all
  Will trace all processes, spawned before or after the call to the debugger.

new

    Will trace all processes spawned after the call to the debugger.

existing

    Will trace all processes existing before the call to the debugger.

Alias

    A registered process alias other than all, new, or existing.

{X,Y,Z}

    Denotes the process represented by the process ID <X.Y.Z>. "<X.Y.Z>" can also be used; it is the result of the pid_to_list/1 BIF.

TraceFlags is a single atom or a list of flags. You can pass any flag accepted by the trace BIFs alongside the following flag abbreviations and aggregations:

s

    Traces sent messages

r

    Traces received messages

m

    Traces sent and received messages

p

    Traces process-related events

c

    Traces global and local calls according to the trace patterns set in the dbg:tp/2 call

sos and sofs

    Denote the set_on_spawn and set_on_first_spawn flags, which we described in the section "Inheritance Flags" on page 360

sol and sofl

    Denote the set_on_link and set_on_first_link flags, which we described in the section "Inheritance Flags" on page 360

all

    Sets all flags

clear

    Clears all set trace flags

You must once again be itching to try out the dbg tracer and generate trace events. In the following example, we will trace all *message*-related events from the ping module that we introduced in "Inheritance Flags" on page 360.

We start our ping process and a separate tracer process. The tracer process will receive and display all of the trace events, including those generated in the shell. In command 5, we enable the trace events for all messages sent and received by the ping process after which we call the ping:send/1 function. In the trace printouts that follow, you can see that upon sending a ping message, the shell process has sent a pong message back in return:

```
3> Pid = ping:start().
<0.41.0>
4> dbg:tracer().
{ok,<0.43.0>}
5> dbg:p(Pid, m).
{ok,[{matched,nonode@nohost,1}]}
6> ping:send(Pid).
pong
(<0.41.0>) << {<0.29.0>,ping}
7> (<0.41.0>) <0.29.0> ! pong
7>
=ERROR REPORT==== 6-Sep-2008::21:40:31 ===
Error in process <0.47.0> with exit value: {undef,[{crash,do_not_exist,[]}]}

7> dbg:stop().
ok
```

You can see from the trace printout right after command 6 that the process within the parentheses is where the trace event originated. Process <0.41.0> receives the message {<0.29.0>, ping}, denoted in the trace by the << symbols, and responds by sending the message pong to process <0.29.0>.

Use this example to experiment with various trace flags, and while doing so, note how the tracer has to be started and stopped. Are the flags you set in dbg:p/2 automatically cleared when you stop the tracer using dbg:stop()?

Using the same ping process, let's now trace all process-related activities using the set_on_spawn flag. Because of this, the process <0.55.0> inherits all of the flags, and as a result, it generates the exit trace event:

```
8> dbg:tracer().
{ok,<0.51.0>}
9> dbg:p(Pid, [p, sos]).
{ok,[{matched,nonode@nohost,1}]}
10> ping:send(Pid).
pong
(<0.41.0>) spawn <0.55.0> as crash:do_not_exist()

=ERROR REPORT==== 6-Sep-2008::21:43:26 ===
Error in process <0.55.0> with exit value: {undef,[{crash,do_not_exist,[]}]}

(<0.55.0>) exit {undef,[{crash,do_not_exist,[]}]}
{ok,[{matched,nonode@nohost,1}]}
11> dbg:stop().
ok
```

Note that in both of the preceding examples, no trace information is logged regarding activities of the tracer process itself. And, if you have not figured it out, dbg:stop/0 will not clear the trace flags. For this purpose, use dbg:stop_clear/0.

Tracing and Profiling Functions

The function dbg:c(Mod, Fun, Args, TraceFlags) is ideal for trace and profile functions executed from the shell. If the TraceFlags argument is omitted, *all* flags are set. In the example that follows, we trace all activity related to an io:format/1 call, in this case, io:format("hello~n").

This tracing shows the inner workings of the input/output mechanism, under which messages are sent to the group leader. The calling process is suspended when it goes into a **receive** statement, and the process is scheduled as soon as a response from the group leader returns:

```
1> dbg:c(io, format, ["Hello World~n"]).
Hello World
(<0.53.0>) <0.23.0> ! {io_request,<0.53.0>,<0.23.0>,
                        {put_chars,io_lib,format,["Hello World~n",[]]}}
(<0.53.0>) out {io,wait_io_mon_reply,2}
(<0.53.0>) << {io_reply,<0.23.0>,ok}
(<0.53.0>) in {io,wait_io_mon_reply,2}
(<0.53.0>) << timeout
ok
```

Using dbg:c/3 is ideal if you want to monitor memory usage and time spent garbage collecting in a particular function, as it isolates the call in a single process. It is not the best way to trace side effects using the set_on_link and set_on_spawn flags, as all flags are cleared as soon as the function returns.

Tracing Local and Global Function Calls

So far, so good, but one of the really powerful features of the trace BIFs, and probably the one used most often is the ability to generate trace events for local and global function calls. In the **dbg** module, you use the following to enable the trace for *global* calls:

```
dbg:tp({Mod, Fun, Arity}, MatchSpec)
```

And you use this for *local* calls:

```
dbg:tpl({Mod, Fun, Arity}, MatchSpec)
```

You use these in conjunction with the dbg:p/2 call, where you specify the traced process together with the c flag. The trace events generated through the intersection of the two sets are the same as those described in the section "Tracing Calls with the trace_pattern BIF" on page 362.

Just as for the trace BIFs, you can define Module, Function, and Arity as '_', where formats of the type {'_', Function,'_'}, in which a wildcard comes before a non-wildcard, are not allowed. We will cover match specifications with dbg soon, but for the time being, we will use [].

The return value of `tp/2` and `tpl/2` has the format `{ok, Matches}`, where `Matches` is a list of tuples reporting the Erlang nodes on which the trace was enabled and the number of functions for which the trace was turned on at each node.

Call tracing is disabled using `dbg:ctp({Mod, Fun, Arity})`. Calling this function will disable function tracing regardless of whether the calls are local or global. This is the call you will probably be using most of the time. Should you want to disable the trace on only a particular global call pattern set with `dbg:tp/2`, however, use `dbg:ctpg({Mod, Fun, Arity})`; local traces are disabled using `dbg:ctpl({Mod, Fun, Arity})`.

All of the calls used to enable and disable function tracing return a tuple of the format `{error, Reason}` or `{ok, MatchDescription}`. `MatchDescription` is a list of tuples of the form `{matched, Node, Number}`, where `Number` denotes how many function calls were enabled or disabled on the particular `Node`. You can clearly see this in the following example.

Also note how we are now receiving trace events generated in the shell. In the examples using the trace BIFs in the section "Tracing Calls with the trace_pattern BIF" on page 362, tracing on the shell could not be enabled because the shell was itself the process receiving the trace events. The messages are now being sent to a tracer process which is not the shell, and so trace events for global calls such as `start/0` and `send/1` have become visible:

```
1> dbg:tracer().
{ok,<0.100.0>}
2> dbg:p(all,[c]).
{ok,[{matched,nonode@nohost,25}]}
3> dbg:tp({ping, '_', '_'}, []).
{ok,[{matched,nonode@nohost,5}]}
4> Pid = ping:start().
<0.105.0>
 (<0.97.0>) call ping:start()
(<0.105.0>) call ping:loop()
5> ping:send(Pid).
(<0.97.0>) call ping:send(<0.105.0>)
pong
=ERROR REPORT==== 7-Sep-2008::12:47:07 ===
Error in process <0.107.0> with exit value: {undef,[{crash,do_not_exist,[]}]}

6> dbg:ctpg({ping, '_', '_'}).
{ok,[{matched,nonode@nohost,5}]}
```

You should spend some time getting acquainted with the tracer and its interface, as this is one of the most powerful tools available with the Erlang runtime distribution. When reading the manual page for dbg, you will discover that there are many variants of the `tp`, `tpl`, `ctp`, `ctpl`, and `ctog` functions. Don't worry about them unless you like using shortcuts. It is probably hard enough to remember the ones we have discussed in this chapter, let alone all of their variants. The good news is that the ones we described are all you need, and they allow you to express all combinations covered in the other variants. It is a trade-off between having a good memory and remembering all of the variants, or typing in a few extra characters. And as a fallback, all you need to revert back to is dbg:h().

 When you are tracing live systems, you should use the dbg tracer with extreme care. If you turn on too many trace events in a busy system, you run the risk of generating so many shell printouts that it becomes impossible to type in the command necessary to stop the trace. As a result, the tracer process becomes unable to keep up with the generated event messages. They queue up, and very quickly, the Erlang runtime system runs out of memory. Support engineers (and one of the authors[§] of this book) have caused service outages in national fixed phone and mobile data networks because of dbg. Use it in a safe test environment as carelessly as you want, but when debugging live systems, make sure you know what you are doing, and exercise extreme care!

Distributed Environments

Tracing can take place in distributed environments, with all of the trace output being redirected to a single tracer. When you enabled or cleared a trace, you must have noticed from the examples that the return value of the calls was of the following format:

```
{ok,[{matched,nonode@nohost,5}]}.
```

As we were not running distributed nodes, all local traces were on the node nonode@nohost. When running the traces in a distributed environment, you would instead get a list of all nodes on which the traces were enabled. If the tracer was unable to connect to a remote node, the result for that particular node would be {matched,Node, 0,RpcError}.

The call dbg:n(Node) adds a distributed Erlang node to the traced list and dbg:cn(Node) removes it from the list. To list the nodes on which you are running a trace, you use dbg:ln(). Once nodes have been added, setting trace items, flags, and function patterns will result in all traced nodes being affected. Process identifiers passed to dbg:p/2 can be on other nodes. The only thing dbg:p/2 does not handle is globally registered names.

Redirecting the Output

In all of the dbg tracer examples we have looked at so far, the trace events have been sent to a tracer process that formatted and printed them out in the shell. This might not always be the best way to handle debugging information. You might instead want to collect statistics, measure garbage collection, or forward the output to a socket or a file.

[§] We have left it as an exercise for you to figure out which one of the authors was responsible.

The good news here is that dbg allows you to define your own fun to handle the trace messages generated by the trace BIFs. If you start the tracer tool using dbg:tracer(process, {HandlerFun, Data}), all events are passed as arguments to HandlerFun, a user-defined fun of arity 2. The trace message is the first argument and the user-defined Data is the second. The fun returns data that is passed back to it in its next iteration, and so this value can be seen as an *accumulator*, keeping track of total resource use, for instance.

In the following example, we will be monitoring the memory usage of the shell process. Our HandlerFun will store the memory usage data when it starts garbage collecting, providing information on whether memory is allocated or released. We'll test our tracer by running some memory-intensive applications on lists. A positive delta value means memory has been freed in the (current) heap and the old heap, and a negative value means memory has been allocated. All sizes are in words. Notice how memory is allocated in the old heap while being freed in the current heap. This will be the longer-lived data that has survived a garbage collection in the current heap and is being moved to the old heap.

The following example is long; any typos will result in you having to type everything from scratch. Be smart; either use an editor to type the example and paste it in the shell, or download it from the book's website:

```
1> HandlerFun =
   fun({trace, Pid, gc_start, Start}, _) ->
         Start;
      ({trace, Pid, gc_end, End}, Start) ->
         {_, {_,OHS}} = lists:keysearch(old_heap_size, 1, Start),
         {_, {_,OHE}} = lists:keysearch(old_heap_size, 1, End),
         io:format("Old heap size delta after gc:~w~n",[OHS-OHE]),
         {_, {_,HS}} = lists:keysearch(heap_size, 1, Start),
         {_, {_,HE}} = lists:keysearch(heap_size, 1, End),
         io:format("Heap size delta after gc:~w~n",[HS-HE])
   end.
#Fun<erl_eval.12.113037538>
2> dbg:tracer(process, {HandlerFun, null}).
{ok,<0.32.0>}
3> dbg:p(self(), [garbage_collection]).
{ok,[{matched,nonode@nohost,1}]}
4> List = lists:seq(1,1000).
[1,2,3,4,5,6,7,8,9,10,11,12,13,14,15,16,17,18,19,20,21,22,
 23,24,25,26,27,28,29|...]
Old heap size delta after gc:0
Heap size delta after gc:6020
5> RevList = lists:reverse(List).
[1000,999,998,997,996,995,994,993,992,991,990,989,988,987,
 986,985,984,983,982,981,980,979,978,977,976,975,974,973,972|...]
Old heap size delta after gc:-676
Heap size delta after gc:3367
```

Redirecting to sockets and binary files

You have seen how to write your own funs to handle trace messages. Using the same mechanism, you can forward trace outputs to a *socket* or a *binary file*. This gives you the advantage of reducing the load on the traced node, particularly as a result of any I/O in the shell. It is this I/O which can get so intensive, making it extremely hard to correctly type in the command needed to stop the tracer!

How do you redirect the output? In the dbg:tracer/2 call, instead of the process argument we demonstrated in the previous sections, you pass the port atom. The call:

    dbg:tracer(port, PortFun)

starts a tracer that will pass the trace messages to an Erlang port. PortFun is a fun encapsulating the opened port and returned by the call dbg:trace_port(ip, Port), where Port is either the port number or the tuple {PortNumber, QueueSize}. QueueSize limits the size of the undelivered message queue, dropping trace events if they are not picked up by the remote socket client. This option will open a listener port, buffer the messages, and send them as soon as a client connects.

To connect the client, call dbg:trace_client(ip, Arg), where Arg is either the PortNumber (if the tracer is running on the same host) or the tuple {HostName, Port Number} if it is running on a different machine, the preferred option if tracing is expected to impose a heavy load. This call will connect to the listener port where the trace is running and retrieve the trace events.

If you want to send your trace events to a binary file, bind the variable PortFun to the return value of dbg:trace_port(file, FileOptions). The FileOptions describes the way in which the trace is to be stored in the filesystem. FileOptions can specify either a filename, or a *wrap files specification*, which will limit the amount of file space used. By default, using {FileName, wrap, FileSuffix} will result in eight wraparound files, each 128 KB in size. To change the default values, use {Filename, wrap, FileSuffix, WrapTrigger, WrapCount}, where WrapTrigger is the size in kilobytes or the tuple {time, Milliseconds}. Suffix is the file suffix and WrapCount is the number of files created before wrapping around.

You can retrieve the messages with the trace_client/2 call, where the first argument is the atom file, reading all of the trace events written so far, or follow_file, which continually reads and processes trace events written after the call. The second argument is the FileName as passed in the FileOptions argument you passed to the dbg:trace_port/2 call.

If you want to handle your own trace events, received through either a file or a socket, you can use the dbg:trace_client(Type, Arg, {HandlerFun, Data}) call. It starts a client that will apply HandlerFun to each incoming trace event in turn. Type and Arg are the same as in the earlier descriptions of trace_client/2, whereas the {HandlerFun, Data} tuple is the same as in tracer/2. The call dbg:stop_trace_client(Pid), where Pid is the return value of the tracer/2 call, stops the client.

In the following example, we start a local tracer that redirects trace events to IP port 1234 on a different Erlang node running on the same host. A tracer client will pick up the events and process them. When dealing with systems carrying live traffic, it is a good practice to redirect your trace events either to a file, which you process on a different machine, or to a socket, where a client on a remote machine picks up the message traffic:

```
1> PortFun = dbg:trace_port(ip, 1234).
#Fun<dbg.12.21848437>
2> dbg:tracer(port, PortFun).
{ok,<0.33.0>}
3>  dbg:p(all, [c]).
{ok,[{matched,nonode@nohost,25}]}
4> dbg:tp({ping, '_','_'}, []).
{ok,[{matched,nonode@nohost,5}]}
5> dbg:tpl({ping, '_','_'}, []).
{ok,[{matched,nonode@nohost,5}]}
6> Pid = ping:start().
<0.39.0>
7> ping:send(Pid).
pong
=ERROR REPORT==== 14-Sep-2008::12:25:23 ===
Error in process <0.41.0> with exit value: {undef,[{crash,do_not_exist,[]}]}
```

The output is received by a tracer client process in a different Erlang shell:

```
1> Pid = dbg:trace_client(ip, 1234).
<0.40.0>
2> <0.30.0>) call ping:start()
(<0.39.0>) call ping:loop()
(<0.30.0>) call ping:send(<0.39.0>)
(<0.39.0>) call ping:loop()
2> dbg:stop_trace_client(Pid).
ok
```

Match Specifications: The fun Syntax

We introduced the powerful (but ugly) match specifications in Chapter 10. As you might recall, a match specification consists of an Erlang term describing a small program that expresses a condition to be matched over a set of arguments. Match specifications are limited in functionality, and in the case of the trace BIFs, they mainly deal with the filtering and manipulation of trace events. If they match successfully, a trace event is generated and some predefined actions can be executed. Match specifications are compiled to a format close to the one used by the emulator, making them more efficient than functions. But apart from being more efficient, the specifications are complex to write, and at first glance, they look incomprehensible.

Luckily, you can generate match specifications covering a majority of simple but useful cases using the dbg:fun2ms/1 call. It converts specifications that are described using fun syntax into match specifications. The results are as efficient as writing the match specifications by hand, but they are much easier to read, write, modify, and debug. We are using `dbg:fun2ms/1` to turn an anonymous fun into a match specification, which is used when setting trace flags on local and global calls. We start by introducing this higher-level approach, but for those who need to harness the full power of the trace BIFs, we follow this by looking at the match specifications themselves.

Generating Specifications Using fun2ms

Remember the bug we described in the beginning of the chapter? The one where we corrupted the ETS table with the tuple {error, unknown_msg}? We can re-create the error in the following way: first, we use the `dp:fill()` call to create and corrupt the ETS table. The crash in the system that occurred when reading the corrupted table will then be generated by calling the `dp:process_msg()` function, which takes the first element in the ETS table and processes it. In the real world, we would not know either of these functions, and would have to rely on higher-level tests that would eventually lead to these calls being made by a worker processes. In this example, we are calling them from the shell, as that will facilitate demonstrating the step-by-step debugging procedures that have to be undertaken to solve the bug using match specifications:

```
1> dp:fill().
true
2> dp:process_msg().
** exception error: no case clause matching [{2,{error,unknown_msg}}]
      in function  dp:process_msg/0
```

The error message tells us immediately that there is a `case` clause error in the `process_msg/0` call. Following the call flow in the `dp:process_msg()` function discovered by the exception error printed in the shell, we immediately notice that the pattern {_,{error, unknown_msg}} is missing in the `case` clause:

```
-module(dp).
-compile(export_all).

process_msg() ->
  case ets:first(msgQ) of
    '$end_of_table' ->
      ok;
    Key ->
      case ets:lookup(msgQ, Key) of
        [{_, {event, Sender, Msg}}] ->
          event(Sender, Msg);
        [{_, {ping,  Sender}}] ->
            ping(Sender)
      end,
      ets:delete(msgQ, Key),
      Key
  end.

event(_,_) -> ok.
ping(_)    -> ok.
```

```
fill() ->
  catch ets:new(msgQ, [named_table, ordered_set]),
  dp:handle_msg(<<2,3,0,2,0>>).
```

The entry {2, {error, unknown_msg}} should not have been inserted in the table in the first place, so we can assume that the case clause is correct and that adding the {error, unknown_msg} clause will not solve the bug. What we need to find out is who inserted it in the table, and stop it from happening again. The corruption must have originated in an ets:insert/2 call, but as there are many in the system, we need to pinpoint which one it is by creating a match specification that generates a trace message only if the tuple element being inserted is {error, unknown_msg}. This match specification, when used together with the trace patterns of the modules, functions, and arguments, will trigger a trace event whenever the call with those arguments is made.

The function dbg:fun2ms/1 takes a *literal fun* as an argument and returns a match specification describing the properties in the fun. By "literal fun," we mean a fun that is typed in and passed as a parameter, not a fun that is bound to a variable which is then passed as an argument, or a fun that results from applying a higher-order function. We could also call this an "explicit" or a "manifest" fun. The fun takes one parameter, which is either a variable or a list of variables, all of which can be pattern-matched or used in guards. You can use the variables in the fun body, but the last expression has to be either a predefined call that is converted to an action, or a term that is ignored.

The Erlang precompiler takes the literal fun and translates it to a match specification. You can type match specifications directly in the shell, but if you include them in your modules, you must include the header to the *ms_transform.hrl* file. It is part of the standard library application, and is most easily included as follows:

```
-include_lib("stdlib/include/ms_transform.hrl").
```

Armed with this knowledge, let's generate a match specification which when passed to the dbg:tp/2 function will trigger an event if the second argument to a call is of the form {_, {error, unknown_msg}}. This would include the call ets:insert(msgQ, {1, {error, unknown_msg}}):

```
dbg:fun2ms(fun([_, {_,{error, unknown_msg}}]) -> true end).
```

Do you see how we are pattern matching in the fun? The atom true returned by the literal fun is ignored, but we have used the pattern-matching facilities of the fun syntax to express the required pattern.

Let's try it out in the shell. Remember, you cannot bind the fun to a variable; it is a *literal fun* expression that is handled by the precompiler, and as such, it has to be entered explicitly as an argument to the call. Note also that the call db:fill/0 puts only one incorrect entry in the ETS tables, but it could have put in thousands of entries, most of which could have been correct. As a result of our tests, however, we want to generate only one trace event that is triggered when the second element of the second argument is the tuple that corrupts our table:

```
3> dbg:tracer().
{ok,<0.58.0>}
4> Match1 = dbg:fun2ms(fun([_,{_,{error, unknown_msg}}]) -> true end).
[{['_',{'_',{error,unknown_msg}}],[],[true]}]
5> dbg:tp({ets, insert, 2}, Match1).
{ok,[{matched,nonode@nohost,1},{saved,1}]}
6> dbg:p(all,[c]).
{ok,[{matched,nonode@nohost,25}]}
7> dp:fill().
true
(<0.54.0>) call ets:insert(msgQ,{2,{error,unknown_msg}})
```

We now know the pid of the calling process, but it did not make us any wiser. As processes are created for every incoming message and are terminated as soon as the message has been queued, there would be no point in tracing it. What we really need is the function that calls this insertion (the caller function).

Luckily, we can request this information by telling the match specification to generate an event that contains information about the caller function. We do this by making our fun return one of a set of predefined literal functions, all of which are also handled by the precompiler. In our case, we would use message(caller()):

```
dbg:fun2ms(fun([_,{_,{error, unknown_msg}}]) -> message(caller()) end).
```

The literal call message(Data) sends a message with the Data to the tracer process; to display it in the shell instead, use display(Data). In our example, we passed the literal function caller() as Data. We will cover other valid data options shortly. Right now, let's focus on identifying the caller and see whether it helps us solve the bug:

```
8> Match2 = dbg:fun2ms(fun([_,{_,{error, unknown_msg}}]) ->
8>                          message(caller())
8>                     end).
[{['_',{'_',{error,unknown_msg}}],[],[{message,{caller}}]}]
9> dbg:tp({ets, insert, 2}, Match2).
{ok,[{matched,nonode@nohost,1},{saved,2}]}
10> dp:fill().
true
(<0.34.0>) call ets:insert(msgQ,{2,{error,unknown_msg}}) ({dp,handle_msg,1})
```

Note how after the ets:insert/2 trace message, the calling function ({dp,handle_msg, 1}) is now also displayed by the tracer. This is as a result of the message(caller()) call in our literal fun. We now know that ets:insert/2 with the wrong data is being called by the dp:handle_msg/1 function. We look at the code, follow the call flow, and immediately find that the {error, unknown_msg} tuple is created in the dp:handle/3 function as a result of the MsgType argument being anything other than the integer 1 or 2:

```
handle_msg(<<MsgId, MsgType, Sender:16, MsgLen, Msg:MsgLen/binary>>) ->
  Element = handle(MsgType, Sender, Msg),
  ets:insert(msgQ, {MsgId, Element}).

handle(1,   Sender, Msg)  -> {event, Sender, Msg};
handle(2,   Sender, _Msg) -> {ping,  Sender};
handle(_Id, _Sender, _Msg) -> {error, unknown_msg}.
```

We can now run a trace on the `handle/3` function, generating events only if the `MsgType` is not the integer 1 or 2. We bind `MsgType` in the `fun` head and test it in a *guard*. Binding of variables is allowed only in the function head: using = in your match `fun` will result in an error being returned, and the match specification failing to compile.

We can easily add this condition by adding a guard to our literal `fun`:

```
dbg:fun2ms(fun([Id, Sender, Msg]) when Id /=1, Id /=2 -> true end).
```

Remember that a semicolon between guards means that at least one of the guards is required to succeed, and a comma means they all must succeed. The guards allowed in literal `fun`s are the same as in conventional guards. They include:

BIFs used in type tests
> is_atom, is_constant, is_float, is_integer, is_list, is_number, is_pid, is_port, is_reference, is_tuple, is_binary, is_function, is_record

Boolean operators
> not, and, or, andalso, orelse

Relational operators
> >, >=, <, =<, =:=, ==, =/=, /=

Arithmetic operators
> +, -, *, div, rem

Bitwise operators
> band, bor, bxor, bnot, bsl, bsr

Other BIFs allowed in guards
> abs/1, element/2, hd/1, length/1, node/1,2, round/1, size/1, tl/1, trunc/1, self/0

If a runtime error occurs as a result of a bad argument when pattern matching in the function head, or because of invalid arguments used in the guard operations, the match fails. If a runtime error occurs in the match specification body, the match specification simply returns the atom `'EXIT'`. The trace event is generated and any return value is ignored.

By using the not equals (/=) conditional guard to test for invalid integers in the shell, we will get the final clue to resolve the bug!

```
11> Match3 = dbg:fun2ms(fun([Id, Sender, Msg]) when Id /=1, Id /=2 -> true end).
[{['$1','$2','$3'],[{'/=','$1',1},{'/=','$1',2}],[true]}]
12> dbg:tpl({dp, handle, 3}, Match3).
{ok,[{matched,nonode@nohost,1},{saved,3}]}
13> dp:fill().

(<0.44.0>) call dp:handle(3,2,<<>>)
(<0.44.0>) call ets:insert(msgQ,{2,{error,unknown_msg}}) ({dp,handle_msg,1})
true
```

Looking at the preceding trace events, we now know that `dp:handle/3` is called with message type 3, and the hardware ID of the sender is 2. Looking at the software revision on the hardware denoted by the ID 2, we realize it is not supported by our software, resulting in the data being corrupted. Problem solved!

Where Did the Bug Come From?

The crash described in this section happened in our test plant. The trace events we generated and stored in the logfile allowed us to discover which message type was not handled and to track down the hardware where it originated using a strategy similar to the one described in the example. We quickly discovered that the software revision of the originating hardware was a later release than the one supported by our installation. So much for workarounds; the bug in this example was in the connectivity module that allowed hardware with nonsupported software revisions to connect to the system.

On top of fixing the connectivity module, we removed the handling of unknown messages in the `dp:handle` call, where we had originally returned the `{error, unknown_msg}` tuple. This was a case of defensive programming that should not have been included in the first place. By making the process handling the message terminate with a `case` clause error the termination would affect only processes handling messages originating from hardware that should not have been connected to the system. This gave us an early warning of the problem and a direct indication of where the bug occurred (with the hardware ID and message type).

You can include the calls in a literal `fun` and they will be translated to actions during the parse transform in the precompiler. Some of these actions you already saw in the examples, but we include all of them here for reference:

`return_trace()`

> Generates an extra event upon completing the traced call, including the return value of the function. When generating this extra trace event, tail-recursive properties of the traced function calls are lost.

`exception_trace()`

> Behaves in the same way as `return_trace`, but if a runtime error occurs, an `exception_from` message is generated.

`display(Data)`

> Generates a side effect by printing the `Data` passed to it. The `Data` can be either one of the arguments bound in the `fun` head or the return value of one of the other literal functions described in this section.

`message(Data)`

> Generates a trace event with the `Data`. The `Data` can be either one of the arguments bound in the `fun` or the return value of one of the other literal functions described in this section.

`message(false)`

> Is a special case of `message/1`, where no trace event is generated for the `call` and `return_to` trace flags. This is useful if you are interested in side effects, such as those generated by the `display/1` literal.

message(true)

Is another special case of message/1, where the trace on the {Module, Function, Arity} behaves as though no match specifications are associated with it. Its only use is to override the message(Data | false) call.

enable_trace(TraceFlag)

Enables a TraceFlag on the process that triggered the match specification. You can pass only one flag at a time, but nothing is stopping you from having several calls each with their own trace flag in the fun body. The TraceFlag is as defined in the trace BIFs; no abbreviations used by the tracer tool are allowed. Executing this call in a match specification is the equivalent of calling erlang:trace(self(), true, [TraceFlag]).

enable_trace(Pid, TraceFlag)

Is the same as enable_trace/1, except that it will enable the trace flag on the Pid, which is either a process identifier or a registered name.

disable_trace(TraceFlag)

Disables a TraceFlag on the particular process that triggered the match specification. The same restrictions to TraceFlag apply as in the enable_trace/1 literal call.

Executing this call in a match specification is the equivalent of calling erlang:trace(self(), false, [TraceFlag]).

disable_trace(Pid, TraceFlag)

Is the same as disable_trace/1, except the TraceFlag will be disabled on the specified Pid.

trace(Disable, Enable) and trace(Pid, Disable, Enable) allow you to enable and disable many flags at the same time.

silent(true)

Turns off all call trace messages originating until a match specification with silent(false) in its body is matched.

set_tcw(Int)

Sets a unique *trace control word* that you can retrieve using the erlang:system_info(trace_control_word) BIF call. It is a free-to-use status value used by some advanced tools (possibly user-defined), allowing these tools to influence the actions taken based on its value.

You can pass the following calls as arguments to the display/1 and message/1 literal functions:

caller()

Returns the {Module, Function, Arity} tuple, allowing you to identify the calling function.

get_tcw()

Returns the trace control word which was previously set using set_tcw/1.

object()
: Returns a list with all of the arguments passed to the match specification.

bindings()
: Returns a list of all bound variables in the match head.

process_dump()
: Returns the process stack properly formatted as a string in the form of a binary. You can pass it as an argument to message/1, where you have implemented your own tracer process. Passing it to display/1 will *not* work, as it is unable to print binaries.

self()
: Returns the process identifier of the calling process.

Odds and ends with fun2ms

When generating match specifications using fun2ms, you must follow certain restrictions, as the functionality in match specifications is limited. Now that you are armed with the knowledge of how match specifications work, these restrictions are easier to understand.

You cannot call Erlang functions in the fun body. Guards and functions allowed in the match body will be translated as is, so even if they compile correctly, using them might cause the match specification to return an error. All variables defined as arguments to the fun are replaced by match specification variables in the order in which they occur. If you look at some of the match specifications we have generated, a fun head such as that which occurs in the following:

```
fun([Id, Sender, Msg])when Id /=1, Id /=2 -> true end).
```

translates to ['$1','$2','$3']. Every occurrence of these variables is replaced in the match specification conditions and body, so in the preceding example, the guards would translate to a condition of the format [{'/=','$1',1},{'/=','$1',2}] and guards such as is_integer(Id) would be translated to a condition resembling [{is_integer, '$1'}].

Variables bound outside the scope of the fun and not appearing in the head are imported, either from the shell or from the function where the match specification is generated. They are translated to constant expressions of the format {const, Constant}, where Constant is the literal value of the Constant. Generating the match specification using the following function:

```
foo(A) -> dbg:fun2ms(fun([B,C]) when B == A -> A end).
```

and calling foo(10) will generate a specification of the following format:

```
[{['$1','$2'], [{'==','$1',{const,10}}], [{const,10}]}].
```

You can bind variables only in the fun head, and even here, you can do so only restrictively. For example, a fun of the form fun({A,[B|C]} = D), where D is bound at the top level to all of the arguments, is allowed, whereas fun({A,[B|C]=D}) is not. In the first case, if you are interested in printing or returning all of the arguments, you would be better off using the literal function object(), which is translated to '$_', a term that expands to all of the arguments. If you are interested only in the variable bindings and you ignore the "don't care" variables, use the literal function bindings(), which translates to '$*' in the lower-level match specification.

Literal term constructions are translated in order to turn them into valid match specifications; tuples are converted to a one-element tuple containing the tuple itself, and constant expressions are converted to the form {const, Constant}. Records and operations on records are converted to tuples and element calls. The is_record/2 guard is converted to {is_record, Var, Arity}, where Var is replaced by a variable of the format '$0' and Arity is the size of the tuple. List pattern matches are converted to use the {hd, List} and {tl, List} constructs. As conditional constructs such as case, if, and catch are not allowed in match specifications, they are not allowed in funs either.

Remember that when dealing with the fun2ms call in a module, the module must always include the *ms_transform.hrl* header file. If you don't include it, the code might compile without any warnings, but the match specification will not be translated, possibly resulting in a runtime error. Translations to match specifications are done at compile time, so runtime performance is not affected by using these literal functions.

Difference Between ets and dbg Match Specifications

There are differences between the match specifications used by the ets and the dbg modules. If the pseudo function triggering the translation is ets:fun2ms/1, the fun's head must contain a single variable or a single *tuple* with its guards acting as filters. The body of the fun, and as a result, the body of the match specification, constructs terms and returns values that are the result of the select/2 call. This is done without generating any side effects or binding variables. Even though you can use ets match specifications from the shell, you will usually find them embedded in program code.

If the pseudo function is dbg:fun2ms/1, on the other hand, the fun's head must contain a single variable or a single *list*. The specification body will contain imperative commands that result in side effects manifesting themselves as printouts, the manipulation of trace flags, or extra data appended to the trace events. The return value of the generated specification is ignored. Trace BIF specifications are most commonly used from the shell, even if nothing is stopping you from integrating them in your programs.

What unites the two is the need for efficient filtering with low overheads that do not affect the real-time properties of the system. If what you have seen so far does not scare you, let's look at these match specifications in more detail. A warning for the faint of heart: what we will describe in the next section, although powerful, is not pretty.

Match Specifications: The Nuts and Bolts

When covering the help function dbg:fun2ms/1, we went into a high-level explanation of its results, namely match specifications. The specifications are tuples with three elements of the format [Head, Conditions, Body]:

- The *head* is used to bind and match variables and terms.
- In the *conditions*, logical tests are applied on the variables. You can define your own logical tests or use predefined guards.
- In the *body* part of the specification, we list a possibly empty set of predefined actions that have to be taken if the match in the head is successful and the logical conditions are met.

The objective of this section is to ensure that you understand the match specifications that result from the dbg:fun2ms/1 and ets:fun2ms/1 calls, and possibly implement some simpler ones for yourself.

The Head

The head is a list of variables, literals, and composite data types. All variables in the head are of the form '$int()', where int() is replaced by an integer of the format '0' or '1', ranging from 0 to 100,000,000. The atom '_' denotes the "don't care" variable, and can be used if you are not interested in matching parts of a particular argument. If you instead want to match on all values and all arities without any variable bindings, you just use '_', giving a specification of the form ['_', Conditions, Body]. You might be tempted to write [[], Conditions, Body] instead, but beware, as this will only match functions of arity 0.

In the following example, we create a specification that matches on functions of arity 2. We bind the first argument to the variable '$1'. The second argument has to be a tuple where the first element is bound to the variable '$2' and the second element is the tuple {error, unknown_msg}. The match specification for such a case is as follows:

```
[{['$1',{'$2',{error,unknown_msg}}],[],[]}]
```

In the preceding code, the conditions and the body are denoted by empty lists, as no logical checks and side effects are needed. Look at the example in the section "Generating Specifications Using fun2ms" on page 375, where we were looking for debug messages, and see what dbg:fun2ms/1 produced.

In the following example, we use dbg:tp/2 to start a trace on all calls to the ets:insert/2 function. The match specification that we pass as the second argument ensures that trace events are triggered only if the second argument is a tuple of size 2, where the second element is the tuple {error, bad_day}:

```
1> dbg:tracer().
{ok,<0.32.0>}
2> dbg:p(all,[c]).
```

```
{ok,[{matched,nonode@nohost,25}]}
3> dbg:tp({ets,insert,2}, [{['$1',{'$2',{error,bad_day}}],[],[]}] ).
{ok,[{matched,nonode@nohost,1},{saved,1}]}
4> ets:new(foo,[named_table]).
foo
5> ets:insert(foo, {1, monday}).
true
6> ets:insert(foo, {1, {error, bad_day}}).
true
(<0.30.0>) call ets:insert(foo,{1,{error,bad_day}})
```

Instead of '$1' and '$2', we could have used "don't care" variables, as we are not doing anything with the values of the arguments in the conditions or body.

It is possible to repeat variables within specifications. In the specification [{['$1','$1'],[],[]}], we match calls to functions of arity 2 where the arguments are the same. The specification would trigger a trace event if the call to foo(1,1) was being traced. The match specification head [{'$1','$2'}, ['$1','$2'|'_ ']] would match functions of arity 2 where the first argument is a tuple with two elements and the second argument is a list whose first two elements are identical to those in the tuple. For instance, the call foo({1,2},[1,2,3]), if traced, would trigger an event.

Conditions

The conditions list allows several logical tests on variables to be combined. If they are positive, the match is successful and the trace event is triggered. All variables must be bound, but unlike in the head, literal and compound data types must be described using a special syntax. To represent a tuple, it is necessary to use either {const, Tuple} or {Tuple}, where Tuple is a literal or the variable representing the tuple. Lists are deconstructed into a head and tail using {hd, List} and {tl, List}, where List is the representation of the list.

Boolean functions include functionality similar to the BIFs you are allowed to use in guards and in logical, Boolean, relational, and arithmetic operators. As match specifications are Erlang terms, however, all of these operators have to be represented as atoms, and operations themselves are grouped together in tuples.

Guards are of the format {Guard, Variable}, where the Guard can be any one of is_atom, is_constant, is_float, is_integer, is_list, is_number, is_pid, is_port, is_reference, is_tuple, is_binary, is_function, or is_seq_trace.

For example, to test for a list, you would use a guard of the following form:

```
{is_list, '$0'}.
```

The guard to check for a record uses is_record and is of the following format:

```
{is_record, '$1', RecordType, record_info(size, RecordType)}
```

In the preceding code, the RecordType has to be a hardcoded literal giving the record type required.

The conditional expression constructs take the following form:

```
{Construct, Exp1, Exp2, ...}
```

In the preceding code, the first element of the tuple is the logical construct and the remaining ones are (possibly nested) conditional expressions or guards. The logical construct {'not', Expression} evaluates to true if the Expression evaluates to false. For any other result, it evaluates to false.

The 'and', 'or', 'xor', andalso, and orelse constructs all take a tuple of size 3 or more, where the first element is the construct and the remaining ones are the logical expressions. Using 'and' or 'or' will evaluate all expressions in the construct, whereas andalso and orelse will stop evaluating as soon as one expression returns false or true, respectively. For 'xor' to evaluate to true, one of its expressions must evaluate to true and the other to false. Here is an example of a conditional expression where all three expressions have to evaluate to true:

```
{'and', {'not', '$1'}, '$2', {'or', '$3','$4'}}.
```

Comparison operators take tuples with three elements, where the first element is one of the operators '>', '>=', '<', '=<', '=:=', '==', '=/=', or '/=', and the remaining two contain expressions whose result is compared. Combining guards, conditional expressions, and comparison operators could give you something similar to the following format:

```
{'and', {is_integer, '$0'}, {is_integer, '$1'}, {'>=', '$0', '$1'}}.
```

The following BIF operations are also allowed in conditional expressions. They behave like their counterpart BIFs and consist of a tuple of size 2 or 3, depending on how many arguments the BIF requires. The operations abs, hd, tl, length, node, round, size, trunc, 'bnot', 'bsl', and 'bsr' all take one argument. Using element, '+', '-', '*', 'div', 'rem', 'band', 'bor', and 'bxor' require a tuple with the operator and two arguments. The BIF operations {self} and {get_tcw} return the process identifier of the calling process or the trace control word. Even if they are most commonly used in the match specification body, they are also allowed as arguments to conditional expressions.

Enough with the syntax and semantics, as it is becoming too much to digest in one go. The best way to tackle it all is to look at some examples and try out some of your own. Can you figure out what triggers the following match specification? Try to do this before you read on:

```
[{['$1', '$2', '$3'],
  [{orelse,
     {'=:=','$1', '$2'},
     {'and',
         {'=:=','$1', {hd, '$3'}},
         {'=:=','$2', {hd, {tl, '$3'}}}}
   }
  ],
  []
}]
```

Let's break up the preceding code bit by bit. The head tells us the function is of arity 3. The condition is an orelse; the first condition here is `'$1' '=: =' '$2'`, requiring that the first argument be exactly equal to the second. If this fails, we try the conjunction (and) of `{'=:=','$1', {hd, '$3'}}`and `{'=:=', '$2', {hd, {tl,'$3'}}}`. From this we can deduce that `'$3'` has to be a list whose first element `{hd, '$3'}` is equal to the first argument `'$1'` and whose second element (the head of the tail) is equal to `'$2'`, the second argument. In Erlang terms, the third argument would be a list where `'$3'` = `['$1','$2'|'_']`. Traced calls that would trigger a match would include `foo(1,2, [1,2,3])` and `foo(true, true, false)`. As an exercise you could try to define a fun that will be transformed to this using `dbg:fun2ms/1`.

As you can see from the example, quite a few operators and constructs are allowed. They all work in the same way as the corresponding guards, BIFs, and operators. The rule of thumb is that if they are allowed in guards, they will be allowed in match conditions.

In case of bad arguments, where the data of the wrong type is bound to the variables that are passed to these operations, the condition fails. The process executing the match specification does not crash.

Let's look at a few more examples. The first one takes two arguments and will match if `'$1'` is greater than or equal to 0 and `'$2'` is less than or equal to 10. Calling the traced function `foo(0,10)` or `foo(5,5)` would successfully trigger the trace event:

```
[{['$1', '$2'],
  [{'and', {'>=', '$1', 0},
           {'=<', '$2', 10}}],
  []
}]
```

The next example will match if the first argument to the function of arity 2 is the atom enable or disable and the second argument is the tuple where the first element is another tuple, {slot, 1, 3}. Note how we have used the {const, Term} construct in the example to denote the atoms enable and disable, and then encapsulated these tuples in another tuple.

As conditions and the body of match specifications consist of tuples, we have to either use the {const, Term} construct, or insert the tuple in a tuple (as in {{slot, 1,3}}) to differentiate a "real" tuple from a conditional construct and the BIF and comparison operators:

```
[{['$1', '$2'],
  [{'and', {'orelse',{'=:=', '$1', {const, enable}},
                     {'=:=', '$1', {const, disable}}},
           {'=:=', {element, 1, '$2'}, {{slot, 1, 3}}}
  }],
  []
}]
```

We would use the preceding specification if we were tracing specific operations on boards in slot 1 placed in subrack 3. Examples of functions that would trigger a trace event include the function calls board:action(enable, {{slot, 1,3}, disabled}) and board:action(disable, {{slot, 1,3}, unknown}).

As you will see in the next section, upon triggering the specification, we could turn on a particular trace flag in the body part of the specification. Before looking at how to do it, let's try to clean up the preceding example, adding a second match specification to the list. In a live environment, both match specifications would yield the same result, but the example that follows allows only tuples of size 2. Using the equivalent of the element BIF, we are stating that the second argument has to be a tuple with at least two elements:

```
[{[enable, {{slot, 1, 3}, '_'}],
  [],
  []},
 {[disable, {{slot, 1, 3}, '_'}],
  [],
  []}
]
```

The Specification Body

If the match is successful and the conditions evaluate to true, we can specify a set of actions, including sending trace events, printing terms, enabling and disabling trace flags, and returning trace information. The same rules as those given in the previous section apply here to variables and literals; actions must always be defined in tuples, even if they take no arguments. The same rules as earlier also apply to '$_' and '$$': namely, '$_' returns the whole parameter list, and '$$' returns the list of all bound variables.

In the following example, if the first match specification is successful, the procs flag will be added to the list of trace flags; if the second match specification is successful, it will remove that flag from the list:

```
[{[enable, {{slot, 1, 3}, '_'}],
  [],
  [{enable_trace, procs}]},
 {[disable, {{slot, 1, 3}, '_'}],
  [],
  [{disable_trace, procs}]}
]
```

The syntax of the specification body to enable and disable calls is of the form {enable_trace, TraceFlag} and {disable_trace, TraceFlag} where TraceFlag is a single flag as defined in the section "The Trace BIFs" on page 357. The enable_trace and disable_trace actions have the same effect of calling the following BIF:

```
erlang:trace(self(), Flag, [TraceFlag])
```

where Flag is replaced by the atom true or false, respectively enabling or disabling the TraceFlag.

The following actions can be executed in the match specification body:

{message, Args}
: Appends the Args to the trace events. The Args could be variables, arguments, or return values of other actions such as process_dump.

{message, false | true}
: Is a special case where messages of the call and return_to trace flags are not sent if the flag is false. Passing true results in a regular trace message and is the same as not having a body at all.

{return_trace}
: Causes a trace message to be sent upon the return of this function. If the function is tail-recursive, this property is lost.

{exception_trace}
: Behaves in the same way as return_trace, with the exception that if the traced function exits because of a runtime error, an exception_from trace message is generated.

{silent, true | false}
: Turns trace messages on and off for that process, but not the tracing itself.

{display, Term}
: Displays a single Term on stdout, and should be used only for debugging purposes!

{set_tcw, Value}
: Sets the trace control word of a node, returning the previous value. Calling erlang:system_flag(trace_control_word, Value) will have the same effect as executing the command in the specification body.

{enable_trace, TraceFlag}
: Enables any of the trace flags accepted by the trace BIF. For every TraceFlag that you want to enable, you need a separate enable_trace call. This would have the same effect as calling erlang:trace(self(), true, [TraceFlag]).

{disable_trace, TraceFlag}
: Disables the TraceFlag. Executing this function in the match specification would be the equivalent of calling the function erlang:trace(self(), false, [TraceFlag]).

{trace, DisableList, EnableList}
: Will disable the flags defined in the DisableList while atomically enabling the ones defined in the EnableList. The flags are the same as those that can be used in the erlang:trace/3 BIF, but you may not include the cpu_timestamp flag. You may, however, include the {tracer, Pid} specifying which Pid should receive the trace events. If defined in both lists, flags (including the tracer directive) in the EnableList take precedence over those in the DisableList.

`{trace, Pid, DisableList, EnableList}`
> Is the same as the function with two arguments, but also allows you to define a process identifier or a registered alias on which the flags should be enabled or disabled. The aggregation of the flags is atomic.

The following calls are used to return values, most commonly used as an argument to the `message` and `display` actions just described:

`{process_dump}`
> Returns textual information on the process, formatted as a string and stored as a binary.

`{caller}`
> Returns the calling function in the format `{Module, Function, Arity}`. The trace event will return `undefined` if the function cannot be determined.

`{get_tcw}`
> Returns the trace control word previously set. This call has the same effect as calling the BIF `erlang:system_info(trace_control_word)`.

`{pid}`
> Returns the pid of the process executing the match specification.

In the following example, we append the process dump to the trace message, pick it up in the tracer fun we have defined ourselves, convert it from a binary to a string, and print it out. Note that the process stack of the shell is deeper. We have removed some of the trailing functions. Also, note how in the head section of the match specification we pass in `'_'`, denoting a function of any arity:

```
1> DbgFun = fun({trace, _Pid, _event, _data, Msg}, _Acc) ->
1>                 io:format("~s~n",[binary_to_list(Msg)])
1>             end.
#Fun<erl_eval.12.113037538>
2> dbg:tracer(process, {DbgFun, null}).
{ok,<0.33.0>}
3> dbg:tp({io,format,1}, [{'_',[],[{message,{process_dump}}]}]).
{ok,[{matched,nonode@nohost,1},{saved,1}]}
4> dbg:p(all,[c]).
{ok,[{matched,nonode@nohost,25}]}
5> io:format("Hello~n").
Hello
=proc:<0.30.0>
State: Running
Spawned as: erlang:apply/2
Spawned by: <0.24.0>
Started: Tue Oct 07 13:17:07 2008
Message queue length: 0
Number of heap fragments: 0
Heap fragment data: 0
Link list: []
Reductions: 8879
Stack+heap: 2584
OldHeap: 2584
```

```
Heap unused: 1271
OldHeap unused: 2584
Stack dump:
Program counter: 0x01b0e904 (shell:eval_loop/3 + 44)
CP: 0x01b02388 (erl_eval:do_apply/5 + 1304)

0x01473450 Return addr 0x01b0f15c (shell:exprs/6 + 368)
y(0)     [{'DbgFun',#Fun<erl_eval.12.113037538>}]
y(1)     []
y(2)     none

0x01473460 Return addr 0x01b0ec40 (shell:eval_exprs/6 + 80)
y(0)     []
y(1)     []
y(2)     [{'DbgFun',#Fun<erl_eval.12.113037538>}]
y(3)     {value,#Fun<shell.7.51306786>}
y(4)     {eval,#Fun<shell.24.79061235>}
y(5)     13
y(6)     []
y(7)     []
y(8)     []
```

In this last example, we are trying to find the caller of the io:format/2 function in the shell. We see the output formatted as the tuple {erl_eval, do_apply, 5} right after the call in command 4:

```
1> dbg:tracer().
{ok,<0.32.0>}
2> dbg:p(all, [call]).
{ok,[{matched,nonode@nohost,25}]}
3> dbg:tp({io, format, 2}, [{'_', [], [{display, {caller}}] }]).
undefined
{ok,[{matched,nonode@nohost,1}, {saved,1}]}
4> io:format("Hello~n",[]).
{erl_eval,do_apply,5}
Hello
(<0.30.0>) call io:format("Hello~n",[])
ok
```

Saving Match Specifications

Look at the preceding example, more specifically the return value of the dbg:tp/2 call. It returns {ok,[{matched,nonode@nohost,1},{saved,1}]}. You already know about the first inner component, {matched, nonode@nohost, 1}; it tells you that on this particular node, one function was matched. The {saved, 1}, however, tells you the match specification in the call was saved with this identifier, that is, the number 1. The dbg tracer assigns integers to all match specifications, allowing you to use this integer instead of the specification itself.

The help functions for retrieving and manipulating these calls include the following:

dbg:ltp()
  Recalls all match specifications used in the session

---

dbg:dtp() and dbg:dtp(Id)
> Delete the stored match specifications (with a particular identifier)

dbg:wtp(FileName) *and* dbg:rtp(Filename)
> Write and read match specifications from a file:

```
5> dbg:tp({io,format,2},[{['_'],[],[{enable_trace, procs}]}]).
{ok,[{matched,nonode@nohost,1},{saved,2}]}
6> dbg:ltp().
1: [{'_',[],[{display,{caller}}]}]
2: [{['_'],[],[{enable_trace,procs}]}]
exception_trace: x
x: [{'_',[],[{exception_trace}]}]
ok
```

Starting with the R13 release of Erlang/OTP, there is a presaved match specification with the identifier exception_trace alias x. It is presaved as the match specification [{'_',[],[{exception_trace}]}] and can be used in calls such as dbg:tp({M, F, A}, x). It deserves a special mention, as it is probably one of the most commonly used specifications when debugging.

Further Reading

As you might have realized, the functionality for tracing live Erlang programs, although powerful, can be complex at times. In this chapter, we provided you with a comprehensive overview of what is available, which should enable you to handle the vast majority of cases you come across. But as you will probably be dealing with live systems running nonstop under high load, you never know what to expect.

There are two functions, set_seq_token/2 and get_seq_token/0, in the dbg:fun2ms/1 fun body call, which we have not covered. They are translated to the sequential trace actions set_seq_token and get_seq_token in the match specification body. Together with the API exported by the seq_trace module, they allow users to follow message propagation paths across processes. If process A sends a message to process B, which as a result sends messages to processes C and D, this propagated message call chain is traced. If you are interested in learning more about them, read the manual page for seq_trace. This topic alone would deserve an additional chapter.

If you want to read more on the trace BIFs, the first point of call should be the erlang module manual page, where all the BIFs are described. If you are interested in exploring match specifications, the *ERTS User's Guide* dedicates a whole chapter to them, including a complete grammar tutorial. If you instead prefer to stay clear from match specifications and opt for match specification transforms using the fun2ms functions, the ms_transform manual page is probably your best bet. The match specification functions you pass to fun2ms, however, are documented in the *ERTS User's Guide*. And finally, the dbg manual page covers the dbg tracer tool.

Exercises

Exercise 17-1: Measuring Garbage Collection Times

Using the trace BIFs, write a program that monitors the number of microseconds a process spends garbage collecting in a specific function. Once the function has completed its execution, print out the time. To ensure that the readings are not affected by previously allocated memory, spawn a new process for every reading.

Test your programs with a tail-recursive function and a non-tail-recursive function, which do the same thing. This will allow you to monitor the impact that garbage collection has on the time of non-tail-recursive functions handling lists of different sizes.

Hint: use `timer:now_diff/2` to calculate the time difference.

You could use the following as a tail-recursive function:

```
average(List) -> sum(List) / len(List).

sum([]) -> 0;
sum([Head | Tail]) -> Head + sum(Tail);

len([]) -> 0;
len({_ | Tail]) -> 1 + len(Tail).
```

And, you could use this as a non-tail-recursive function:

```
average(List) -> average_acc(List, 0,0).

average_acc([], Sum, Length) -> Sum / Length;
average_acc([H | T], Sum, Length) -> average_acc(T, Sum + H, Length + 1).
```

When you execute your code, it might look something like this:

```
1> List = lists:seq(1,1000).
[1,2,3,4,5,6,7,8,9,10,11,12,13,14,15,16,17,18,19,20,21,22,
 23,24,25,26,27,28,29|...]
2> gc_mon:measure(gc_test, average, List).
Gc monitoring terminated
Microseconds:8
ok
```

Exercise 17-2: Garbage Collection Using dbg

Rewrite the solution to Exercise 17-1 using dbg and your own tracer fun. When doing so, add counters to measure how much memory is reclaimed during execution of the process. Pay special attention to the different memory types.

Exercise 17-3: Tracing ETS Table Entries

A crash report was logged. Further investigation shows that an ETS table called `coun tries` gets corrupted by the entry {'EXIT', Reason}. The tuple is written instead of a record of type `countries`. The table is updated thousands of times per day, and it would thus not be appropriate to trace all table entries as the overhead and amount of data to be filtered would be too large.

Create a match specification that is triggered every time the tuple {'EXIT', Reason} is passed as the second argument to the `ets:insert/2` function. When you get the match specification to work, start a separate Erlang node on which your tracer receives and prints all trace messages.

When it works, rewrite the match specification that triggers on both {'EXIT', Reason} and {'EXIT', Pid, Reason} using conditional clauses.

Exercise 17-4: Who Is the Culprit?

Determine which of the two authors of this book managed to single-handedly cause a nationwide outage of a mobile data network through the careless use of the dbg tracer tool.

CHAPTER 18

Types and Documentation

The basic types in Erlang—integers, floating-point numbers, atoms, strings, tuples, and lists—were introduced in Chapter 2; records were covered in Chapter 7; and further types—binaries and references—in Chapter 9. When we have declared functions and other definitions, we have also given an informal description of the types of their inputs and outputs.

This chapter shows how you can write down the types of functions as a part of their formal documentation in Erlang, using the EDoc documentation framework, written by Richard Carlsson. What you write down as the type of a function can be checked for consistency against the function definition using the TypEr tool, built by the implementers of Dialyzer. TypEr will infer types without any user input, and so it can be an essential tool for program understanding. TypEr and Dialyzer are the result of the High Performance Erlang (HiPE) team's research at Uppsala University. All of these tools are part of the standard Erlang distribution.

Types in Erlang

Let's start this chapter with an example and follow it with an overview of the type notation for Erlang.

An Example: Records with Typed Fields

We discussed record definitions earlier in the book. In the example of the mobile user database in Chapter 10, you saw a declaration of a record to hold information about a particular user of the mobile phone system:

```
-record(usr, {msisdn,          %int()
              id,              %term()
              status = enabled %atom(), enabled | disabled
              plan,            %atom(), prepay | postpay
              services = []}). %[atom()], service flag list
```

The usr record has five fields, and in the comments that follow, each field type is indicated.

You can take that a step further, however, and make these comments a part of the program itself. First, you introduce type declarations defining types for the kind of plan, user status, and service:

```
-type(plan()     :: prepay | postpay).
-type(status()   :: enabled | disabled).
-type(service()  :: atom()).
```

The three type declarations define the plan, status, and service types. As with constant functions, they are followed by a set of parentheses, ():

plan()

Has two elements, the atoms prepay and postpay. You use the | symbol to indicate alternatives (as you would in regular expressions and grammars). Here, the alternatives are the two possible members of the type.

status()

Also has two elements, the atoms enabled and disabled.

service()

Is a synonym for atom(), but makes it clear that when it is used, the *intention* is for an atom in this position to represent a service of some kind.

With these definitions in place, you can explicitly give types to the record fields in this *record with typed fields*:

```
-record(usr, {msisdn              ::integer(),
              id                  ::integer(),
              status = enabled    ::status(),
              plan                ::plan(),
              services = []       ::[service()]
}).
```

This more clearly indicates what the elements of the usr record type should be and how they should be used by programs that manipulate them. But most importantly, it provides tools with information that will help you detect type errors in the code. This short example provided you with a taste of the Erlang type notation. Now let's cover it in more detail.

Erlang Type Notation

A number of Erlang types are predefined, and they include the following:

any()

Includes *all* Erlang data values (as does its synonym, term()).

atom()

Includes all Erlang atoms.

binary()

Consists of all binary values.

`boolean()`
> Contains the atoms true and false.

`byte()`
> Contains the numbers 0–255.

`char()`
> Is the Unicode subset of the `integer()` type.

`deep_string()`
> Is a recursive type, equal to `[char+deep_string()]`.

`float()`
> Consists of all floating-point numbers.

`function()`
> Contains all funs

`integer()`
> Consists of all integer values; recall that these are "big integers."

`list(T)`
> Is the type of a list of type T; it can also be written `[T]`.

`nil()`
> Has one element, the empty list `[]`.

`none()`
> Is a void type with no elements; it is used as the return type of functions that never return.

`number()`
> Consists of the union of the `float()` and `integer()` types.

`pid(), port(), reference()`
> Are all self-explanatory.

`string()`
> Is a synonym for the list of characters, `[char()]`.

`tuple()`
> Contains all tuples.

You can find details of other predefined types in the EDoc and Dialyzer documentation. In addition to the aforementioned predefined types, you can define your own types using the following notation:

atom
> Any atom can be used as a type; for example, the type ok has the element ok.

| or +
> You can use these to form the union of two types, as in `true|false`.

#rec{}
> This is the record type named *rec*.

{T1,T2,...}

> This is the type of a tuple whose first element comes from type T1, second element from type T2, and so on. So, for example, {error,atom()} consists of all pairs where the first element is the atom error and the second is any atom.

[T]

> This is the type of a list whose elements come from the type T.

L..U

> This is the range of integers from lower bound L to upper bound U. This is used in defining byte() and char() and a number of other built-ins, including pos_integer().

A function type has the following form:

```
(Argument_types) -> Result_type
```

where the argument types may also contain the names of the corresponding parameters.

You use a -spec statement to specify a type (or prototype) for a function in a program. To be clear, -spec is used to *specify the type* of a function, whereas –type is used to *define a type*. Revisiting the usr_db example from Chapter 10, you can say:

```
-spec(create_tables(FileName::string()) -> {ok, ref()} | {error, atom()}).

create_tables(FileName) ->
    ... .

-spec(close_tables() -> ok | {error, atom()}).

close_tables() ->
    ... .
```

The type of create_tables here indicates that the FileName argument is a string, and that there are two alternatives for the return type:

- If the operation is successful, this is signaled by returning a tuple with the first element ok, and the second element a reference to the tables created.
- If it fails, a tuple with the first element error will be returned. Its second element is an atom, presumably indicating the nature of the error.

The use of FileName here is optional. Writing the following would have the same effect:

```
-spec(create_tables(string()) -> {ok, ref()} | {error, atom()}).
```

However, including the parameter name—assuming it is chosen to reflect its purpose—improves the documentation.[*]

---

[*] The EDoc system described later in the chapter will automatically include parameter names in its generated type documentation even if they do not appear in the -spec statement.

TypEr: Success Types and Type Inference

The TypEr system, built by Tobias Lindahl and Kostis Sagonas,[†] is used to check the validity of -spec annotations, as well as to infer the types of functions in modules without type annotations.

You use TypEr from the command line. You can see the full range of options by typing:

```
typer --help
```

Taking the example of the mobile user database from Chapter 10, the following command:

```
typer --show usr.erl usr_db.erl
```

gives the following output (shortened for brevity):

```
Unknown functions: [{ets,safefixtable,2}]

%% File: "usr.erl"
%% --------------
-spec start() -> 'ok' | {'error','starting'}.
-spec start(_) -> 'ok' | {'error','starting'}.
-spec stop() -> any().
-spec add_usr(_,_,_) -> any().
-spec delete_usr(_) -> any().
 ...

%% File: "usr_db.erl"
%% ------------------
-spec create_tables(_) -> any().
-spec close_tables() -> any().
-spec add_usr(#usr{}) -> 'ok'.
-spec update_usr([tuple()] | tuple()) -> 'ok'.
-spec delete_usr(_) -> 'ok' | {'error','instance'}.
 ...
```

In a *statically typed* language such as Haskell, the type of a function inferred by the type checker will provide a guarantee that the function will not fail if applied to arguments of the input type. Erlang is a *dynamically typed* language, and so the TypEr tool takes a different approach.

> TypEr infers *success types*, which encapsulate all the ways in which a function can be applied successfully. In general, this cannot be accurate, but it will always be an *overapproximation*, so using the function in any other way will be guaranteed to fail.
>
> To make this clear, if a function f is given a success typing (S) -> T, and E is any Erlang expression so that f(E) successfully evaluates to V, then E must be of type S, and V of type T.

[†] TypEr is described in the papers from the Erlang Workshops in 2005 (Tallin) and 2007 (Freiburg) (*http://doi .acm.org/10.1145/1088361.1088366* and *http://doi.acm.org/10.1145/1292520.1292523*).

In the usr_db.erl example earlier, no useful information could be inferred for create_tables/1, as its input and output types are any(). Just to be clear, TypEr was applied to a version of usr_db *without* type annotations; if you add these, the result of applying TypEr may be different:

```erlang
-spec(create_tables(string()) -> {ok, ref()} | {error, atom()}).

create_tables(FileName) ->
  ets:new(subRam, [named_table, {keypos, #usr.msisdn}]),
  ets:new(subIndex, [named_table]),
  dets:open_file(subDisk, [{file, FileName}, {keypos, #usr.msisdn}]).

-spec(close_tables() -> ok | {error, atom()}).

close_tables() ->
  ets:delete(subRam),
  ets:delete(subIndex),
  dets:close(subDisk).

-spec(add_usr(#usr{}) -> ok).

add_usr(#usr{msisdn=PhoneNo, id=CustId} = Usr) ->
  ets:insert(subIndex, {CustId, PhoneNo}),
  update_usr(Usr).

-spec(update_usr(#usr{}) -> ok).

update_usr(Usr) ->
  ets:insert(subRam, Usr),
  dets:insert(subDisk, Usr),
  ok.

-spec(delete_usr(integer()) -> ok|{error,atom()}).

delete_usr(CustId) ->
  case get_index(CustId) of
    {ok,PhoneNo} ->
      delete_usr(PhoneNo, CustId);
    {error, instance} ->
      {error, instance}
  end.

-spec(delete_usr(integer(),integer()) -> ok|{error,atom()}).

delete_usr(PhoneNo, CustId) ->
  dets:delete(subDisk, PhoneNo),
  ets:delete(subRam, PhoneNo),
  ets:delete(subIndex, CustId),
  ok.

...
```

Running TypEr on the annotated file will check the specified types, and will give this result:

```
%% File: "usr_db.erl"
%% -----------------
-spec create_tables(string()) -> {'ok',ref()} | {'error',atom()}.
-spec close_tables() -> 'ok' | {'error',atom()}.
-spec add_usr(#usr{}) -> 'ok'.
-spec update_usr(#usr{}) -> 'ok'.
-spec delete_usr(integer()) -> 'ok' | {'error',atom()}.
-spec delete_usr(integer(),integer()) -> 'ok' | {'error',atom()}.
  ...
```

In doing this, TypEr will check the specified type against the inferred type, and report any inconsistencies. For example, if you change the –spec for add_usr to the following:

```
-spec(add_usr(#usr{}) -> integer()).
```

TypEr will report this:

```
typer: Error in contract of function usr_db:add_usr/1
    The contract is: (#usr{}) -> integer()
    but the inferred signature is: (#usr{}) -> 'ok'
```

On the other hand, if you change the spec of create_tables/1, no error will be reported, since the inferred type for this function is consistent with *any* one-argument function type.

Dialyzer: A DIscrepancy AnaLYZer for ERlang Programs

TypEr gives an analysis of types in Erlang programs. Dialyzer extends this to perform static analysis on Erlang programs to identify *software discrepancies*, including redundant tests and unreachable code, as well as obvious type errors.

To speed up its operation, Dialyzer can create a Persistent Lookup Table (PLT), using the --build_plt option. When you include the kernel, standard libraries, and Mnesia, as shown here:

```
dialyzer --build_plt -r <erl-lib>/kernel-2.12.5/ebin <erl-lib>/
stdlib-1.15.5/ebin <erl-lib>/mnesia-4.4.7/ebin
```

it takes some minutes to generate the PLT and produce this report:

```
Creating PLT /Users/simonthompson/.dialyzer_plt ...
re.erl:41: Call to missing or unexported function unicode:characters_to_binary/2
re.erl:134: Call to missing or unexported function unicode:characters_to_list/2
re.erl:200: Call to missing or unexported function re:compile/2
re.erl:226: Call to missing or unexported function unicode:characters_to_binary/2
re.erl:245: Call to missing or unexported function unicode:characters_to_list/2
re.erl:505: Call to missing or unexported function unicode:characters_to_list/2
re.erl:545: Call to missing or unexported function unicode:characters_to_binary/2
Unknown functions:
  compile:file/2
  compile:forms/2
  compile:noenv_forms/2
  compile:output_generated/1
  crypto:des3_cbc_decrypt/5
  crypto:start/0
```

```
  done in 16m43.44s
done (warnings were emitted)
```

Subsequently calling Dialyzer on the files for the running example gives this report (in less than a second):

```
dialyzer -c usr.erl usr_db.erl
  Checking whether the PLT /Users/simonthompson/.dialyzer_plt is up-to-date... yes
  Proceeding with analysis...
usr.erl:110: The pattern [] can never match the type {'error','instance'}
usr_db.erl:69: Call to missing or unexported function ets:safefixtable/2
  done in 0m0.33s
done (warnings were emitted)
```

You can find more information about Dialyzer in the online documentation.

Documentation with EDoc

It is sometimes said that functional programs are self-documenting. Sadly, although functional programming languages may produce programs that are more readable, complex programs are not self-documenting, let alone obvious to understand. Free text comments in program modules, if kept up-to-date, are a first step in describing a program. However, they have the disadvantage of lacking structure as well as being difficult to scan, search, and read independently of the program text.

EDoc provides a *documentation framework* for Erlang that overcomes these disadvantages and generates documentation from information you have inserted in your modules:

- EDoc provides a *structure* for comments, including **type** and **spec** information as well as textual comments on functions.

- EDoc is a *documentation generator*: a standard style of HTML document is generated from the structured information in each module.

- EDoc provides a framework for adding information covering a whole set of modules (e.g., in an application or a package), providing an *overview* of the larger-scale structure or assumptions for the whole system.

EDoc is part of the standard Erlang distribution, and it has a lot in common with similar systems such as Haddock (for Haskell), Javadoc, pydoc, and RDoc (for Ruby). In this section, we will introduce you to many of EDoc's features by documenting the mobile user database example from Chapter 10.

 It is intended that in later releases of Erlang/OTP, EDoc will share information with TypEr, and so will use the type information in the -type and -spec declarations. In the meantime, this typing information is given in a different format, described in this section.

Documenting usr_db.erl

EDoc generates documentation from tags of the form `@tag text`, embedded within comments. Each tag can continue over multiple lines, until the next tag or noncomment line. You can use different kinds of tags in different ways:

Module tags

Provide module-level documentation. Occurrences must precede the module declaration itself.

Function tags

Are associated with the function that follows them, and give information about that particular function.

Generic tags

Might include "to do" information or a type definition, but can occur anywhere within a file.

We will look at these in turn for the usr_db example.

Module tags

In the example, the module tags are as follows:

```
%% @author Francesco Cesarini <support@erlang-consulting.com>
%% @author Simon Thompson [http://www.cs.kent.ac.uk/~sjt/]
%% @doc Back end for the mobile subscriber database.
%% The module provides an example of using ETS and DETS tables.
%% @reference <a href="http://oreilly.com/catalog/9780596518189/">
   Erlang Programming</a>,
%%   <em> Francesco Cesarini and Simon Thompson</em>,
%%   O'Reilly, 2009.
%% @copyright 2009 Francesco Cesarini and Simon Thompson
```

The tags give information about the following:

`@author`

The author(s) of the module and optional contact information (email or HTML).

`@copyright`

A copyright statement.

`@doc`

A description of the module, in well-formed XHTML text. The first sentence of this text is used as a summary of the module.

`@reference`

A reference giving further information, which can include XHTML links.

You can use other tags to give information about version number (`@version`), when the module was introduced into the system (`@since`), and whether the documentation should be visible or not (`@hidden` or `@private`).

Function tags

The principal documentation for a function is given by its type and a description of what it does:

@spec

> Gives the type for a function. The form of the type description was given earlier. In future versions, it is expected that this will be replaced by use of information in a -spec declaration.

@doc

> Indicates the general documentation for a function, using XHTML markup.

Other available tags include a cross-reference to another object's documentation (@see), a description of which types of exceptions can be thrown (@throws), and whether a function is deprecated (@deprecated); @hidden, @private, and @since, as described earlier, can also be used.

For a fragment of the usr_db module, the documentation will be as follows:

```
%% @doc Create the ETS and DETS tables which implement the database. The
%% argument gives the filename which is used to hold the DETS table.
%% If the table can be created, an 'ok' tuple containing a
%% reference to the created table is returned; if not, it returns an 'error'
%% tuple with an atom describing the error.

%% @spec create_tables(string()) -> {ok, reference()} | {error, atom()}

-spec(create_tables(string()) -> {ok, ref()} | {error, atom()}).

create_tables(FileName) ->
  ets:new(subRam, [named_table, {keypos, #usr.msisdn}]),
  ets:new(subIndex, [named_table]),
  dets:open_file(subDisk, [{file, FileName}, {keypos, #usr.msisdn}]).

%% @doc Close the ETS and DETS tables implementing the database.
%% Returns either 'ok' or an 'error'
%% tuple with the reason for the failure to close the DETS table.

%% @spec close_tables() -> ok | {error, atom()}

-spec(close_tables() -> ok | {error, atom()}).

close_tables() ->
  ets:delete(subRam),
  ets:delete(subIndex),
  dets:close(subDisk).

%% @doc Add a user (of the 'usr' record type) to the database.

%% @spec add_usr(#usr{}) -> ok

-spec(add_usr(#usr{}) -> ok).
```

```
add_usr(#usr{msisdn=PhoneNo, id=CustId} = Usr) ->
    ets:insert(subIndex, {CustId, PhoneNo}),
    update_usr(Usr).
```

In the preceding code, we provided the types using `@spec`; it is expected that in future releases of Erlang, `-spec` declarations will be used instead, making `@spec` tags superfluous.

Generic tags

Generic tags can appear anywhere in a module:

`@type`

Will give a definition of a type and is picked up by EDoc to include in the generated documentation. It is expected that in future releases this will be replaced by the use of `-type` declarations.

`@todo`

Used to indicate "to do" notes; these will not appear in the generated documentation unless the `todo` option is activated.

Running EDoc

The main EDoc functions are in the `edoc` module. A call to `edoc:application/1` will generate the documentation for an application, and a call to `edoc:files/1` will generate the documentation for a set of files; these functions use the default EDoc options. Two-argument versions of the functions allow a set of option choices to be passed in the second argument.

Module pages

Figure 18-1 shows the page generated for the `usr_db` module that results from running `edoc:files(["usr_db.erl", "usr.erl"])`. This shows the structure of a typical EDoc page for an Erlang module.

The page begins with links to its major sections, and is followed by the one-sentence summary of the module and other module tags; the full description of the module follows.

In the function index, functions are listed in alphabetical order, rather than in their order in the file. Each function is hyperlinked to its details, and is given a one-sentence overview: the first sentence of its `@doc` tag.

Each function has details regarding its type. Note that the information contains not only the types of the arguments and results, but also the names of the arguments that are automatically extracted from the source code. For instance, the `@spec` for `create_tables/1` says the following:

```
%% @spec create_tables(string()) -> {ok, reference()} | {error, atom()}
```

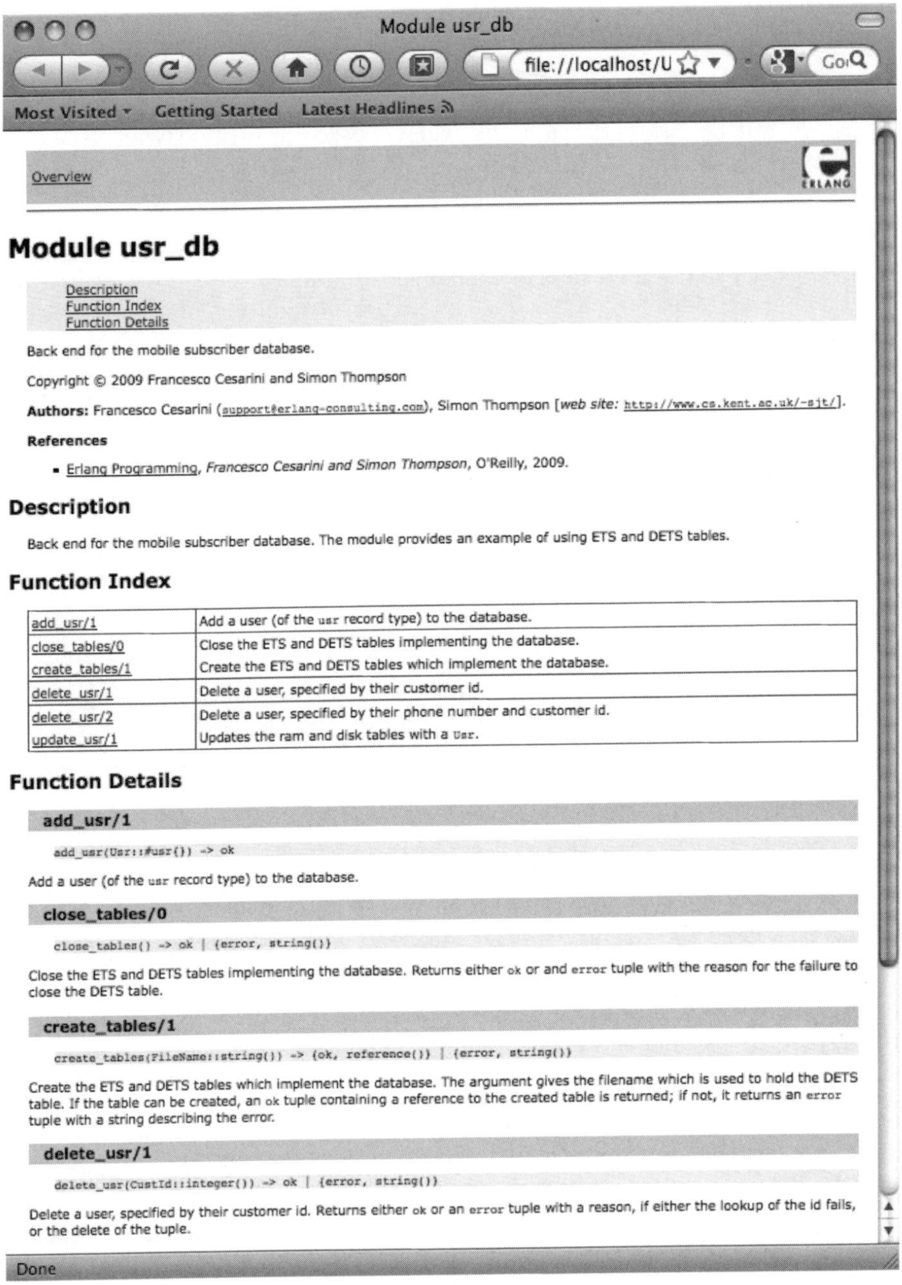

file://localhost/U   GoQ

Most Visited ▾   Getting Started   Latest Headlines ⟲

Overview

Module usr_db

Description
Function Index
Function Details

Back end for the mobile subscriber database.

Copyright © 2009 Francesco Cesarini and Simon Thompson

Authors: Francesco Cesarini (support@erlang-consulting.com), Simon Thompson [*web site:* http://www.cs.kent.ac.uk/~sjt/].

References

- Erlang Programming, *Francesco Cesarini and Simon Thompson*, O'Reilly, 2009.

Description

Back end for the mobile subscriber database. The module provides an example of using ETS and DETS tables.

Function Index

add_usr/1	Add a user (of the usr record type) to the database.
close_tables/0	Close the ETS and DETS tables implementing the database.
create_tables/1	Create the ETS and DETS tables which implement the database.
delete_usr/1	Delete a user, specified by their customer id.
delete_usr/2	Delete a user, specified by their phone number and customer id.
update_usr/1	Updates the ram and disk tables with a Usr.

Function Details

add_usr/1

`add_usr(Usr::#usr{}) -> ok`

Add a user (of the usr record type) to the database.

close_tables/0

`close_tables() -> ok | {error, string()}`

Close the ETS and DETS tables implementing the database. Returns either ok or and error tuple with the reason for the failure to close the DETS table.

create_tables/1

`create_tables(FileName::string()) -> {ok, reference()} | {error, string()}`

Create the ETS and DETS tables which implement the database. The argument gives the filename which is used to hold the DETS table. If the table can be created, an ok tuple containing a reference to the created table is returned; if not, it returns an error tuple with a string describing the error.

delete_usr/1

`delete_usr(CustId::integer()) -> ok | {error, string()}`

Delete a user, specified by their customer id. Returns either ok or an error tuple with a reason, if either the lookup of the id fails, or the delete of the tuple.

Done

Figure 18-1. EDoc page for (a fragment of) usr_db.erl

Here is the documentation that is produced:

```
create_tables(FileName::string()) -> {ok, reference()} | {error, atom()}
```

This gives you the useful extra information that the string argument represents a filename. Without any @spec information, EDoc will still include the names of the parameters in the generated information.

Overview page

For each project, an overview page will be generated, providing an index for the modules in the project. Further information can be provided for this in an *overview.edoc* file, which should typically appear in a *doc* subdirectory where the other documentation will be placed.

The *overview.edoc* file has the same content tags as the header for a module, but it is not necessary to enclose each line in a comment. The results of this for the running example are shown in Figure 18-2, as generated from a file that begins like this:

```
@author Francesco Cesarini <support@erlang-consulting.com>
@author Simon Thompson [http://www.cs.kent.ac.uk/~sjt/]
@reference <a href="http://oreilly.com/catalog/9780596518189/">Erlang Programming</a>,
  <em> Francesco Cesarini and Simon Thompson</em>,
  O'Reilly, 2009.
```

Types in EDoc

The usr_db.erl module contains no type definitions, but usr.hrl contains a number of these, and they are referenced in usr.erl. Types are documented like this: the definition is given first, and this can be followed by an optional description:

```
%% @type plan() = prepay|postpay. The two payment types for mobile subscribers.
%% @type status() = enabled | disabled. The status of a customer can be enabled
%% or disabled.
%% @type service() = atom(). Services are specified by atoms, including
%% (but not limited to) 'data', 'lbs' and 'sms'. 'Data' confirms the user
%% has subscribed to a data plan, 'sms' allows the user to send and receive
%% premium rated smses, while 'lbs' would allow third parties to execute
%% location lookups on this particular user.
```

This is extracted in the documentation for the usr module, and if a defined type is used in the @spec for a function, this is hyperlinked to its definition. The information about types appears in the documentation after the general description and before the function index. Figure 18-3 shows a fragment of the documentation for usr.erl, with the data type definitions appearing after the module description.

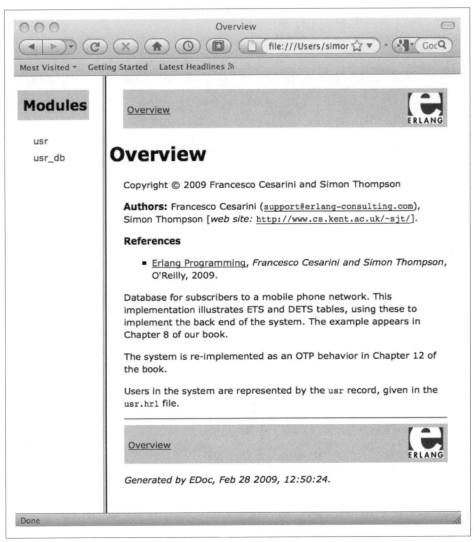

Figure 18-2. The EDoc overview page for the database

Going Further with EDoc

If you want to generate comprehensive documentation for applications and other projects, EDoc provides a number of facilities for formatting, cross-referencing, and preprocessing.

EDoc comes with a set of *predefined macros*, called by enclosing them in braces, thus: {@name} and {@name argument}. These include @date, @time, @module, and @version as well as the {@link reference. description} form. This creates a *link* to the object identified by the reference; the description gives the text for the anchor of the link.

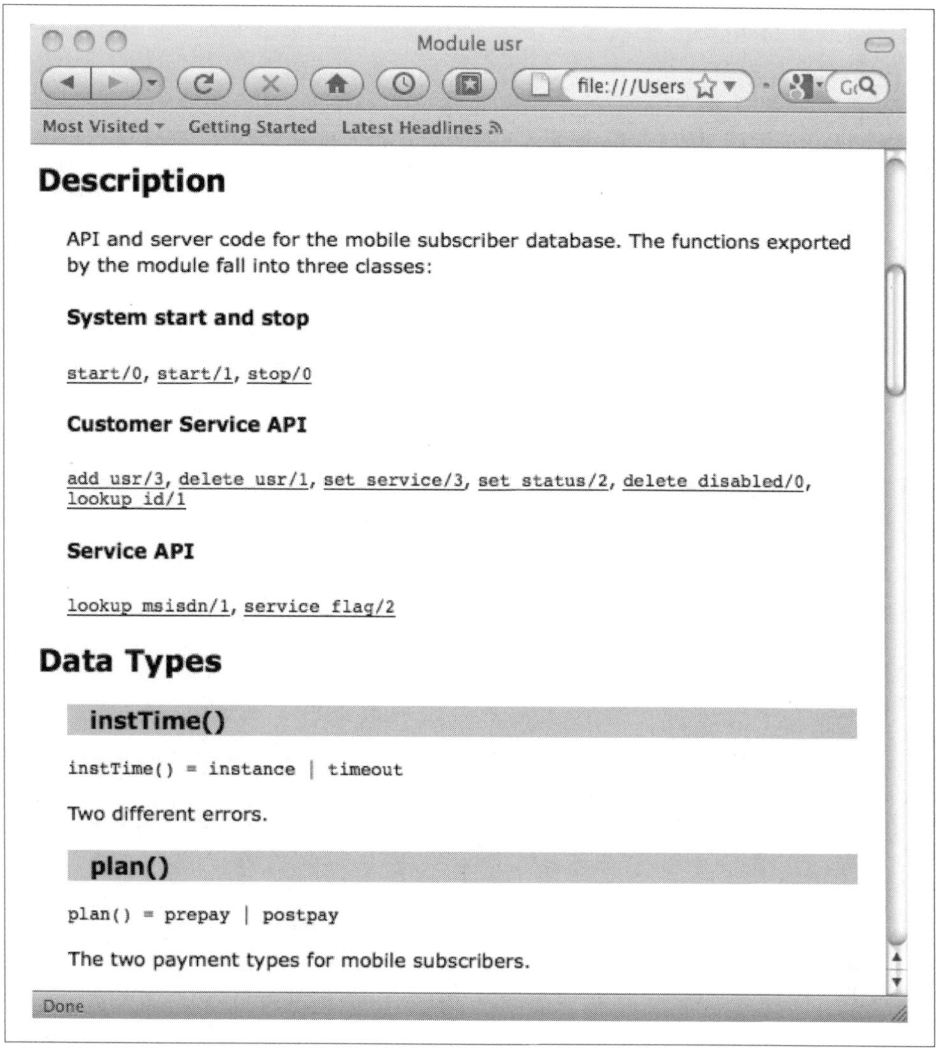

Figure 18-3. A fragment of the web page for usr.erl

References include name/arity for functions within the module, and Mod:name/arity for functions in the module Mod. A typename() refers to a type within the same module, and Mod:typename() in another. So, for example, the following link:

```
{@link set_status/2. 'set_status/2'}
```

gives a link to the documentation for the set_status/2 function in the same module.

In this example, the link is also labeled set_status/2: you use another EDoc facility for *verbatim quotation*. The quotation of the text '...' puts the enclosed text into verbatim (code) form, properly escaping any XHTML-significant characters in the text, such as <,

=, and >. y shows an example using the `@link` construct, where the description part of the documentation gives an index for the functions divided according to function, rather than alphabetically.

For author convenience, EDoc supports *wiki-style formatting* instead of XHTML. Blank lines are taken to delimit paragraphs (i.e., `<p>` ... `</p>` elements). Headings can be generated automatically from `==Heading==`, and similarly from `===`, however, `=` should *not* be used. This header markup will automatically generate an anchor (equivalent to `<a name="Heading">`), which can then be referred to by an EDoc macro, `{@section Heading}`. For further details on formatting and cross-referencing in EDoc, see the online documentation.

Exercises

Rather than give formal exercises for this chapter, we encourage you to try TypEr, Dialyzer, and EDoc on the solutions you wrote for exercises earlier in the book.

CHAPTER 19

EUnit and Test-Driven Development

As you are writing a program, how do you understand how the program will behave? You might have a model in your mind of what the program will do, but you can be sure of it only when you *exercise* or *interact* with your program in some way. Chapter 18 showed you how you can use -spec to express what you think the input and output types of a function should be; TypEr can check whether this is consistent with the code itself.

Types don't tell you how a program behaves, however, and *testing* is one of the best ways to understand how your code will function. We have been doing this informally throughout the book; each time we have given some definitions, we have immediately gone to the Erlang shell and tried them out in practice. When you're developing in Erlang your coding and test cycles tend to be small. You write a few functions, and you test them. You add a few more, and you test them again. Repeating all of the tests in the shell every time becomes both time-consuming and error-prone.

In this chapter, we'll introduce the EUnit tool, which gives you a *framework for unit testing in Erlang.* We'll show how it is used in practice, as well as discuss how it supports what is known in software engineering circles as test-driven development.

Test-Driven Development

The waterfall model of software development saw a software system being developed in a series of steps: first, the requirements of the system would be elucidated; on the basis of this, the system would be designed, and only once this was complete would the implementation begin. No wonder so many software projects failed to deliver what the customer wanted! Needless to say, the waterfall model is as disastrous in Erlang as it is in any other programming language.

A *test-driven* approach turns this on its head. Implementation begins on day one, but with a focus: each feature that is to be implemented is first characterized by a set of tests. This increment is added to the code and is accepted only if the tests pass. At the same time, this should not invalidate any earlier test; *regression tests* must also pass.

The implementer benefits from such an approach because she has a clear target at each stage: pass the tests. The customer is also in the happy position of being able to interact with or *exercise* the system continually: each time the evolving system builds, he can see whether it behaves as it ought to.

Test-driven development (TDD) has been associated with *agile programming*, but it would be wrong to identify the two. An informal test-driven approach has characterized functional programming since the early LISP systems: most functional programming systems have at their top level a read-evaluate-print loop, which encourages a test- or example-based approach. Erlang comes out of that tradition, with its roots in functional programming and Prolog, and the examples of interactions in the Erlang shell show this informal test-driven approach in practice.

Test-driven development is not confined to the single-person project. There is anecdotal evidence from industry that TDD can be effective in larger software projects, too, principally because it means the development team engages with potential problems earlier, and in the process develops more comprehensive test suites than it would with a more traditional approach. In turn, this leads to better software being delivered. It is needless to say that Erlang and TDD go hand in hand. In the remainder of this chapter, we will cover the EUnit system, which provides support for a more formal test-driven approach, and which is included in the standard Erlang distribution.

EUnit

EUnit provides a framework for defining and executing unit tests, which test that a particular program unit—in Erlang, a function or collection of functions—behaves as expected. The framework gives a representation of tests of a variety of different kinds, and a set of macros which simplify the way EUnit tests can be written.

For a function without side effects, a test will typically look at whether the input/output behavior is as expected. Additionally, it may test whether the function raises an exception (only) when required.

Functions that have side effects require a more complex infrastructure. Side effects include operations that might affect persistent data structures such as ETS or Dets tables, or indeed operating system structures such as files and I/O, as well as operations that contain message passing among concurrent processes. The infrastructure needed to test these programs includes the following:

- Testing *side-effecting programs* typically requires some setup and initial modification of the data before checking the behavior of a particular operation. This needs to be followed by a cleanup of the program state.

- Testing units within a *concurrent program* typically requires a test rig in which some mock objects or stubs are written to stand in the place of other components expected to interact with the unit under test.

EUnit supports testing side-effecting and concurrent programs. Let's start with the simplest case of functional testing.

How to Use EUnit

EUnit provides a detailed representation of tests, as well as a layer of macros making tests easier to write. It also provides a mechanism for gathering all the tests in a module, executing them all, and providing a report on the results. For a module to use EUnit it needs to include the `eunit` library:

```
-include_lib("eunit/include/eunit.hrl").
```

In EUnit, a single test is identified by a function named *name*_test; a test-generating function will be named *name*_test_ (i.e., with an _ at the end). We define test-generating functions in the section "The EUnit Infrastructure" on page 416.

With the EUnit header file included, compilation of the module (*mod*, say) will result in a function `test/0` being exported, and all the tests within the module are run by calling *mod*:`test()`.

It is possible to separate the test functions from the module (*mod*, say) into the module *mod*_tests.erl, which should also include the *eunit.hrl* header file. The tests are invoked in exactly the same way, using *mod*:`test()`.

It may be that you want to use EUnit on your code—for example, to use the `?assert` macro—but you don't want to generate tests. In this case, you need to insert the following code before the line that includes the `eunit.hrl` library:

```
-define(NOTEST, true)
```

If this macro appears in a header file included by all modules in an application, it gives a single point controlling whether testing is enabled. Now we will illustrate how tests are written, using an example you've seen before.

Functional Testing, an Example: Tree Serialization

In the section "Serialization" on page 208 in Chapter 9, we looked at how a binary tree could be serialized as a list, and then reconstructed from the list. We presented an optimized version of this, but left the original version as an exercise. In this unoptimized version, the first element of the representation gives its size recursively through the tree?

```
treeToList({leaf,N}) ->
  [2,N];

treeToList({node,T1,T2}) ->
  TTL1 = treeToList(T1),
  [Size1|_] = TTL1,
  TTL2 = treeToList(T2),
  [Size2|_] = TTL2,
  [Size1+Size2+1|TTL1++TTL2].
```

```
listToTree([2,N]) ->
  {leaf,N};

listToTree([_|Code]) ->
  case Code of
    [M|_] ->
      {Code1,Code2} = lists:split(M,Code),
      {node,
        listToTree(Code1),
        listToTree(Code2)
      }
  end.
```

The function `treeToList/1` converts the tree to a list, and `listToTree/1` should convert that representation back to the original tree. We can now test the functions with some example trees:

```
tree0() ->
  {leaf, ant}.

tree1() ->
  {node,
    {node,
      {leaf,cat},
      {node,
        {leaf,dog},
        {leaf,emu}
      }
    },
    {leaf,fish}
  }.
```

We require that the two functions applied one after the other return the original value:

```
leaf_test() ->
  ?assertEqual(tree0() , listToTree(treeToList(tree0()))).
node_test() ->
  ?assertEqual(tree1() , listToTree(treeToList(tree1()))).
```

where we use the `eunit` macro `assertEqual` to test for equality between the value of two Erlang terms.

We can also test for particular values of the `treeToList` function:

```
leaf_value_test() ->
  ?assertEqual([2,ant] , treeToList(tree0())).
node_value_test() ->
  ?assertEqual([11,8,2,cat,5,2,dog,2,emu,2,fish] , treeToList(tree1())).
```

In testing `listToTree`, we do something different. This function is partial, and we can check that it is undefined outside the range of `treeToList`:

```
leaf_negative_test() ->
  ?assertError(badarg, listToTree([1,ant])).
node_negative_test() ->
  ?assertError(badarg, listToTree([8,6,2,cat,2,dog,emu,fish])).
```

These tests use the `assertError` macro, which captures a raised exception and checks that it is of the form specified by the first argument (here, a `badarg`).

With these tests included in the file, running the `test/0` function in the shell gives the brief report that all tests are successful:

```
2> serial:test().
  All 6 tests successful.
ok
```

There is nothing more to say when tests are successful. Alas, not all tests are successful. What does EUnit report when tests fail? In true test-driven development style, as an illustration, we can *regression-test* the optimized version of serialization with the same test set and see what happens:

```
11> serial2:test().
serial2:leaf_negative_test...*failed*
::error:{assertException_failed,[{module,serial2},
                                 {line,66},
                                 {expression,"listToTree ( [ 1 , ant ] )"},
                                 {expected,"{ error , badarg , [...] }"},
                                 {unexpected_success,{leaf,ant}}]}
  in function serial2:'-leaf_negative_test/0-fun-0-'/0

serial2:node_value_test...*failed*
::error:{assertEqual_failed,[{module,serial2},
                             {line,72},
                             {expression,"treeToList ( tree1 ( ) )"},
                             {expected,[11,8,2,cat,5,2|...]},
                             {value,[8,6,2,cat,2|...]}]}
  in function serial2:'-node_value_test/0-fun-0-'/1

serial2:node_negative_test...*failed*
 ... details similar ...


=========================================================
  Failed: 3.  Aborted: 0.  Skipped: 0.  Succeeded: 3.
error
```

This report shows that three of the six tests have failed, and gives detailed feedback on the cause of failure. In the case of `leaf_negative_test`, it is that the particular function application succeeds unexpectedly, instead of raising a `badarg` exception. In the second case, the actual result was different from the actual value, both of which are printed in the report.

The failed tests either cover values for which the original functions failed, or are sensitive to changes in the new implementation, where the details of the list representation have changed. The first two tests that check that applying the functions one after the other returns the original argument, however, confirm that the crucial property still holds.

If you want to put the test functions into a separate `serial_tests` module, you can use an `import` directive to include the tests without making any changes from the `serial` module:

```erlang
-module(serial_tests).
-include_lib("eunit/include/eunit.hrl").
-import(serial,
        [treeToList/1, listToTree/1
         tree0/0, tree1/0,]).

leaf_test() ->
  ?assertEqual(tree0() ,
               listToTree(treeToList(tree0()))).
... etc ...
```

The EUnit Infrastructure

In this section, you will learn about the foundations of the EUnit system, with which you can build tests and test sets.

Assert Macros

The basic building block of EUnit is a single test, given by a `..._test()` function. On the right side of the earlier code examples, we used `assertEqual` and `assertError` to check values and exceptions. Other assert macros include the following:

`assert(BoolExpr)`
: Can be used not only in tests, but anywhere in a program to check the value of a Boolean expression at that point

`assertNot(BoolExpr)`
: Is equivalent to `assert(not(BoolExpr))`

`assertMatch(GuardedPattern, Expr)`
: Will evaluate the `Expr`, and if it fails to match the guarded pattern, an exception is reported on `test()`

`assertExit(TermPattern, Expr)` and `assertThrow(TermPattern, Expr)`
: Will test for a program exit or a throw of an exception, similar to `assertError`

In the example, we used `assertEqual(E, F)` instead of `assert(E =:= F)` because `assertEqual` generates more informative messages when the test fails.

Test-Generating Functions

Beyond a single test, you can define test-generating functions that combine a number of tests into a single function. A test generator returns a representation of a set of tests to be executed by EUnit.

The simplest way to represent a test is as a `fun` expression that takes no arguments:

```
leaf_value_test_() ->
    fun () -> ?assertEqual([2,ant] , treeToList(tree0())) end.
```

In the preceding code, the differences from the definition of leaf_value_test are high-lighted. The macro library allows you to write this more succinctly:

```
leaf_value_test_() ->
    ?_assertEqual([2,ant] , treeToList(tree0())).
```

In this code, the _assertEqual macro plays the same role as assertEqual, but for test representations rather than tests.

A test-generating function will, in general, return a *set* of tests. For instance, the following code encapsulates two tests into a single function:

```
tree_test_() ->
    [?_assertEqual(tree0() , listToTree(treeToList(tree0()))),
     ?_assertEqual(tree1() , listToTree(treeToList(tree1())))].
```

When EUnit runs this test, *all* the tests in the list are performed.

EUnit Test Representation

EUnit represents tests and test sets in a variety of different ways. Here is a list of the most useful; a full description is in the EUnit documentation. A test representation TestRep is *run* by calling eunit:test(TestRep):

Simple test objects
> The simplest test object is a nullary fun, that is, a fun that takes no arguments. A simple test object is also given by a pair of atoms, {Module, Function}, referring to a function within a module.

Test sets
> Test sets are given by lists and deep lists. A module name (atom) is also used to represent the tests within the module.

Primitives
> The primitives do not contain embedded test sets as arguments, but instead are descriptions of tests that lie within a module, as in {module, Module}; within a directory, as in {dir, Path::string()}; and within an application, a file, and so forth. Generators are also embedded as {generator, GenFun::(() -> Tests)}.

Control
> It is possible to control how and where tests are to be executed:

{spawn, Tests}
> Will run the tests in a separate subprocess, with the test process waiting until the tests finish

{timeout, Time::number(), Tests}
> Will run the tests for Time seconds, terminating any unfinished tests at that time

```
{inorder, Tests}
```
Will run the tests in strict order

```
{inparallel, Tests}
```
Will run the tests in parallel where possible

Fixtures
Fixtures support the setup and cleanup for a particular set of tests to run; we discuss them in more detail in the next section.

Testing State-Based Systems

To explain this topic, we will go back to the example of the mobile user database. When we introduced the example in Chapter 10, we tested it from the Erlang shell (see the section "A Mobile Subscriber Database Example" on page 231). This section shows how you can incorporate this style of testing into EUnit. We will also look in more detail at the way in which tests are represented.

Fixtures: Setup and Cleanup

It is characteristic of state-based systems that you can test particular properties only after you have set up the right configuration; once you've done this, the test can take place, but after that, you need to clean up the system to prepare for any further tests.

The first test we'll write will test a lookup on an empty database after the tables are created with `create_tables("UsrTabFile")`:

```
?_assertMatch({error,instance}, lookup_id(1))
```

After the test is run, we clean up by removing the file *UsrTabFile*. This is implemented by writing a fixture: a test description that allows setup and cleanup. The simplest fixtures have the following form:

```
{setup, Setup, Tests}
{setup, Setup, Cleanup, Tests}
```

where, for some type T:

```
Setup   :: (() -> T)
Cleanup :: ((T) -> any())
```

the `Setup` function is executed before the `Tests`, returning a value X of type T. After the tests, `Cleanup(X)` is performed; this allows information from the setup—for example, about pids or tables—to be communicated to the cleanup phase.

For our example, we can write the following:

```
setup1_test_() ->
    {spawn,
     {setup,
      fun ()  -> create_tables("UsrTabFile") end,        % setup
      fun (_) -> ?cmd("rm UsrTabFile") end,               % cleanup
```

```
        ?_assertMatch({error,instance}, lookup_id(1))
      }
    }.
```

Note in the cleanup that we can call an external Unix command to remove the file using the ?cmd macro, and that the test is executed by a call to eunit:test/1. To test that the database is functioning correctly, we need a rather more elaborate setup; on termination of the spawned process, the ETS tables constructed by the program will be destroyed:

```
setup2_test_() ->
  {spawn,
    {setup,
      fun () ->
        create_tables("UsrTabFile"),
        Seq = lists:seq(1,100000),
        Add = fun(Id) -> add_usr(#usr{msisdn = 700000000 + Id,
                                       id = Id,
                                       plan = prepay,
                                       services = [data, sms, lbs]})
              end,
        lists:foreach(Add, Seq)
      end,
      fun (_) -> ?cmd("rm UsrTabFile") end,
      ?_assertMatch({ok, #usr{status = enabled}} , lookup_msisdn(700000001) )
    }
  }.
```

We can run these two tests by calling Mod:test() or eunit:test(Mod), where Mod is the name of the module containing the tests.

Testing Concurrent Programs in Erlang

When a program consists of a number of objects evolving concurrently, it is harder to see how a unit-testing framework such as EUnit can directly help. Although EUnit provides the scaffolding for an expression to be applied at a particular point during system evolution, it is harder to monitor the evolution of the system itself. The biggest challenges in testing concurrent systems are race conditions. You run your test, and you can easily reproduce your error. You turn on the trace on these processes, and all of a sudden everything works as expected, as the extra I/O causes your processes to execute in a different order.

Some EUnit facilities can be useful; it is possible to ?assert a property at any point in the program, and so to monitor when a pre- or post-condition or system invariant is broken. EUnit also provides support for debugging, with messages reported to the Erlang console rather than to standard output. These macros include the following:

debugVal(Expr)

> Will print the source code and current value of Expr. The result is always the value of Expr, and so the macro can be written around any expression in the program without affecting its functionality.

```
debugTime(Text, Expr)
```
Will print the Text followed by the (elapsed) execution time of the Expr.

If the NODEBUG macro is set (to true) before the eunit header file is included in the module, the macros have no effect in that module.

Other systems that support the testing of concurrent programs include Quviq Quick-Check,[*] which implements property-based random testing for Erlang; McErlang,[†] a model checker for Erlang written in Erlang; and Common Test, a systems-testing framework based on the OTP Test Server application, and part of the standard Erlang distribution.

Exercises

Exercise 19-1: Testing Sequential Functions

Revisit the exercises in Chapter 3, and devise EUnit tests for your solutions: do your solutions pass all the tests? Is that because of faults in the solutions or in the way you have defined the tests?

Exercise 19-2: Testing Concurrent Systems

Devise tests within EUnit for the echo process introduced in Chapter 4. How do you have to change the tests when the implementation is modified to register the process?

Exercise 19-3: Software Upgrade

How would you define EUnit tests for a software upgrade in the db_server example given in Chapter 8?

Exercise 19-4: Testing OTP Behaviors

Incorporate the test examples given in Chapter 12 into the EUnit framework.

Exercise 19-5: Devising Tests for OTP Behaviors

Devise EUnit tests for the solutions to the exercises in Chapter 12.

---

[*] *http://www.quviq.com/*; an earlier version of the tool, and its application in testing telecom software, is discussed at *http://doi.acm.org/10.1145/1159789.1159792*.

[†] *https://babel.ls.fi.upm.es/trac/McErlang/*; an early version of the tool is described at *http://doi.acm.org/10 .1145/1291220.1291171*.

CHAPTER 20

Style and Efficiency

Throughout this book, we have covered the do and don'ts of Erlang programming. We have introduced good practices and efficient constructs while pointing out bad practices, inefficiencies, and bottlenecks. Some of these guidelines you will probably recognize as being relevant to computing in general; others will be Erlang-related, and some will be virtual-machine-dependent. Learning to write efficient and elegant Erlang code will not happen overnight. In this chapter, we summarize design guidelines and programming strategies to use when developing Erlang systems. We cover common mistakes and inefficiencies and look at memory handling and profiling.

Applications and Modules

A collection of tightly interacting modules in Erlang is called an *application*. Erlang systems consist of a set of loosely coupled applications. It is good practice to design an application to provide a single point of entry for calls originating from other applications. Collecting all externally exported functions into one module provides flexibility in maintaining the code base, since modifications to all "internal" modules are not visible to external users. The documentation for this module gives a complete description of the *interface* to the application. This single point of entry also facilitates tracing and debugging of the call flow into the application. In large systems, it is good practice to prefix the modules in a particular application with a short acronym. This ensures that the choices of module names in different applications will never overlap, which would cause problems if both applications were used together in any larger system.

Modules are your basic building blocks. When designing your modules, you should try to export as few functions as possible. As far as the user of a module is concerned, the complexity of a module is proportional to the number of exported functions, since the user of a module needs to understand only the exported functions. Having as small an interface as possible also gives the maintainer of the application greater flexibility, as it makes it much easier to refactor the internal code.

You should try to reduce intermodule dependencies as much as you can. It is harder to maintain a module with calls to many modules instead of calls to just a few modules.

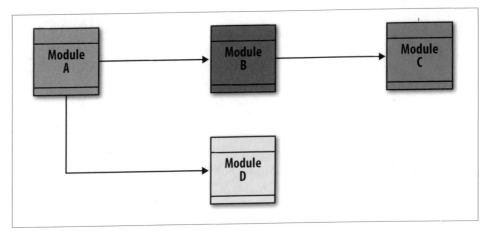

Figure 20-1. Intermodule dependencies

Reducing interdependencies not only facilitates the maintenance of the modules being called, but also makes it easier to refactor them as a smaller number of external calls have to be maintained.

Intermodule dependencies should form an acyclic graph, as shown in Figure 20-1; that is, there should be no module X, say, that depends on another module that (through a chain of dependencies) depends on module X.

Within a particular module, place related functions close to each other.[*] For example, you should place functions such as **start** and **stop** and **init** and **terminate** next to each other, and place all of the message-handling functions in the same part of a module. Keeping related functions close together makes your code easier to follow and inspect, especially for cases where you do or open something in one function and undo or close it in another.

Libraries

You should put commonly used code into libraries. Libraries should be collections of related functions, possibly those that manipulate values of the same type. Where possible, ensure that library functions are free of side effects, as doing so will enhance their reusability. When there are functions with side effects such as message passing, destructive database operations, or I/O, you should try to ensure that all the operations with side effects that are related—for instance, all those manipulating a particular ETS table—are contained in a single library module.

---

[*] Research in Chris Ryder's Ph.D. thesis, "Software Measurement for Functional Programming," University of Kent, 2004, suggests that this is, in fact, what programmers do as a matter of course.

Dirty Code

Dirty code consists of anything you "should not" do but are forced to do anyhow. Getting away from writing dirty code can sometimes be hard and other times impossible.

Dirty code includes the use of the process dictionary, the use of the `process_info/2` BIFs or code that makes assumptions about the internal structure of data types, or other internal constructs. For example, you could use the `process_info/2` BIF to view the length of the message queue or to peek at the first message in the mailbox. You might want to store a global variable in the process dictionary, or have a dynamic function to look up a record type and generically use it to forward a request to a callback module.

It is important to isolate tricky or dirty code into separate modules, and do everything possible to minimize and avoid such code. It is also crucial to document dirty code, and to say in what way it is dirty so that anyone modifying your code has a clear view of the assumptions you made when writing this dirty module.

Interfaces

Document all your exported interfaces. Documentation should include parameter values, possible ranges, and return values. If your function has side effects such as message passing, database updates, or I/O, you should include information on these. Provide references to specifications and protocols and document all principal data structures. Comments should be clear, informative, and not superfluous; you want to describe what the function does, not how it does it. An example of this is provided by the EDoc documentation in Chapter 18, a full version of which is available on this book's website.

Decide why a function is exported. It is a good practice to divide export directives into categories based on the following function types:

- A user interface function
- An intermodule function (used in the same application, but not exported to others)
- An internal export (used by the same module in BIFs such as `apply/3` and `spawn/3`)
- A standard behavior callback function (`init`, `handle_call`, etc.)

It obviously depends on how many functions you are exporting; a rule of thumb is that if your export clause spans more than one line, it is time to break it up. You might include client functions such as `start/0`, `stop/0`, `read/1`, and `write/2` in one export directive, and callback functions such as `init/1`, `terminate/2`, and `handle_call/3` in another. This allows anyone who is reading the code instead of the documentation to distinguish between exported interfaces and internal exports or callback functions.

Use the `-compile(export_all)` directive only when developing your code (and even then, only if you must). When the development work is done, don't forget to remove the directive; more often than not, this does not happen.[†] An alternative to including this as a module directive is to make it an option on compilation, like so:

```
compile:file(foo, [compile_all,...]).
```

Return Values

If a function might not always succeed, tag its return values, because if you don't a positive result might be interpreted as a negative one. In the following example, what happens if you do a key search on key 1 using the list [{0,true}, {1,false}, {2,false}]? How do you distinguish the return value of the key with the atom false returned as the base case when the entry is not found?

```
keysearch(Key, [{Key, Value}|_]) -> Value;
keysearch(Key, [_|Tail])         -> keysearch(Key, Tail);
keysearch(_key, [])              -> false.
```

The Erlang approach is to use standard return values of the form ok, {ok, Result}, or {error, Reason}. In our case, we would distinguish between a successful lookup using {ok, false} and a failed lookup using the atom false. The correct implementation of keysearch/2 would thus be:

```
keysearch(Key, [{Key, Value}|_]) -> {ok, Value};
keysearch(Key, [_|Tail])         -> keysearch(Key, Tail);
keysearch(_key, [])              -> false.
```

An alternative approach to this is to raise an exception in the case of no corresponding key value being found. With this approach, each use of keysearch will need to be in the context of a try ... catch construct to handle the exceptional case.

If you know that the function will always be successful, return a single value such as true or false, or just an integer, atom, or composite data type. This will allow the return value of this function to be passed as a parameter to another function without having to check for success or failure.

You should always pick return values that will simplify the caller's task. Doing so will make the code more compact and readable. For example, if you know that get_status/1 will always succeed, why write a tagged return value, from which you have to extract the return value:

```
{ok, Status} = get_status(BladeId),
NewStatus = reset(BladeId, Status)
```

when all you need to do is return the Status? Doing so will allow the function call to be passed as an argument to the reset/2 call:

```
NewStatus = reset(BladeId, get_status(BladeId))
```

[†] Perhaps even in this book!

Make no assumptions about what the caller will do with the result of a function. In the following example, we are assuming that the caller of the function wants to print out an error message stating that the person for whom we want to raise taxes is not Swedish:

```erlang
tax_to_death(Person) ->
  case is_swede(Person) of
    true ->
      {ok, raise_taxes(Person)};
    {error, Nationality} ->
      io:format("Person not Swedish:~p~n",[Nationality]),
      error
  end.
```

Let's not make that assumption, and instead allow the caller of this function to make the decision based on the return value of the call:

```erlang
tax_to_death(Person) ->
  case is_swede(Person) of
    true ->
      {ok, raise_taxes(Person)};
    {error, Nationality} ->
      {error, Nationality}
  end.
```

Another way to look at this example is that, in general, it is best for a function to do one thing: in this case, perform the update or print an error message of a particular form. If you find you have a function doing two things, you can refactor it into two separate functions, each of which can be reused separately.

Internal Data Structures

Do not allow private data to leak out. All details of private data structures should be abstracted out of the interface. In abstracting your interfaces, encapsulate information that the users of the function do not need. Design with flexibility so that any changes introduced to your internal data representations will not influence the exported functional interface. The following queue module:

```erlang
-module(q).
-export([add/2, fetch/1]).

add(Item, Q) -> lists:append(Q, [Item]).

fetch([H|T]) -> {ok, H, T};
fetch([])    -> {error, empty}.
```

could be used as follows:

```erlang
NewQ = [],
Queue1 = q:add(joe, NewQ),
Queue2 = q:add(klacke, Queue1).
```

In the preceding code, we are leaking out the fact that the queue data structure is a list when we bind the variable NewQ to the empty queue. This is not good. The q module should have exported the function `empty() -> []`, which we use to create the empty queue we bind to the variable `NewQ`:

```
Newq = q:empty(),
Queue1 = q:add(joe, Newq), ...
```

Now, it would be possible to rewrite the queue implementation, for instance, as a pair of lists keeping the front and rear separately, without having to rewrite any code using the queue module. One exception to this is if you want to determine whether two queues are equal; in this case, you will need to define an equality function over queue representations, because the same queue can be represented in different ways in this new implementation.

Processes and Concurrency

Processes are the basic structuring elements of a system in Erlang. A fundamental principle of design in Erlang is to create a one-to-one mapping between the parallel processes in your Erlang program and the set of parallel activities in the system you are modeling. This could add up to quite a few simultaneous processes, so where possible, avoid unnecessary process interaction and unnecessary concurrency. Ensure that you have a process for every concurrent activity, not for only two or three! If you are coming from an object-oriented background, this will not equate to a process for every object or every method applied to an object. Or, if you are dealing with users logged in to a system, this will probably not equate to a process for every session. Instead, you will have a process for every event entering the system. This will equate to massive numbers of transient processes and a very limited number of persistent ones.

You should always implement a process loop and its peripheral functions in one module. Avoid placing code for more than one process loop in the same module, as it becomes very confusing when you try to understand who is executing what. You might have many process instances executing the same code base, but ensure that the loop in the module is unique.

Hide all message passing in a *functional interface* for greater flexibility and information hiding. Instead of having the following expression in your client code:

```
resource_server ! {free, Resource}
```

replace it with this function call:

```
resource_server:free(Resource)
```

Place the client functions in the same module as the process. Client functions are functions called by other processes that result in a message being sent to the process defined in the module. This makes it easy to follow the message flow without having to jump between modules, let alone having to find out in which module the clause receiving the messages is located. From your end, it reduces and simplifies the documentation you have to produce, as you need to describe only the functional API, and not the message flow. You will also get more flexible code, as you are hiding the following:

- That the resource server is a process
- That it is registered with the alias `resource_server`
- The message protocol between the client and the server
- The fact that the call is asynchronous

If you use a functional interface, you have the flexibility to change all of this without affecting the client code.

Registered processes should, where possible, have the same name as the module in which they are implemented. This facilitates the troubleshooting of live systems and improves code readability and maintainability. Registered processes should have a long life span; you should never toggle between registering and deregistering them. As the space used by atoms is not garbage-collected, avoid dynamically creating atoms to register processes, as this might result in a memory leak.

Processes should have well-defined behaviors and roles in the system. You should seriously consider using the OTP behaviors described in Chapter 12, including servers, event handlers, finite state machines, supervisors, and applications.

When working with message passing, all messages should be tagged. It makes the order of clauses in the `receive` statement unimportant, and as a result, it facilitates the addition of new messages without changing the existing behavior, thus reducing the risk of bugs.

Avoid pattern matching only on unbound variables in `receive` clauses. In the following example, what if you want your process to also handle the message `{get, Pid, Variable}`?

```
loop(State) ->
  receive
    {Mod, Fun, Args} ->
      NewState = apply(Mod, Func, Args),
      loop(NewState)
  end.
```

It is easy to see that you would have to match the message above the `{Mod, Fun, Args}` pattern, but what if your `receive` statement was matching on many more messages? It would not be that obvious. Or, what would happen if you want to apply a function in the module `get`? By tagging messages, you get full flexibility in rearranging the order of your messages and avoid incorrectly pattern matching messages in the wrong clause:

```
loop(State) ->
  receive
    {apply, Mod, Fun, Args} ->
      NewState = apply(Mod, Func, Args),
      loop(NewState);
```

```
    {get, Pid, Variable} ->
      Pid ! get_variable(Variable),
      loop(State)
  end.
```

When using **receive** clauses, you should not be receiving unknown messages. If you do, they should be treated as bugs and you should allow your system to crash, just as an OTP application will do. If you don't let the system crash, make sure you log the unexpected messages. This allows you to ascertain where they are coming from, and to diagnose how to deal with them. You will probably find that these messages originate from ports or sockets or from bugs that should be picked up during a unit test.

If you do not handle unknown messages, you will notice that the CPU usage of your system will start to increase and that the response time will decrease, until the Erlang runtime system runs out of memory and crashes. The increase in CPU usage and response time is explained by the fact that whenever a process receives a message, it has to traverse the potentially hundreds or thousands of messages in the mailbox before reaching one that matches. The more messages that are not matched, the more memory is leaked.

When you are sending and receiving messages, you should hide the internal message protocol from the client. Always tag the messages using the client function name, as it facilitates the hop from the client function to the location in your process loop where you handle the message:

```
free(Resource) ->
    resource_server ! {free, Resource}.
```

Use references. In complex systems where similar responses and requests might originate from different processes, use references to uniquely identify responses from specific requests:

```
call(Message) ->
  Ref = make_ref(),
  resource ! {request, {Ref, self()}, Message},
  receive
    {reply, Ref, Reply} -> Reply
  end.

reply({Ref, Pid}, Message) ->
  Pid ! {reply, Ref, Message}.
```

Be careful with timeouts. If you use them, always flush messages that arrive late. If you don't, your next request will result in the response that previously timed out. You can avoid this problem by using references and assuming that no other processes will send messages of the format {reply, Ref, Reply}. Just flush messages you do not recognize:

```
call(Message) ->
  Ref = make_ref(),
  resource ! {request, {Ref, self()}, Message},
  wait_reply(Ref).
```

```
wait_reply(Ref) ->
  receive
    {reply, Ref, Reply} -> Reply;
    {reply, _,   Reply} -> wait_reply(Ref)
  end.
```

If you are using timeouts to deal with the case when a process might have terminated, it is much better to use a link or a monitor.

Be very restrictive in trapping exits, and when doing so, do not toggle. Make sure you remember to flush exit signals, and when linking and unlinking processes, be aware of race conditions. When you link to a process, it might have already crashed. When you unlink, it might have terminated, and its EXIT signal will be waiting to be retrieved from the process mailbox.

Separate error recovery and normal code, as combining them will increase the complexity of your code. It will also ensure that crashes and recovery strategies are handled consistently. Never try to fix an error that should not have occurred and then continue. If an unexpected error occurs, make your process crash and let a supervisor process deal with it. It is more likely that in trying to handle errors, you will generate more bugs than you think you are solving. In the following example, what would you do if the variable List is not a list?

```
bump(List) when is_list(List) ->
  lists:map(fun(X) -> X+1 end, List);
bump(_) ->
  {error, no_list}.
```

Everywhere you call bump/1, you would have to cater for two return values in a case clause, but still not make the person reading the code any wiser as to why List got corrupted. Don't be defensive, and instead write:

```
bump(List) ->
  lists:map(fun(X) -> X+1 end, List).
```

If List does not contain a list, a runtime error will occur in lists:map/2 and will result in your process terminating. Let your supervisor handle the exit signal and let it decide the recovery strategy consistently with all of the other processes it is supervising. Make sure the crash is recorded and can be used for post-mortem debugging; log these errors and crashes in a separate error logger process.

As we should be hiding processes and message passing behind functional interfaces, ensuring that the dependencies form an acyclic graph also ensures that there are no deadlocks between these processes. Deadlocks are extremely rare in Erlang, but they do occur if the concurrency is not well thought out and properly designed. Ensuring that there are no cycles in your module dependencies is a step in the right direction.

Stylistic Conventions

Programs are written not just to be executed by a computer, but also to be *read* and *understood* by their authors and other programmers. Writing your programs in a way that makes them easier to read and understand will help you remember what your program does when you come back to it six months down the road, or will help another programmer who has to use or modify your program. It will also make it easier for someone to spot errors or other problems, or to interpret debugging information. So, being consistent in the way that you write programs will help everyone. In this section, we give a set of commonly used conventions for style in writing Erlang programs.

First, avoid writing deeply nested code.[‡] In your **case**, **if**, **receive**, and **fun** clauses, you should never have more than two levels of nesting in your code. Here is how *not* to do it:

```
reset(BladeId, AdminState, OperState) ->
  case AdminState of
    enabled ->
      case OperState of
        disabled ->
          enable(BladeId);
        enabled ->
          disable(BladeId),
          enable(BladeId)
      end;
    disabled ->
      {error, admin_disabled}
  end.
```

A common trick to reduce indentation is to create temporary composite data types. If you have nested **if** and **case** statements, join them together in a tuple and pattern-match them in one clause:

```
reset(BladeId, AdminState, OperState) ->
  case {AdminState, OperState} of
    {enabled, disabled} ->
      enable(BladeId);
    {enabled, enabled} ->
      disable(BladeId),
      enable(BladeId);
    {disabled, _OperState} ->
      {error, admin_disabled}
  end.
```

You can reduce indentation by introducing pattern matching in your function heads. By pattern matching on the **AdminState** and the **OperState** in the function clause, not only do we make the code more readable, but we also reduce the level of nested clauses to zero and reduce the overall code size:

[‡] The Wrangler refactoring tool will indicate this and a number of other bad smells in Erlang code. Wrangler is available from *http://www.cs.kent.ac.uk/projects/protest/*.

```
reset(BladeId, enabled, disabled) ->
  enable(BladeId);
reset(BladeId, enabled, enabled) ->
  disable(BladeId),
  enable(BladeId);
reset(_BladeId, disabled, _OperState) ->
  {error, admin_disabled}.
```

Avoid using `if` clauses when `case` clauses are a better fit. This is especially common with programmers coming from an imperative background who still do not feel at ease with pattern matching. Always ask yourself whether you can rewrite your `if` clause into a `case` clause using pattern matching and guards that will avoid pattern matching on the atoms `true` and `false`. If so, is it more readable and compact?

```
get_status(A,B,C) ->
  if
    A == enabled ->
      if
        B == enabled ->
          if
            C == enabled ->
              enabled;
            true ->
              disabled
          end;
        true ->
          disabled
      end;
    true ->
      disabled
  end.
```

The preceding example is an extreme occurrence of the misuse of the `if` statement. By creating a composite data type with the variables A, B, and C, and replacing the `if` with a `case` statement, we reduce the level of indentation and make the code more readable and compact:

```
get_status(A,B,C) ->
  case {A,B,C} of
    {enabled,  enabled,   enabled } -> enabled;
    {_status1, _status2, _status}   -> disabled
  end.
```

Aim to keep your modules to a manageable size. Short modules facilitate maintenance and debugging as well as the understanding of your code. A manageable module should have no more than 400 lines of code, comments excluded. Split long modules in a logically coherent way and remember that long lines will not solve your problem.

Lines of code should not be too long, or drift across the page. A line of code should never be more than 80 characters long. Too many times developers have tried to convince us that their long lines were readable by widening their editor window to cover the whole screen. That will not solve your problem.

If your code is drifting across the page, spending a few minutes on reformatting will save the next person from spending hours trying to maintain and debug it. The following code style is a typical example of what happens when someone writes code, tests it, but never goes back to review it:

```
name(First, Second) ->
   case person_exists(First, Second) of
      true ->
         Y = atom_to_list(First) ++ [$ |atom_to_list(Second)] ++
                                    [$ |get_nickname(First,
                                        Second)],
         io:format("true person:~s~n",[Y]);
      false ->
            ok
   end.
```

You can easily rewrite this to the following:

```
name(First, Second) ->
   case person_exists(First, Second) of
      true ->
         Y = atom_to_list(First) ++
            [$ |atom_to_list(Second)] ++
            [$ |get_nickname(First, Second)],
         io:format("true person:~s~n",[Y]);
      false ->
         ok
   end.
```

After a second iteration and a bit of thinking, the preceding code would convert to the following:

```
name(First, Second) ->
   case person_exists(First, Second) of
      true ->
         NickName = get_nickname(First, Second),
         io:format("true person:~w ~w ~s~n",[First, Second, NickName]);
      false ->
         ok
   end.
```

If your code drifts across the page, solve the problem by:

- Picking shorter variable and function names
- Using temporary variables to divide your statement or expression across several lines of code
- Reducing the level of indentation in your case, if, and receive statements
- Moving possibly duplicated code into separate functions

Choose meaningful function and variable names. If the names consist of several words, separate them with either capital letters or an underscore. Some people will pick one style for variables and the other for functions, without mixing them together. Whatever the case, when you pick a style, stick to it throughout your code; this is true of all the conventions discussed here: consistency makes code readable, whereas chopping and changing between styles will distract a reader away from understanding what is going on.

Avoid long names, as they are a prime reason for code drifting across a page. A long name might look all right in the function head, but imagine calling it in the second level of a clause together with long variable names. If you are using longer names, it makes more sense to use them in functions—which will potentially be used across your code base—rather than for variables whose scope will be limited to a particular function.

Use abbreviations and acronyms to shorten your names, but ensure that they make sense and are easy to understand. A common pitfall is to use names that are very similar, and so can easily be confused: `Name`, `Names`, and `Named` all look pretty similar! If you use prefixes, pick ones that will provide hints regarding the variable types or function return values. Finding meaningful function and variable names takes practice and patience, so don't underestimate the task.

Avoid the underscore on its own, when using "don't care" variables. Even if you are not using the variable, its contents might be of interest to whoever is reading the code. Not using an underscore in front of unused variables will result in compiler warnings. But be warned that variables of the format `_name` are regular single-assignment variables, so once they have been bound, they cannot be reused as "don't care" variables.

In the following example, if the variable `AdminState` is bound to the atom `disabled`, `_State` will also be bound to `disabled`. If `OperState` is bound to `enabled`, the second `case` clause will fail with a `case` clause error, as none of the patterns match:

```
restart(BladeId, AdminState, OperState) ->
  case AdminState of
    enabled ->
      disable(BladeId);
    _State ->
      ok
  end,
  case OperState of
    disabled ->
      ok;
    _State ->
      stop(BladeId)
  end.
```

Use records as the principal data structure to store compound data types. If records are used by one module only and are not exported, the definition should be placed at the beginning of the module to stop others from using it in their modules. Records used by many modules should be placed in an include file and properly documented.

Use record selectors to access a field and pattern matching to access values in more than one field:

```
Cat = #cat{name = "tobby", owner ="lelle"},
#cat{name = Name, owner = Owner} = Cat,
Name2 = Cat#cat.name.
```

Never, *ever* use the internal tuple representation of records, as this defies the flexibility that records bring to the table:

```
Cat = #cat{name = "tobby", owner = "lelle"},
{cat, Name, _owner} = Cat
```

If you write code such as this and then add a field to your record, you will have to update the code that uses the tuple representation even if the function is not affected by the field.

Use variables only when you need them. There are two reasons for using a variable. The first is when you need to use the value they are bound to more than once. The second is to improve clarity in your code and shorten the length of a line. You might be tempted to write the following:

```
Sin = sin(X),
Cos = cos(X),
Tan = Sin / Cos
```

But passing the return values of the function directly to another function will make your code much clearer and more compact:

```
Tan = sin(X) / cos(X)
```

Always ask yourself whether the variable really makes the code clearer; the more you program in Erlang, the less you will find yourself using variables.

Be careful when using catch and throw. It is nearly always preferable to use the try ... catch construct than to use catch and throw. Try to minimize their usage, and ensure that any throw is caught in the same module as a catch. As you saw in Chapter 3, it makes sense to use catch only when you can safely ignore the return value of a function.

If you are using catch and throw, always make sure you also handle runtime errors that might be caught. For instance, the following code appeared in code that was about to be sent into production:

```
Value = (catch get_value(Key)),
ets:insert(myTable, {Key, Value})
```

What would have happened if a runtime error occurred in the get_value/1 call? You would have caught the runtime error {'EXIT', Reason}, bound it to the variable Value, and stored it in the ETS table.

Be very careful with the process dictionary; in fact, avoid it like the plague. The BIFs put and get are destructive operations. They will make your functions nondeterministic and hard to debug, as determining the contents of the process dictionary after the process has crashed will be very difficult, if at all possible, to do. Instead of the process dictionary, introduce new arguments in your functions.

If you do not know what the process dictionary is,[§] move along; there is nothing to see here, move along.

Use the `-import(mod, [fun/arity...])` directive with care. It can make the code hard to follow and will initially confuse the most experienced of Erlang programmers. A temptation might be to use it when reducing the length of lines, but this comes at the expense of comprehension. There are some cases where it makes sense:

- Common functions from the `lists` module, such as `map`, `foldl`, and `reverse`, are going to be comprehensible to anyone reading your code.
- If you want to move EUnit tests from the module `foo` to the module `foo_tests` without changing them, you will need to import all the definitions from `foo` into `foo_tests`.

Developing your own Erlang programming style might take years. Make sure you are consistent with yourself in any single system. This includes your use of indentation and spaces as well as your choice of variable, function, and module names. In large projects, style guidelines are often given to you.

Coding Strategies

The user should be able to predict what will happen when using the system. The foundation of this determinism originates in predictable and consistent results being returned by your functions. A consistent system in which modules do similar things is easier to understand and maintain than a system in which modules do similar things in quite different ways.

Minimize the number of functions with side effects, collecting them in specific modules. These modules could handle files or encapsulate database operations. Functions with side effects will cause permanent changes in system state. Knowledge of these states is imperative for these functions to be used and debugged, making reusability and maintainability of the code difficult. Try to reduce the number of side effects by maximizing the number of pure functions, separating side effects into atomic functions rather than combining them with functional transformations.

Some of the most challenging bugs are those caused by race conditions, especially in the advent of multicore systems in which code is truly running in parallel. Here's a common scenario: you run a test and are able to re-create the same bug every time. As soon as you turn on trace printouts, the overhead causes the process to run more slowly, changing the order of execution. All of a sudden, your bug is not reproducible anymore. What can you do to avoid this? Make your code as deterministic as possible.

A *deterministic* program is one that will always run in the same manner and provide the same result or manifest the same bug, regardless of the order of execution. What

---

[§] We barely mention it in this book for this reason.

makes a solution deterministic? Assume a supervisor has to start five workers in parallel and that the start order of these processes does not matter. A nondeterministic solution would spawn all of them and check that they have started correctly. A deterministic solution would spawn the processes one at a time, ensuring that each one has started correctly before proceeding to the next one. Although both *might* provide the same result, the nondeterministic solution may make start errors hard to reproduce, providing different results based on the order in which the processes have been spawned. Although determinism is not guaranteed to eliminate race conditions, it will certainly reduce the number of them and will make your system more predictable, as well as making debugging much more straightforward.

Abstract out common design patterns. If you have the same code in two places, isolate it in a common function.‖ If you have similar patterns of code, try to define the difference in terms of a variable or a separate function, combining the pattern in a common call. Where appropriate, replace your recursive functions iterating on lists with a call to one of the higher-order functions in the `lists` module. You can encapsulate your functionality on the list elements in a `fun` and choose a recursive pattern using one of the higher-order functions provided in the `lists` module. To see what is happening to the elements in the list, all you need to do is inspect the `fun`. To see what recursive pattern is being applied, examine which function in the `lists` module is being called.

When you are designing a library of your own, you will find that many of the functions follow similar patterns, such as `fold` or `map` on lists. You can then write your own higher-order functions to encapsulate these patterns, allowing you and other users of the module to abstract their functions. With higher-order functions encapsulating common computation patterns, many recursive functions with multiple clauses and base cases become much more compact, readable, and maintainable.

Avoid defensive programming. Trust your input data if it originates from within the system, testing it only if it enters through an external interface. Once the data has entered the runtime system, it is the responsibility of the calling function to ensure that the input data is correct, and not of the function that was called. If your function is called with erroneous input data, the advice is to *make it crash*.

An example of defensive programming would be a function converting atoms denoting months to their respective numbers:

```
month('January') -> 1;
month('February') -> 2;
...
month('December') -> 12;
month(_other) -> {error, badmonth}.
```

It might be tempting to add the last clause acting as a catchall. If `month/1` is called with an incorrect atom, it would return the tuple `{error, badmonth}`. This return value would either force the user of the function to test for and handle this error value, or would

---

‖ The Wrangler system can help you to find duplicate code and to factor it into a common function.

cause a runtime error somewhere in the system, where a function expecting an integer receives the tuple. Removing the last defensive clause would instead cause a function clause error, providing the call chain and the misspelled atom in the crash report.

Do not future-proof your code. Don't try to write code that will be able to deal with every possible eventuality as the system evolves. It will make your code harder to understand and maintain, adding unnecessary complexity. And best of all, you will probably not end up using what you have added anyhow. Be kind to Erlang consultants supporting and maintaining your code in the years to come. Do not try to predict what will happen to your code after it has gone into production. Just implement what is needed.

Here are two final pieces of advice that are not necessarily related to Erlang, but to programming in general: avoid cut-and-paste programming and do not comment out dead code, just delete it, as you can always get it back from your repository if you need to sometime in the future.#

Efficiency

The Erlang virtual machine is constantly being optimized and improved. What might have been inefficient constructs or necessary workarounds in earlier releases are not necessarily a problem in the current version. So, beware when reading about efficiency, workarounds, and optimizations in old performance guides, blog entries, and especially old posts in newsgroups and mailing list archives. If in doubt, always refer to the release notes and documentation of the runtime system you are using. And most importantly, benchmark, stress test, and profile your systems accordingly.

Sequential Programming

The most common misconceptions regarding efficiency concern funs and list comprehensions. List comprehensions allow you to generate lists and filter elements, and funs allow you to bind a functional argument to a variable. Today, the compiler translates list comprehensions to ordinary recursive functions, and funs were optimized a long time ago and have gone from being highly inefficient black magic to having performance between that of a regular function call and using an apply/3.

Strings are not implemented efficiently in Erlang. In the 32-bit representation, every character consists of four bytes, with an additional four bytes pointing to the next character. In the 64-bit representation, this doubles to eight bytes. On the positive side, Unicode is not an issue. On the negative side, if you are dealing with large data sets and memory does become an issue, you will need to convert your strings to binaries and match using the bit syntax.

Assuming, of course, that you are using version control....

Use the re library module to handle regular expressions if speed and throughput are important. This library supports PCRE-style regular expressions in Erlang, with a substantially more efficient implementation than the regexp library module, the use of which is now deprecated.

At one time, you were encouraged to reorder your function clauses as well as receive and case clauses, putting the most common patterns at the top. Today, the compiler will rearrange the clauses for you, and in most cases will use an efficient binary search to jump to the right clause, regardless of the number of clauses. An exception is if you have code of the following form:

```
month('January') -> 1;
month('February') -> 2;
month(String) when is_list(String) -> {error, badmonth};
month('March') -> 3;
...
month('December') -> 12;
month(_other) -> {error, badmonth}.
```

As the variable String will always match and be bound, with the clause possibly failing on the guard, the compiler needs to treat this case separately. It will first try to match 'January' or 'February' using a binary search, after which it binds the argument to the variable String and evaluates the guard. If the guard fails, a new binary search is executed on the remaining months. Moving the guarded clause either to the beginning:

```
month(String) when is_list(String) -> {error, badmonth};
month('January') -> 1;
month('February') -> 2;
...
```

or as the next-to-last clause, immediately before month(_other) -> ..., will improve efficiency slightly.

If you are setting and resetting a lot of timers, avoid using the timer module, as it serializes all requests through a process and can become a bottleneck. Instead, try using one of the Erlang timer BIFs erlang:send_after/3, erlang:start_timer/2, erlang:cancel_timer/1, or erlang:read_timer/1.

Where possible, use tuples instead of lists. The size of the tuple is two words plus the size of each element. Lists will consume one word for every element. As a result, tuples consume less memory and are faster to access.

Keep in mind that atoms are not garbage-collected. If you generate atoms dynamically using the list_to_atom/1 BIF on dynamic data, you might eventually run out of memory or reach the limit of allowed atoms, which is slightly more than 1 million. This BIF, if converting external data to atoms, makes your system open for denial of service attacks. Where this is a possibility, you should instead be using list_to_existing_atom/1.

Lists

Always ask yourself whether you really need to work with flat lists, as the function `lists:flatten/1` is an expensive operation. I/O operations through ports and sockets accept nonflat lists, including those consisting of binary chunks, so there is no need to flatten the list.

The same applies to the BIFs `iolist_to_binary/1` and `list_to_binary/1`. If your list is of depth one, use `lists:append/1` instead:

```
1> Str = [$h,[$e,[$l,$l],$o]].
[104,[101,"ll",111]]
2> io:format("~s~n",[Str]).
hello
ok
3> lists:append([[1,2,3],[4,5,6]]).
[1,2,3,4,5,6]
```

Left-associated concatenation is inefficient. Concatenating strings using the following:

```
lines(Str) ->
    "Hello " ++ Str ++ " World".
```

will result in the strings on the left side of the `++` being traversed multiple times. Instead of using `++`, you can let the compiler do the concatenation for you by writing:

```
lines(Str) ->
    ["Hello ", Str, " World"].
```

Do not append elements to the end of a list using `++` through `List ++ [Element]` or `lists:append(List, [Element])`. Every time you append an element, the list on the lefthand side of the `++` needs to be traversed. Do it once and you might get away with it. Do it recursively, and then for every recursive iteration, you will start getting serious performance problems:

```
double([X|T], Buffer) ->
    double(T, Buffer ++ [X*2]);
double([], Buffer) ->
    Buffer.
```

It is much more efficient to add the element to the head of the list and when the recursive call reaches the base case, reverse it. This way, you traverse the list only twice:

```
double([X|T], Buffer) ->
    double(T, [X*2|Buffer]);
double([], Buffer) ->
    lists:reverse(Buffer).
```

If you are dealing with functions that accept nonflat lists, add your element to the end of the nonflat list by using `[List, Element]`, creating a list of the format `[[[[[1],2], 3],4],5]`. It will save you from having to reverse the list once you're done.

So, you should use `++` when appending lists that you know do not consist of single elements where the result has to be flat. Isn't the following line of code:

```
List1 ++ List2 ++ List3
```

more elegant than this?

```
lists:append([List1, List2, List3])
```

Try to traverse lists only once. With small lists, you might not notice any difference in execution time, but as soon as your code gets into production and the line length increases, performance might become an issue. In the following example, we take a list of integers, extract all of the even numbers, multiply them by a multiple, and add them all together:

```
even_multiple(List, Multiple) ->
  Even = lists:filter(fun(X) -> (X rem 2) == 0 end, List),
  Multiple = lists:map(fun(X) -> X * Multiple end, Even),
  lists:sum(Multiple).
```

We traverse the list three times in this definition of even_multiple. Instead, encapsulate the filtering, and add the integers in a fun, and use a higher-order function to traverse the list. This allows you to implement the same operation more efficiently, by traversing the list only once:

```
even_multiple(List, Multiple) ->
  Fun = fun (X, Sum) when (X rem 2) == 0 ->
              X + Sum;
            (X, Sum) ->
              Sum
        end,
  Multiple * lists:foldl(Fun, 0, List).
```

Tail Recursion and Non-tail Recursion

Non-tail-recursive functions are often more elegant than tail-recursive functions, but when dealing with large data sets, be careful of large bursts of memory usage. These bursts can occasionally result in your Erlang runtime system running out of memory and therefore terminating. Building a large data structure can be efficient in non-tail-recursive form, but other operations on large data structures, necessitating deeply recursive calls, may be more efficient using tail recursion. This is because it makes last-call optimization possible, and as a result, it will allow a function to execute in a constant amount of memory.

In the end, our advice about this is to make sure you measure the performance of your system to understand its memory usage and behavior.

Concurrency

Operations that require a lot of memory will affect performance, as they will trigger the garbage collector more often. When dealing with memory-intensive operations, spawn a separate process, terminating it once you're done. The garbage collection time will be reduced as all its memory area will be deallocated at once.

You can take this one step further and spawn a process using the following:

```
spawn_opt(Module, Function, Args, OptionList)
```

where `OptionList` includes the tuple `{min_heap_size, Size}`. `Size` is an integer denoting the size of the heap (in words) allocated when spawning the process. The default allocated heap size is 233 words, a conservative value set to allow massive concurrency.

When fine-tuning your system, increasing the heap size will reduce the number of garbage collections, potentially speeding up some operations. Use `min_heap_size` with care and make sure to measure your system performance before and after ensuring that the change you made has had the desired effect. If you are not careful, your system might end up using more memory than necessary and run more slowly due to the worsened data locality. You can also set the heap size for all processes when starting the Erlang runtime system by setting the `+h Size` flag to `erl`.

Tune for full garbage collection. If you remember the discussion in Chapter 3, the memory heap is divided into the *new heap* and the *old heap*. Data in the new heap that survives a sweep by the garbage collector is moved to the old heap. The option `{fullsweep_after, Number}` makes it possible to specify the number of garbage collections which should occur before the old heap is swept, regardless of whether the old heap is full.

Setting the value to `0` will force a full sweep every time. This is a useful option in embedded systems where memory is limited. If you are using lots of short-lived data, especially large binaries, see whether setting the value between 10 and 20 will make a difference. Use the `fullsweep_after` option only if you know there are problems with the memory consumption of your process, and ensure that it makes a difference. You can set the `fullsweep_after` and `min_heap_size` flag options globally for all newly spawned processes using the `erlang:system_flag(Flag, Value)` BIF.

When a process is suspended in a **receive** clause waiting for an event, the garbage collector will not be triggered unless a message is received and more memory is required. This is regardless of how long the process waits or of the quantity of unused data in the heaps. To get around this, you can force a garbage collection using the BIF `garbage_collect/0` on the calling process or `garbage_collect/1` on a particular one. You can save even more memory by using `erlang:hibernate/3`; use of this is supported in `gen_server`, `gen_fsm`, and `gen_event`.

Use binaries to encode large messages. Messages sent between processes result in the message being copied from the stack of the sending process to the heap of the receiving one. Avoid unnecessary concurrency, and, where possible, keep messages small.

If you need to send a large message to many processes, convert it to a binary. Binaries larger than 64 bytes (the reference counted binaries) are passed around as pointers, stopping large amounts of data from being copied among processes. So, if you have many recipients or are forwarding the unchanged message to many processes, the cost of converting your data type to a binary will be less than copying it from the stack of the sending process to the heap of every receiving process.

It is also more efficient to use binaries when sending large amounts of data to ports and sockets. In particular, there is no need to convert your output to lists of integers.

And Finally...

Let others review your code, as they will provide you with feedback and comments on style and optimizations. Always try to write clean and understandable code; type-specify your interfaces and model your data structures. Put effort in choosing algorithms and constructs that scale effectively. You have to think about real-time efficiency from the start, as it is not something that you can easily resolve later. And finally, never optimize your code in your first stage of development; instead, always program with maintainability and readability in mind. When you have completed your application, profile it and optimize your code only where necessary.

Common mistakes often made by beginners include:

- Functions that span many pages
- Deeply nested `if`, `case`, and `receive` statements
- Badly typed and untagged return values
- Badly chosen function and variable names
- Unnecessary or superfluous processes
- Badly and unindented code
- Use of `put` and `get`
- Misuse of `catch` and `throw`
- Bad, superfluous, or missing comments
- Usage of tuple representation of records
- Bad and insufficient use of pattern matching
- Trying to make the program fast, with unnecessary optimizations

A programmer will be able to understand a problem in full only when he has solved it at least once. And when he has found a solution, he will have thought of much better ways he could have solved the problem. Always try to go back and rewrite your code, keeping the aforementioned beginner errors in mind. You will quickly discover that your refactored program will consist of an elegant and efficient code base which, as a result, becomes easier to test, debug, and maintain.[*] Expect code reductions of up to 50% when rewriting your first major Erlang programs. As you become more

---

[*] If you are writing industrial applications, don't believe for one second that you will be the last one to touch your code. Be kind to the others who will take over after you!

experienced and develop your programming style, this reduction will decrease as you start getting things optimal the first time around.

The golden rules when working with Erlang should always be as follows:

First make it work.

Then make it beautiful.

And finally, only if you really have to, make it fast while keeping it beautiful.

You will quickly discover that in the majority of cases, your code will be fast enough. Happy Erlang programming!

Using Erlang

This appendix begins by helping you to get started with Erlang. We then recommend a number of tools to help you to develop Erlang-based systems more effectively. Finally, we tell you where you can find out more about Erlang, particularly from the many web-based resources for Erlang.

Getting Started with Erlang

This section tells you how to get started with Erlang: first, how you install it, and then how to run Erlang programs. We conclude by showing you the various commands in the Erlang shell that help you to be a power user of the shell, using the history mechanism and the line editing commands.

Installing the System

The Erlang distribution is available from the Erlang website, *http://erlang.org/download .html* or one of the many mirror sites. You can also download Erlang using BitTorrent, as well as find it bundled in many of the major Linux distributions. The sources are provided for compilation on Unix operating systems, including Linux and Mac OS X; for Windows, use the binary installer. When building from source, follow the instructions that come with the distribution.

Running the Erlang Shell

In Unix, Linux, and Mac OS X, you can run the Erlang shell from the command line by typing the **erl** command, setting whatever options you require.

To open an Erlang file in Windows, double-click the file icon; this will ensure that the system is opened in the correct directory. There are two variants of the system on Windows, available under "Open With" when you right-click an Erlang source file:

Erl
> Opens the file in an Erlang shell from the Command Prompt program

Werl

> Opens an Erlang shell in a window supporting copy and paste operations more accessibly than the standard Command Prompt program

To run Erlang with options set—for example, the `-smp` option required to run wxErlang—you can run these commands from the Run window, or by typing them as commands to a command prompt. This option allows you to change into the appropriate directory before issuing the command, as in the following:

```
C:\Documents and Settings\Administrator>cd Desktop\programming\wxex
C:\...\...\wxex>"c:\Program Files\erl5.7\bin\erl.exe" -smp miniblog.erl
Eshell V5.7  (abort with ^G)
1> miniblog:start().
```

You can also change these features by right-clicking the file icon and selecting Properties. In the Shortcut tab, you can change your target to include your start options and change the "Start in directory" to point to where your source code is.

In the Erlang shell on Unix and in Werl on Windows, there is a standard set of editing operations on the commands typed:

Up and down arrows

> Fetch the previous and next command line; this may be part of a command, since commands can span multiple lines, in general

Ctrl-P and Ctrl-N

> Have the same effect as the up and down arrows

Left and right arrows

> Move the cursor one character to the left and right

Ctrl-B and Ctrl-F

> Have the same effect as the left and right arrows

Ctrl-A

> Takes the cursor to the start of the line

Ctrl-E

> Takes the cursor to the end of the line

Ctrl-D

> Deletes the character under the cursor

As you've seen in the body of the text, there is a set of commands in the Erlang shell. The most commonly used commands include the following:

`c(File)`

> Compiles and loads the `File`, purging old versions of code

`b()`

> Prints the current variable bindings

`f()`

> "Forgets" all the current variable bindings

`f(X)`
> "Forgets" the binding for the variable `X`

The history of commands and their results are remembered:

`h()`
> Will print the history list (which has a default length of 20, but can be changed)

`e(N)`
> Will repeat command number `N`

`e(-N)`
> Will repeat the nth previous command; for example, `e(-1)` is the previous command

`v(N)`
> The return value of command `N`

`v(-N)`
> The return value of the nth previous command; for example, `v(-1)+v(-2)` will return the sum of the previous two values

Other shell commands, including those for dealing with record definitions in the shell, are described in detail in the documentation for the `shell` module.

Tools for Erlang

If you want to get started writing programs in a new language, the last thing you want to have to do is to learn a new editor, too. Luckily, many common editors and IDEs provide support for writing programs. In this section, we'll discuss these editors and IDEs, as well as some of the other tools we find useful beyond those we've already described. Some tools come as a part of the Erlang distribution, and there's a comprehensive description of those tools in the documentation that accompanies the distribution.

Editors

Erlang programs are contained in text files, and so they can be created within any text editor. However, a number of editors support Erlang-aware operations, and, taking a leaf from the Java and C++ communities, a growing number of fully fledged IDEs support Erlang:

- According to a recent survey of Erlang users,[*] the principal development tool for the dedicated Erlang programmer is *Erlang mode for Emacs*, documented in the tools reference manual from the Erlang online documentation. This gives syntax

---

[*] Available at *http://www.protest-project.eu/publications/survey.pdf*, this was undertaken as part of the ProTest project.

coloring for Erlang, as well as context-sensitive formatting that will help you to lay out your program so that you and others can read it. The mode will also check that your module names and filenames match, as well as providing skeletons for common OTP behaviors.

- *Distel*, or Distributed Emacs Lisp, takes the Emacs support for Erlang to another level, as it allows Emacs to interact with running Erlang nodes, as we described for other programming languages in Chapter 16. Distel provides completion of names of functions and modules; runs Erlang code from within Emacs; offers some limited refactoring support; and features an interactive debugger. Distel is a live Google Code project.

- There is also an Erlang plug-in package for Vim, which supports indentation and syntax highlighting, as well as folding and partial omni-completion. This is available from the Vim website, *http://www.vim.org/scripts/script.php?script_id=1584*.

- Eclipse is the tool of choice for many Java and C++ programmers, and Erlang is now supported in Eclipse through *Erlide*, an Erlang plug-in for Eclipse, available on Sourceforge.net. Erlide implements syntax highlighting and indentation, but also offers evaluation of Erlang expressions within the IDE and automatic compilation on file save. Its structure also defines the scope for Erlang projects, and gives debugging support. Erlide also provides access to the refactorings in Wrangler (discussed shortly).

- Other IDEs and editor support include *ErlyBird*, an Erlang IDE based on NetBeans, and *UltraEdit*, which some Windows developers use to highlight syntax in Erlang programs.

Other Tools

In addition to the tools we described earlier, other tools we find helpful include the following:

- Although EUnit provides the framework for unit testing, *Common Test* gives a framework for complete Erlang-based systems. Common Test is part of the Erlang standard distribution.

- Traditional testing allows you to check the behavior of a system under particular inputs; an alternative approach, embodied in *QuickCheck*, asks the tester to state properties of the system and then tests the properties for randomly generated input values. QuickCheck is also able to test properties of concurrent systems using finite state machines to exercise their behavior, and to shrink any data that fails to satisfy the property to minimal counterexamples. QuickCheck is a product of Quviq AB.

- Beyond simple testing, static analysis can check for dead code as well as type anomalies. *Dialyzer* comes as part of the Erlang standard distribution.

- Refactoring for Erlang is supported by the *Wrangler* tool, embedded in both Emacs and Erlide, and available from the University of Kent in the United Kingdom.

RefactorErl also supports some refactorings, as do Distel and the `syntax_tools` modules that come with the Erlang standard distribution.

- Although testing tools can check the behavior of a system under only a selection of inputs, model checking enables the user to check all possible behaviors of the system under test. *McErlang*, developed at the Universidad Politécnica de Madrid, is a model checker for Erlang written in Erlang. Another approach to model checking translates Erlang into μCRL and then model-checks the results.

Where to Learn More

The best place to start when you want to learn more about Erlang is the Erlang home page, *http://www.erlang.org*. Here, you can find out about upcoming events, books, courses, and jobs, as well as access the system documentation.

The system documentation—which you can access at *http://www.erlang.org/doc/* and download to your computer when you download Erlang—can be a bit overwhelming at first, but it contains a lot of useful information:

- Each Erlang module in the distribution is documented at *http://www.erlang.org/doc/*: click the Modules link in the top-lefthand corner of the home page.
- To get other information about a topic or potential function, you can use the index generated from the documentation, which is accessible from the top-lefthand corner of the home page.

The tabs down the left side of the main page give links to documentation regarding the main Erlang applications and tools. Particularly useful are:

- The installation guide (under the Erlang/OTP tab)
- Getting Started, a mini tutorial (under Erlang Programming)
- The Erlang reference manual (under Erlang Programming)
- The FAQs at *http://www.erlang.org/faq.html*

You can find other Erlang information in the following locations:

- The Erlang community site, Trapexit.org, at *http://www.trapexit.org*.
- The website for this book, *http://www.erlangprogramming.org*, which also has links to all the sites mentioned here, as well as a lot more background information.
- The Erlang mailing lists, accessible from *http://erlang.org/faq.html* and available in archived form at Nabble.com and elsewhere.
- The many Erlang-focused blogs. Just search for "Erlang" on *http://blogsearch.goo gle.com*; many of the blogs are aggregated at *http://planet.trapexit.org* and at Planet Erlang, *http://www.planeterlang.org*.
- The two annual Erlang events: the Erlang Workshop, sponsored by ACM SIGPLAN and collocated with the International Conference on Functional

Programming (*http://www.erlang.org/workshop/*), and the Erlang User Conference, which takes place in Stockholm each year (*http://www.erlang.org/euc/*).

- The Erlang Factory, which runs commercial Erlang conferences and whose website, *http://www.erlang-factory.com/*, contains slides and videos from many of the talks given at the conferences.

Index

Symbols

" (quotation marks) (see quotation marks)
# (hash), 15
$ symbol, 22
$Character notation, 16, 23
( ) parentheses (see parentheses)
* (multiplication) operator, 17, 378
* (unary multiplication) operator, 17
+ (addition) operator, 17, 378
+ (unary addition) operator, 17
++ operator, 26, 27, 201
, (comma), 52, 378
-- operator, 26
.beam file extension, 41
.erl file extension, 40
.erlang file extension, 186
/ (unary division) operator, 17
/= (not equal to) operator, 28, 378
: (colon), 25, 205
; (semicolon), 52, 378
< (less than) operator, 28, 378
<< >> (double angled brackets), 206
<= (leftwards arrow), 206
<= (less than or equal to) operator, 28, 378
=/= (exactly not equal to) operator, 28, 378
=:= (exactly equal to) operator, 28, 378
== (equal to) operator, 28, 378
> (greater than) operator, 28, 378
>= (greater than or equal to) operator, 28, 378
? (question mark), 165
?FILE macro, 167
?LINE macro, 167
?MACHINE macro, 167
?MODULE macro, 167
?MODULE_STRING macro, 167
@ (at) symbol, 19
[ ] (square brackets), 22, 23
_ (underscore), 19, 37
{ } (curly brackets), 21
|| (double vertical bar), 206
~ (tilde), 57
– (subtraction) operator, 17, 378
– (unary subtraction) operator, 17

A

abort function, 299
abs/1 function, 378
accept function, 331
accumulating parameter, 63
add/2 function, 76
addition (+) operator, 17, 378
add_handler function, 132
add_path function, 286
add_patha function, 181, 184
add_pathz function, 181
add_table_index function, 302
add_usr function, 300
agile programming, 412
all flag, 359
all function, 196
allocate function, 119, 123
Amazon.com, 2
AMQP, 2
and logical operator, 20, 378
andalso logical operator, 20, 378
any function, 196
append function
    ++ operator and, 27
    wxErlang function, 311

We'd like to hear your suggestions for improving our indexes. Send email to *index@oreilly.com*.

E

F

f/0 shell command, 84, 446
f/1 shell command, 447
Facebook, 2
fault tolerance
    distributed programming and, 245
    distributed systems and, 245, 247
    layering and, 149
features, Erlang
    concurrency, 5, 6
    distributed computation, 7
    high-level constructs, 4
    integration, 8
    message passing, 5
    robustness, 6
    soft real-time properties, 6
FFI (foreign function interface), 352
file function, 163, 168, 179
file module, 79
file2tab function, 226
filename module, 79
files/1 function, 405
fill/0 function, 375, 376
filter function, 191, 192, 196
finite state machines (see FSMs)
firewalls, 261
first/1 function, 221
float/1 function, 54
floating-point division operator, 17
floats
    defined, 17
    Erlang type notation, 397
    mathematical operations, 17
float_to_list/1 function, 54
flush/0 shell command, 93, 324, 359
foldl/3 function
    lists module, 196
    mnesia module, 305
foreach statement, 193
foreign function interface (FFI), 352
format/1 function, 369
format/2 function, 57, 101, 356
frequency module
    allocate function, 119, 123
    deallocate function, 120, 124
    init function, 121
Fritchie, Scott Lystig, 215
FSMs (finite state machines)
    busy state, 117

chapter exercises, 138
    offline state, 117
    online state, 117
    process design patterns, 117, 126–131, 290
fun2ms/1 function
    dbg module, 375–382, 383–391
    ets module, 223, 225, 382, 383–391
function clause
    components, 38
    conditional evaluations, 38, 46
    guards, 50–52
    runtime errors, 68
    variable scope, 49
function definitions
    case expressions and, 47
    fun expressions, 192
    overview, 38
    pattern matching, 4
functional data types (funs)
    already defined functions, 194
    defined, 189
    Erlang type notation, 397
    example, 190
    fun expressions, 192
    functions and variables, 195
    functions as arguments, 190–192
    functions as results, 193
    lazy evaluation, 197
    predefined higher-order functions, 195–196
    transaction support, 299
functional programming, 9, 45, 189
functional testing, 413–415
functions, 45
    (see also BIFs; higher-order functions)
    already defined, 194
    arguments and, 38, 190–192
    as results, 193
    binding to variables, 5, 30
    callback, 132, 265
    chapter exercises, 44, 83, 86
    client, 122
    coding strategies, 435
    EDoc documentation, 403, 404
    fully qualified function calls, 176
    grouping, 40
    hash, 215
    list comprehensions and, 200
    list supported, 25–27

About the Authors

Francesco Cesarini is the founder of Erlang Training and Consulting (*http://www.erlang-consulting.com*). Having used Erlang on a daily basis since 1995, he started his career as an intern at Ericsson's computer science lab, the birth place of Erlang. He spent four years at Ericsson working with flagship Erlang projects, including the R1 release of the OTP middleware. He has taught Erlang/OTP to all parties involved in the software cycle, including developers, support engineers, testers, and project and technical managers. In 2003, he also started teaching undergraduate students at the IT University of Gothenburg.

Soon after Erlang was released as open source, he founded Erlang Training and Consulting. With offices in the U.K., Sweden, Poland (and soon in the U.S.), the company has become the world leader in Erlang-based consulting, contracting, support, and training and systems development. Francesco is active in the Erlang community not only through regular talks, seminars, and tutorials at conferences worldwide, but also through his involvement in international research projects. He organizes local Erlang user groups and, with the help of his colleagues, runs the trapexit.org Erlang community website.

Simon Thompson is a professor of logic and computation in the computing laboratory of the University of Kent, where he has taught computing at undergraduate and postgraduate levels for the past 25 years, and where he has been department head for the last 6. His research work has centered on functional programming: program verification, type systems, and, most recently, development of software tools for functional programming languages. His team has built the HaRe tool for refactoring Haskell programs and is currently developing Wrangler to do the same for Erlang.

Simon's research has been funded by various agencies, including EPSRC and the European Framework programme. His training is as a mathematician: he has an M.A. in mathematics from Cambridge and a D.Phil. in mathematical logic from Oxford. He has written three books in his field of interest: *Type Theory and Functional Programming*; *Miranda: The Craft of Functional Programming*, and *Haskell: The Craft of Functional Programming*, Second Edition (all books published by Addison-Wesley).

Colophon

The animal on the cover of *Erlang Programming* is a brush-tailed rat kangaroo (*Bettongia penicillata*). The brush-tailed rat kangaroo is a small mammal found in western and southern Australia. It is a cross between a rat and a small wallaby, and although some of its features are reminiscent of a rat, it is not a rodent and is instead classified as a marsupial. In south Australia, they are found in semi-arid scrublands and grasslands; in western Australia, they prefer eucalyptus forests containing a vegetative layer of tussock grass, low woody scrub, and occasional bare patches of ground. They once inhabited more than 60% of the Australian mainland, but now they inhabit less than 1%.

Brush-tailed rat kangaroos have an unusual mammalian diet that consists of bulbs, tubers, seeds, insects, resins, and underground fungi; they do not drink water or eat green plants. Although fungi are not considered a good food source for mammals in general, they provide the nutrients necessary for the brush-tailed rat-kangaroo's health.

The kangaroos' coats are yellowish-gray in color, their feet are pale brown and have hairs that bristle, and their long tails have a prominent black crest. Their tails are also useful: brush-tailed rat kangaroos are able to curl their tails to carry bundles of material to build their nests. They are relatively slow-moving creatures, but are able to hop away quickly when disturbed.

Brush-tailed rat kangaroos are extremely nocturnal. During the day they rest in well-constructed, hidden nests made up of grass and shredded bark. They appear to be solitary except when ready to mate.

Mating occurs year round, and females give birth to one young after a gestation period of 21 days. The newborn remains in the mother's pouch for about 98 days, and then stays in a nest until a new infant is born. As with many other kangaroos, the brush-tailed rat kangaroo mates shortly after giving birth and can keep embryos in a state of dormancy until they are needed.

The cover image is from Cassell's *Natural History*. The cover font is Adobe ITC Garamond. The text font is Linotype Birka; the heading font is Adobe Myriad Condensed; and the code font is LucasFont's TheSansMonoCondensed.